University Forum for Disaster Risk Reduction Conference 2021

IOP Conference Series: Earth and Environmental Science
Volume 989

Manokwari, Indonesia
29 September 2021

ISBN: 978-1-7138-4970-4
ISSN: 1755-1307

Printed from e-media with permission by:

Curran Associates, Inc.
57 Morehouse Lane
Red Hook, NY 12571

Some format issues inherent in the e-media version may also appear in this print version.

Institute of Physics
Dirac House, Temple Back
Bristol BS1 6BE UK

Phone: 44 1 17 929 7481
Fax: 44 1 17 920 0979

techtracking@iop.org

Additional copies of this publication are available from:

Curran Associates, Inc.
57 Morehouse Lane
Red Hook, NY 12571 USA
Phone: 845-758-0400
Fax: 845-758-2633
Email: curran@proceedings.com
Web: www.proceedings.com

TABLE OF CONTENTS

IOP Conf. Series: Earth and Environmental Science **989** (2022) 011001 doi:10.1088/1755-1315/989/1/011001

PREFACE

The University Forum for Disaster Risk Reduction (UF DRR) in collaboration with the University of Papua has successfully held an "UF DRR Conference 2021" via the online zoom system on September 29th, 2021. UF DRR Conference 2021 aims to share their knowledge and experience, thoughts and opinions on how to mitigate, adapt, prepare, and build resilience to climate-related hazards in line with sustainable development. The theme of the UF DRR Conference 2021 is "Disaster Risk Management for Climate-related Hazards in Tropical Countries: Progress and Challenges".

UF DRR Conference 2021 invited the Keynote Speakers who are very dedicated and competent in the field of disaster and the climate, and they are Dr. Raditya Jati, S.Si., M.Si. (Deputy of System and Strategy, National Disaster Management Agency) and Dr. Ardhasena Sopaheluwakan (Head of Center for Applied Climate Information Services, Meteorology, Climatology, and Geophysics Agency). Besides that, we also invited Expert Speeches from University of Hawaii (Prof. Dr. Karl Kim), University of Hiroshima (Assoc. Prof. Dr. Lee Han Soo), Chungbuk National University (Prof. Dr. Sungsu Lee), and University of the Phillipines Los Banos (Prof. Dr. Juan Pulphin). Furthermore, we requested several invited speakers who are leading in disaster and climate experts such as Dr. Nelly Rima (BMKG), Dr. Andreas Dipi Patria (Ministry of Coordinator Maritime and Investment) and Dr. Rahmawati Husein (Muhammadiyah Disaster Management Center, MDMC).

Topic issues for this seminar are climate-related hazards management (flood and landslide, flash flood, whirling, wind, drought and land/forest/bush fire, and tropical cyclone) in the archipelago and coastal and small island; infrastructure resilience; community-based approach; early warning system, preparedness and awareness building; emergency response and relief management; build back-better post-disaster recovery; and Sendai Framework and Sustainable Development Goals perspective. There were 51 articles submitted according to the topic above, and after being selected, there were 32 articles chosen to be presented and submitted to the IOP-EES Proceedings. Article authors consist of lecturers, practitioners, teachers and students. The participants who joined at the time of the seminar was 110 participants.

Many thanks to all of the speakers, authors and participants. See you at the next conference of UF DRR.

Best Regards

Dr. Eng. Hendri, S.Si., M.Si

Chair of the committee

IOP Conf. Series: Earth and Environmental Science **989** (2022) 011001 doi:10.1088/1755-1315/989/1/011001

Committee
University Forum for Disaster Risk Reduction (UF-DRR) Conference 2021

Advisory Board :

1. Prof. Krishna S Pribadi (Institut Teknologi Bandung, Indonesia)
2. Prof. Euis Sunarti (Universitas IPB, Indonesia)
3. Prof. Ricardo Tapilatu (Universitas Papua, Indonesia)
4. Prof. Waonho Yi (Kwangwoon University, South Korea)
5. Prof. Dr. Ramasamy R. Krishnamurthy (University of Madras, India)
6. Prof. Dr. Budy P. Resosudarmo (Australian National University, Australia)

Conference Chair :

Dr. Eng. Hendri, S.Si., M.Si (Universitas Papua, Indonesia)

Conference Vice Chair :

Sorja Koesuma, S.Si., M.Si (Universitas Sebelas Maret, Indonesia)

Secretary :

Dr. Nuraini Rahma Hanifa (Lembaga Ilmu Pengetahuan Indonesia, Indonesia)

Treasure :

Dr. Tukino (Politeknik Kesejahteraan Sosial Bandung, Indonesia)

Editor

1. Dr. Endang Hilmi (Universitas Jenderal Soedirman, Indonesia)
2. Dr. Dicky Peluppesy (Universitas Indonesia, Indonesia)
3. Dr. Ananto Aji (Universitas Negeri Semarang, Indonesia)
4. Dr. Ferad Puturuhu (Universitas Pattimura, Indonesia)
5. Dr. Jamhir Safani (Universitas Haluoleo, Indonesia)
6. Dr. Aviasti (Universitas Islam Bandung, Indonesia)
7. Dr. Gusti Zulkifli Mulki (Universitas Tanjungpura, Indonesia)

Technical Members :

1. Dr. Dedi I. Inan, ST., MT., Ph.D. (University of Technology Sydney, Australia)
2. Dr. Syafrudin Raharjo, S.Pi., MT. (Universitas Papua, Indonesia)
3. Dr. Rima HS Siburian, S.Hut., M.Si. (Universitas Papua, Indonesia)
4. Gandi YS Purba, S.IK., M.Si. (Universitas Papua, Indonesia)
5. Dr. Rina A. Mogea, S.Pi., M.Si. (Universitas Papua, Indonesia)

All papers published in this volume have been reviewed through processes administered by the Editors. Reviews were conducted by expert referees to the professional and scientific standards expected of a proceedings journal published by IOP Publishing.

1. Type of peer review: Single Anonymous
2. Conference submission management system: Easychair
3. Number of submissions received: 51
4. Number of submissions sent for review: 51
5. Number of submissions accepted: 33
6. Acceptance Rate (Submissions Accepted / Submissions Received × 100): 64.7 %
7. Average number of reviews per paper: 4.94
8. Total number of reviewers involved: 11
9. Contact person for queries:
   Name: Hendri
   Affiliation: Universitas Papua, Indonesia
   Email: h.hendri@unipa.ac.id

The potential of high tidal flooding disaster in North Jakarta using mapping and mangrove relationship approach

Endang Hilmi [1, *], Amron [2], and Dandhy Christianto [3]

[1]Program of Aquatic Resources Management, Faculty of Fisheries and Marine Sciences, Universitas Jenderal Soedirman. Jl. Dr Soeparno, Purwokerto Utara, Banyumas 53122, Central Java, Indonesia. Tel.: +62-281-6596700

[2]Marine Science Proram, Faculty of Fisheries and Marine Sciences, Universitas Jenderal Soedirman. Jl. Dr Soeparno, Purwokerto Utara, Banyumas 53122, Central Java, Indonesia. Tel.: +62-281-6596700

[3]Under graduade of Marine Science Proram, Faculty of Fisheries and Marine Sciences, Universitas Jenderal Soedirman. Jl. Dr Soeparno, Purwokerto Utara, Banyumas 53122, Central Java, Indonesia. Tel.: +62-281-6596700

*Email: dr.endanghilmi@gmail.com

Abstract. North Jakarta is widely used for various activities including industry, settlement, transportation, tourism and others. North Coast Jakarta also has high probability of coastal disaster including high tidal flooding which is caused by sea level rise, mangrove degradation and geomorphologicly area. This paper aimed to analysis the potential of high tidal flood using mapping analysis and mangrove density relationship. Mapping analysis and mangrove ecosystem relationships were used to analyzing the potential for tidal flooding. Data showed the density of mangrove ecosystems reaches 220-1100 trees ha^{-1}, a single daily tide, sea level reached 2.75m, sea level rise reached 2.74m (the rate reached 0.71 cm/year), tidal flooding to reach an area of 1014.69-1548.91 ha and land subsidence reach 12 cm/year

1. Introduction

North Jakarta is an area in Jakarta which located on the northern of the Java Sea [1], [2] which has many ecosystem that are the estuary ecosystem [2], [3], swamp ecosystem, coral reef and mangrove ecosystem.[2], [4]. The North Coast of Jakarta is an area to buffer and support development center and market building. In addition, based on economic activities, the North Coast of Jakarta is a main area in Jakarta to support many economic activities, like as industry, ports, settlements, markets and villages. Mangrove ecosystem has important fuction to support economic activity in North Jakarta

The mangrove ecosystem is a green growth and natural resources which have many important functions like as the function as a green belt and buffer to reduce coastal disasters. Some of the coastal disasters in the north coast of Jakarta are tidal waves, ROB, abrasion and land subsidence [5]–[8]

Tidal Flood or ROB is a natural disaster phenomena in the north coast of Jakarta which occurs during the full moon or new moon and will known as the Spring Tide process. Tidal flooding is a serious disaster to cities in coastal areas [9] for example in La Briere, Nantes, France and Rotterdam, the Netherlands [10]. North Jakarta has high potential of tidal waves which caused by sea level rise and relatively low and flat topography. In 1925-2010, North Jakarta has increasing of sea level with average of 0.5 cm/year, and rate of land subsidence reached 5 cm to 12 cm/year (in some locations reached 20-

IOP Conf. Series: Earth and Environmental Science **989** (2022) 012001 doi:10.1088/1755-1315/989/1/012001

28 cm/year) [11]. This conditions are the influencing factor of tidal flooding[9], [11]–[13], including in North Jakarta .

The phenomenon of tidal flooding give negative impacts on coastal areas like as change the physical environment, high pressure for communities, buildings, and settlement infrastructure, and ecosystems and landscapes damages. And the mangrove ecosystem in North Coast of Jakarta has function to reduce impact of tidal flooding and land subsidence [2], [7], because mangrove has strengthen and elasitic of mangrove wood and a distinctive root system to reduce impact of high tidal waves [7]. This paper aimed to analysis the potential of high tidal flood using mapping analysis and mangrove density relationship

2. Methods

2.1. Research Site

This research was conducted on the North Coast of Jakarta from 2016 – 2021. The location determination is determined from the possibility of potential tidal flooding and the presence of mangrove ecosystems (Figure 1).

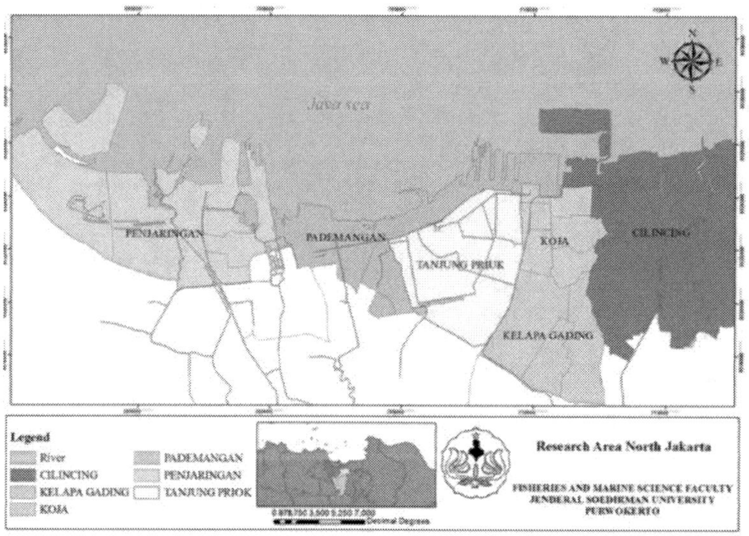

Figure 1. Research site

2.2. Research Procedure

2.2.1. Vegetation Analysis

The vegetation samples for each sampling plot were collected using a 10 m x 10 m transect with 3 replications [14]–[16] (Figure 2). For each sampling plot was measured the number of mangrove species, the number of mangrove tress, diameter of breast high (for mangrove diameter > 4 cm) (Figure 2)[15] And then the vegetation from line transect will be calculated the mangrove density following [14], [15], [17] with equation :

$$mangrove\ density = \frac{the\ number\ of individual\ for\ mangrove\ trees}{area\ of\ sampling\ plot} \quad (1)$$

2.2.2. High Tidal flooding (ROB) analysis

The ROB analysis was built by analyzing

2.3. Data Analysis
Data analysis was carried out using a geographic information system which was integrated with mean sea level (MSL), DEM, highest sea level value (HHWL) and Interferometric Synthetic Aperture Radar (InSAR).

3. Result and Discussions

3.1. Mangrove density in North Jakarta
The potential of mangrove density in North Coast Jakarta can be seen in Table 1. The density data showed the potential density of mangroves (for diameter > 4cm) between 220 trees/ha-1565 trees/ha. According to Hilmi, 2019, the potential density of mangroves is classified as very rare - dense. This condition will lossing of the mangrove ecosystem as a buffer and preservation to reduce coastal disasters in North Jakarta [2], [4], [7]

Table 1. Mangrove Density in North Jakarta

Indicator	Mangrove density (trees/ha)								
	North Jakarta (station)								
	Ekowisata	Lindung 1	Lindung 2	Lindung 3	Arboretum	galatama	Muara gembong	Cilincing	Tol Sedyatmo
Total	1100	311	330	220	600	738	1575	335	1025
SD	255	32	66	54	113	278	480	60	556
Class	rare	very rarely	very rarely	very rarely	rare	rare	dense	Very rare	rare

3.2. Tidal and Sea level Rise
The trend projections of annual sea level rise between 1984 - 2030 were shown in Figure 2. Base on Fromzahl's number = 12.91 (or F > 3), had meaning that the tidal tyoe in the North Coast was a single daily tidal type. These result is similiar with Indriani's research (2010) writes that the tidal type around Tanjung Priok Port is a single daily with a value of 3.44 fromzahl. This data also showed that North Jakarta [19] had one high tide and one low tide/day. The monthly tidal data and an average every year from Tanjung Priok station were processed by least square linear regression on the bivariate distribution value for 27 years of observation (1984 – 2011) showed increasing trend of sea level rise with the equation y = 0.0588x + 212.83, with increasing of sea level rise reached 0.71 cm/year.

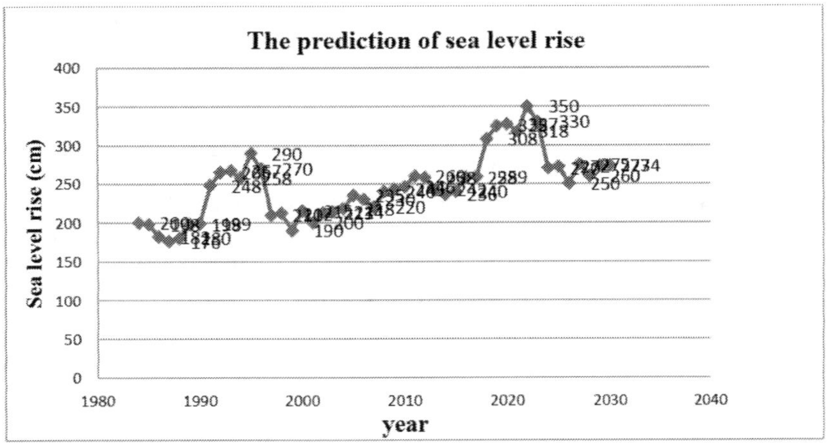

Figure 2. Sea level rise in North Jakarta

3.3. Affected area of ROB or high tidal flooding and land subsidence

The potential of affected area for tidal flooding due to tides, sea level rise and land subsidence can be seen in Figures 3 and Tables 2.

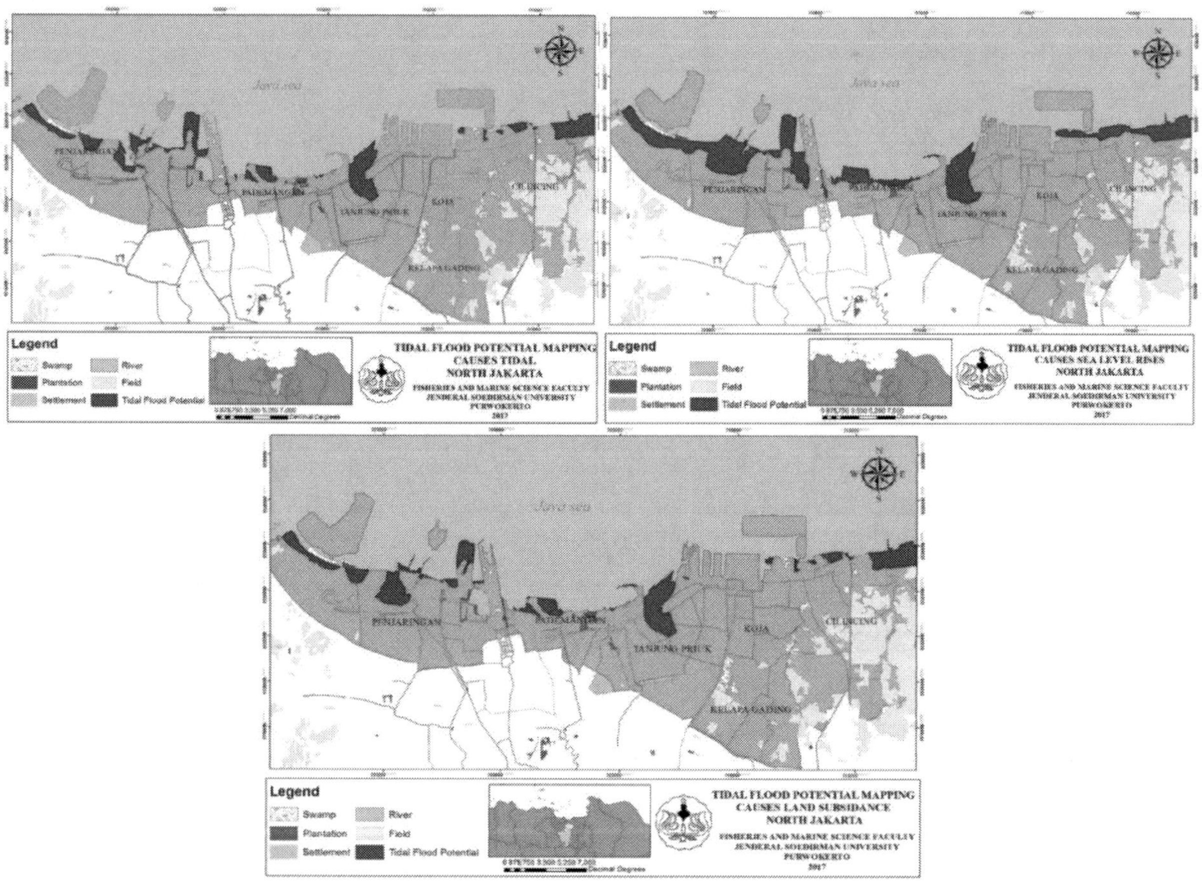

Figure 3. Potential of affected area of ROB or Tidal flooding and land subsinece

The potential area on the North Coast of Jakarta to be affected by ROB between 110.81 ha/district - 385.09 ha/district with a total area reached 1014.69 ha. Potential tidal flooding due to sea level rise reached 1548.91 ha with potential affected area between 110.76 ha/district– 826.09 ha/district. The potential ROB also due to land subsidence reaches 1197.02 ha with the distribution of each sub-district between 152.85 – 409.08 ha.

Based on the data between period 2000-2015 showed that an average of tidal height was 2.05 m with the highest potential in 2030 approximately 2.75 m and with rate of sea level rise reached 0.71 cm/ year. [20] writes that the sea level rise in North Jakarta is 0.575 cm/yr and land subsidence varies between of 1-15 cm/year (in other area can reached 20-28 cm/year). Using InSAR showing Land subsidence in the coastal area of North Jakarta in the period June 2006 - February 2007 reached 12 cm/year. [9], [18], [21]write that the trend of land subsidence in Jakarta is 80 cm (1982-1991), 160 cm (1991-1997) and 20 cm (1997-1999).

Table 2. The potential of affected area of ROB, sea level rise and land subsidence

District	District area (ha)	Affected area of ROB	Affected area of sea leve rise	Affected area of land subsidence
Penjaringan	3478,57	385,09	826,09	409,08
Pademangan	1773,76	231,86	323,87	242,31
Tanjung Priok	1988,10	110,81	110,76	152,85
Cilincing	4194,40	286,93	287,79	392,78
Jumlah	11434,83	1014,69	1548,91	1197,02

3.4. The total affected area of tidal flooding (ROB) and Land subsidence

The total potential of affected area of ROB due to tides, sea level rise and land subsidence could be seen in **Figures 4** and **Table 3**. The integration data between tides, sea level rise and land subsidence showed the total potential of affected area of ROB floading in North Jakarta reached 3581.83 ha.

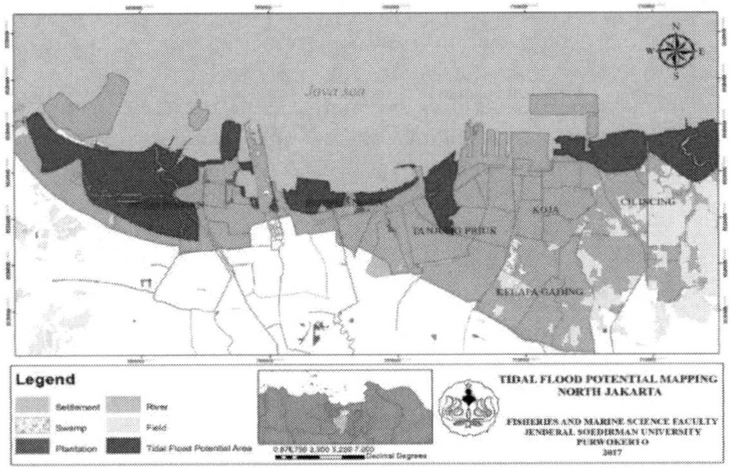

Figure 4. The total of affected area of ROB and Land Subsidence

The total area of tidal flooding in North Jakarta using overlaying DEM data (digital elevation model) with data from tides, sea level rise, and land subsidence. The data's showed that in Penjaringan being inundated between 100 – 150 cm and Pademangan between 30-100 cm give high negative impact for settlements and developments, groundwater exploitation and land subsidence [2], [9], [18], agriculture, industry, housing and others [21]

Table 3. The total of affected area of ROB and Land Subsidence

No	districts	Districts area (Ha)	Area of tidal flooding and land subsidnece (Ha)
1	Penjaringan	3478,57	1872,21
2	Pademangan	1773,76	505,24
3	Tanjung Priok	1988,10	166,58
4	Cilincing	4194,40	1037,80
	Jumlah	11434,83	3581,83

4. Conclussion
The potential of tidal flooding (ROB) in North Jakarta reached 3581,83 ha give negative impact for settlement, industry, market, ground water and land subsidence. The factors influence tidal flooding are sea tide and degradation of mangrove ecosystem. The mangrove ecosystem has function to reduce impact high tidal because the mangrove ability to preserve and protect from ROB disaster.

Aknowledgement
We would like to thank for Dr. Ir. Isdy Sulystio, DEA (Dean of Fisheries and Marine Science Faculty Unsoed), LPPM Unsoed to support grant with Riset Terapan (2021) and researcher colleague for providing his advice for this research. We would also like to thank anonymous reviewers for their helpful and constructive comments which greatly helped us improve our manuscript

References
[1] Onrizal, Rugayah, and Suhardjono, "Flora Mangrove Berhabitus Pohon di Hutan Lindung Angke-Kapuk," *Jurnal Biodiversitas*, vol. 6, no. 1, pp. 34–39, 2005.
[2] E. Hilmi, C. Kusmana, E. Suhendang, and Iskandar, "Correlation Analysis Between Seawater Intrusion and Mangrove Greenbelt," *Indonesian Journal of Forestry Research*, vol. 4, no. 2, pp. 151–168, 2017, doi: 10.20886/ijfr.2017.4.2.151-168.
[3] K. . Kusumahadi, A. Yusuf, and R. . Maulana, "Analisis Keanekaragaman Jenis Vegetasi Mangrove di Kawasan Hutan Lindung Angke-Kapuk dan Taman Wisata Alam Angke-Kapuk Muara Angke Kota Jakarta Utara," *Jurnal Ilmu dan Budaya*, vol. 41, no. 69, p. 8123, 2020.
[4] E. Hilmi, L. K. Sari, T. N. Cahyo, A. Mahdiana, and S. R. Samudra, "The affinity of mangrove species using Association and Cluster Index in North Coast of Jakarta and Segara Anakan of Cilacap , Indonesia," vol. 22, no. 7, pp. 2907–2918, 2021, doi: 10.13057/biodiv/d220743.
[5] D. M. Alongi, "Mangrove forests: Resilience, protection from tsunamis, and responses to global climate change," *Estuarine, Coastal and Shelf Science*, vol. 76, no. 1, pp. 1–13, 2008, doi: 10.1016/j.ecss.2007.08.024.
[6] R. Cochard, *Coastal Water Pollution and its Potential Mitigation by Vegetated Wetlands: An Overview of Issues in Southeast Asia*, vol. 1. Elsevier Inc., 2017.
[7] E. Hilmi, "Mangrove landscaping using the modulus of elasticity and rupture properties to reduce coastal disaster risk," *Ocean and Coastal Management*, vol. 165, no. July, pp. 71–79, 2018, doi: 10.1016/j.ocecoaman.2018.08.002.
[8] B. K. van Wesenbeeck *et al.*, "Aquaculture induced erosion of tropical coastlines throws coastal communities back into poverty," *Ocean and Coastal Management*, vol. 116, pp. 466–469, 2015, doi: 10.1016/j.ocecoaman.2015.09.004.
[9] E. J. Bomer, C. A. Wilson, R. P. Hale, A. N. M. Hossain, and F. M. A. Rahman, "Surface elevation and sedimentation dynamics in the Ganges-Brahmaputra tidal delta plain, Bangladesh: Evidence for mangrove adaptation to human-induced tidal amplification," *Catena*, vol. 187, no. September, p. 104312, 2020, doi: 10.1016/j.catena.2019.104312.
[10] Wahyudi, "Perbandingan Penanganan Banjir Rob di La Briere (Prancis), Rotterdam (Belandadan Perspektif di Semarang (INDONESIA).," *Riptek*, vol. 4, no. 11, pp. 29–35, 2010.
[11] Indonesia Climate Change Sectoral Roadmap (ICCSR), *Scientific basis: Analysis and Projection of Sea level rise and Extreme Weather Events*. 2010.
[12] H. Takagi, M. Esteban, T. Mikami, M. B. Pratama, V. P. B. Valenzuela, and J. E. Avelino, "People's perception of land subsidence, floods, and their connection: A note based on recent surveys in a sinking coastal community in Jakarta," *Ocean and Coastal Management*, vol. 211, no. March, p. 105753, 2021, doi: 10.1016/j.ocecoaman.2021.105753.
[13] A. Cao *et al.*, "Future of Asian Deltaic Megacities under sea level rise and land subsidence: current adaptation pathways for Tokyo, Jakarta, Manila, and Ho Chi Minh City," *Current Opinion in Environmental Sustainability*, vol. 50, no. February, pp. 87–97, 2021, doi: 10.1016/j.cosust.2021.02.010.

[14] C. Kusmana *et al.*, *Teknik Rehabilitasi*. Fakultas Kehutanan, Institut Pertanian Bogor, Bogor, 2005.

[15] E. Hilmi, L. K. Sari, and Amron, "Distribusi Sebaran Mangrove Dan Faktor Lingkungan Pada Ekosistem Mangrove Segara Anakan Cilacap," *Prosiding Seminar Nasional "Pengembangan Sumber Daya Perdesaan dan Kearifan Lokal Berkelanjutan IX" 19-20 November 2019*, 2020.

[16] S. E. Hilmi Endang, Sari Lilik kartika, Cahyo Tri Nur, Kusmana Cecep, "Carbon sequestration of mangrove ecosystem in Segara Anakan Lagoon, Indonesia," *BIOTROPIA : the Southeast Asian Journal of Tropical Biology*, vol. 26, no. 3, pp. 181–190, 2019, doi: DOI : 10.11598/bcb.2019.26.3.1099.

[17] F. Dangan-Galon, R. G. Dolorosa, J. S. Sespeñe, and N. I. Mendoza, "Diversity and structural complexity of mangrove forest along Puerto Princesa Bay, Palawan Island, Philippines," *Journal of Marine and Island Cultures*, vol. 5, no. 2, pp. 118–125, 2016, doi: 10.1016/j.imic.2016.09.001.

[18] L. M. Bott, T. Schöne, J. Illigner, M. Haghshenas Haghighi, K. Gisevius, and B. Braun, "Land subsidence in Jakarta and Semarang Bay – The relationship between physical processes, risk perception, and household adaptation," *Ocean and Coastal Management*, vol. 211, p. 105775, 2021, doi: 10.1016/j.ocecoaman.2021.105775.

[19] Indriani, N. Kurniawati, and M. Hendri, "Simulasi Pemodelan Arus Pasang Surut di Luar Kolam Pelabuhan Tanjung Priok Menggunakan Perangkat Lunak," *Maspari Journal*, vol. 01, pp. 79–83, 2010, [Online]. Available: http://download.portalgaruda.org/article.php?article=159487&val=5974&title=Simulasi Pemodelan Arus Pasang Surut di Luar Kolam Pelabuhan Tanjung Priok Menggunakan Perangkat Lunak SMS 8.1.

[20] N. Plamonia, . "Kajian Pengaruh Kenaikan Muka Air Laut , Reklamasi Pantai dan Degradasi Lahan di DAS Hulu Terhadap Banjir di pesisir Terbangun DKI Jakarta – DAS Ciliwung," ITB, 2010.

[21] P. J. Ward, M. A. Marfai, F. Yulianto, D. R. Hizbaron, and J. C. J. H. Aerts, "Coastal inundation and damage exposure estimation: A case study for Jakarta," *Natural Hazards*, vol. 56, no. 3, pp. 899–916, 2011, doi: 10.1007/s11069-010-9599-1.

IOP Conf. Series: Earth and Environmental Science **989** (2022) 012002 doi:10.1088/1755-1315/989/1/012002

Groundwater Vulnerability Zones Mapping Using DRASTIC and GOD Methods in Krendowahono Village, Karanganyar Regency

Sorja Koesuma*, Umi Rosidah, Ari Handono Ramelan

Department of Physics, Sebelas Maret University
Jl. Ir. Sutami 36A Kentingan Surakarta 57126. INDONESIA

*Email: sorja@staff.uns.ac.id

Abstract. Krendowahono Village is located in Karanganyar Regency, Central Java, Indonesia. During the dry season, several hamlets in Krendowahono Village experience a shortage of clean water. So an attempt must be made to maintain the quality and protect groundwater sources in areas of Krendowahono Village with fresh groundwater supplies. Analysing the vulnerability of groundwater is one of the attempts that can be made. The DRASTIC and GOD methods were used to conduct a vulnerability analysis of groundwater. The DRASTIC and GOD vulnerability assessments were then validated using water samples taken from resident wells. Groundwater vulnerability assessment utilizing the DRASTIC method resulted in the identification of two vulnerability classes: low vulnerability and medium vulnerability. While the GOD method generates three vulnerability classes for groundwater, they are low vulnerability class, medium vulnerability class, and high vulnerability class. The maximum correlation value between the results of the GOD method and the test of water samples with parameters of free ammonia, total dissolved solids (TDS), and salinity was obtained when the results of the GOD method and the test of water samples were validated using Spearman's correlation. The correlation value associated with the pH parameter was the highest in the DRASTIC method. Likewise, the GOD method produced favourable performance in regional studies. This is demonstrated by the significant correlation between the GOD susceptibility test findings and the water sample test results.

1. Introduction

During the dry season, several hamlets in Krendowahono Village encounter a clean water shortage. On the other hand, the majority of Krendowahono Village residents depend on well water to meet their daily water needs. An anomaly in Krendowahono Village, as in the form of connate water, caused multiple wells to discharge brackish water [1]. Additionally, Krendowahono Village is adjacent to a densely populated area and industry, making it concerned about development's consequences.

This research was taken to preserve the quality and protection of groundwater resources in regions of Krendowahono Village that still have access to fresh groundwater. Analyzing the vulnerability of groundwater is one of the measures that can be taken. There are several methods for determining the vulnerability of groundwater. DRASTIC and GOD methods are used in this research. Numerous researchs on groundwater vulnerability analysis utilizing the DRASTIC and GOD methods have been conducted, with the GOD method demonstrating a higher correlation with the DRASTIC method and producing comparable vulnerability maps [2] [3].

IOP Conf. Series: Earth and Environmental Science **989** (2022) 012002 doi:10.1088/1755-1315/989/1/012002

The DRASTIC and GOD methods produce a vulnerability score based solely on an area's geological, hydrological, and hydrogeological features. As a result, it is required to evaluate the content of groundwater samples collected at the research site to ascertain the quality of the water produces. Therefore, based on the problems in Krendowahono Village and the findings of earlier studies, an evaluation of aquifer vulnerability was conducted using the DRASTIC and GOD methods, as well as correlation with the results of a groundwater content test.

2. Methods

2.1. Study Site

Krendowahono Village is located between 7⁰ 27' 21" S - 7⁰ 28' 12" S and 110⁰ 48' 38" E - 110⁰ 50' 45" E (Figure 1). Based on the Geological Map Sheet Salatiga, Java [4], Krendowahono Village is composed of the Kalibening Formation, Notopuro Formation, and Kabuh Formation. The Kalibening Formation occupies the western part of Krendowahono Village, consisting of solid marl with tuffaceous sandstones and limestone piles. The Notopuro Formation dominates in the middle to the east, consisting of intercalated lava breccias with interspersed tuff and tuffaceous sandstones. While the Kabuh Formation, occupying the middle part and located between the Kalibening formation and the Notopuro formation, consists of alternating conglomerates, sandstones and tuff.

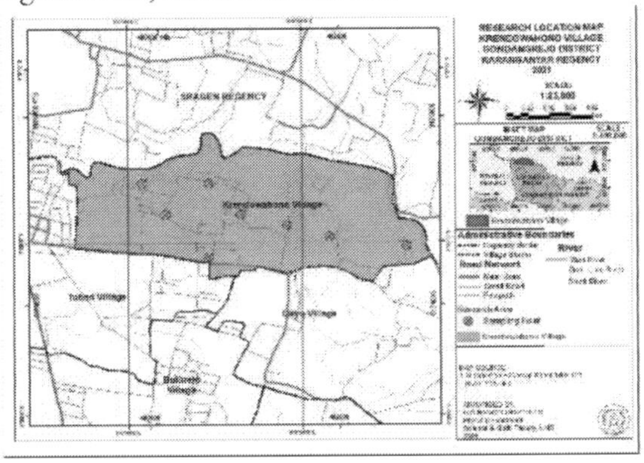

Figure 1. Research location

2.2. Description of the DRASTIC Method

DRASTIC method was designed by Aller *et al.*, (1985), belongs to the overlay and index categories and is believed to be the most common vulnerability assessment method [5]. There are seven hydrogeological parameters considered in the DRASTIC method. These parameters are represented by each letter that composes DRASTIC [6].

To generate a groundwater vulnerability map, seven individual parameter maps need to be prepared according to the conditions of each parameter [6] [7].Determination of the level of groundwater vulnerability using the DRASTIC method by using weighting and ranking of each parameter used. Each parameter has a constant weight and cannot be changed [8] [9]. In addition to weight, each assessment parameter also has a rating that connotes vulnerability results [5] [8]. The level of groundwater vulnerability with the DRASTIC method was determined using the DRASTIC index value (Equation 1) [10].

$$DRASTIC\ INDEX = D_r D_w + R_r R_w + A_r A_w + S_r S_w + T_r T_w + I_r I_w + C_r C_w \quad (1)$$

in which, r represents the rating of each parameter, w represents the weight of each parameter, and D, R, A, S, T, I, and C is the seven parameters of the vulnerability.

IOP Conf. Series: Earth and Environmental Science **989** (2022) 012002 doi:10.1088/1755-1315/989/1/012002

2.3. Description of the GOD Method

GOD method devised in England by Foster in 1987 and reformed in 1998 [10] [11]. The name of the GOD method is an acronym for three parameters, namely: G = groundwater occurrence, O = overlying lithology, and D = depth table of the groundwater specific. This method is carried out by grouping the vulnerability index weight values with three parameters, in which each parameter has a specific weight [12].

The area vulnerability index is calculated by determining the aquifer type parameter rating then multiplied by the lithological type rating of the upper layer of the aquifer and the groundwater table depth parameter rating [13]. In this case, the calculation of the GOD index refers to equation 2 [14].

$$\text{GOD INDEX} = C_A \times C_L \times C_D \tag{2}$$

in which, C_A is the type of aquifer, C_L is the lithological type of the top layer of the aquifer, and $C_D$is the depth of the groundwater table [14]. The level of groundwater vulnerability of the GOD method can be seen from the resulting index value [15] [16].

2.4. Water Sample-test

Water quality is obviously a vital role because it is key to maintaining the environmental balance [17]. The composition of recharge water develops as it passes through the unsaturated zone and enters and flows through the aquifer [18]. Mineral ions in water become a determining parameter of water quality.

The condition of the water is known by testing the water quality in the laboratory. The test results are then compared with water quality standards for certain uses. The representation of water quality standards is the Regulation of the Minister of Health of the Republic of Indonesia number 492 of 2010 concerning the requirements for drinking water quality. Based on water conditions in Krendowahono Village, several parameters used in the laboratory test are the amount of dissolved solids, pH, salinity, ammonia, and manganese (Mn).

2.5. Data Processing

This research uses several data, such as, geoelectric data, groundwater table depth surveys data, rainfall data, Digital Elevation Model (DEM) Nasional data, and Indonesian Topographic Map data. Geoelectric is a method used to identify the medium on earth based on its resistivity [19] [20]. The data used is the result of data acquisition using the Vertical Electrical Sounding method with the Schlumberger configuration. Through the Schlumberger configuration, the geological structure below the vertical surface can be known [21]. This configuration is also suitable for deep penetration in groundwater search [22]. Groundwater depth values were taken directly through a field survey conducted in January and May 2021. Rainfall data were obtained from the Badan Meteorologi, Klimatologi, dan Geofisika. Meanwhile, the National DEM data and the RBI Map were obtained through the Ina-Geoportal website.

The level of groundwater vulnerability by DRASTIC and GOD methods is generated in the form of a map of the distribution of each parameter. Groundwater vulnerability zone maps were obtained by interpolating the index values of each method. The resulting vulnerability map is then used as a reference to compare the two methods. Validation of the results of the groundwater vulnerability map is carried out by looking at the correlation between the map and the results of the groundwater sample test.

3. Result and Discussion

3.1. DRASTIC Method Vulnerability Map

The DRASTIC vulnerability map in Krendowahono Village was generated based on the conditions of the parameters of groundwater table depth, net recharge, aquifer media, soil media, topography, vadose zone, and hydraulic conductivity. By determining the condition of each parameter, the appropriate score can be determined based on the DRASTIC method. The individual parameter condition map is presented in Figure 2.

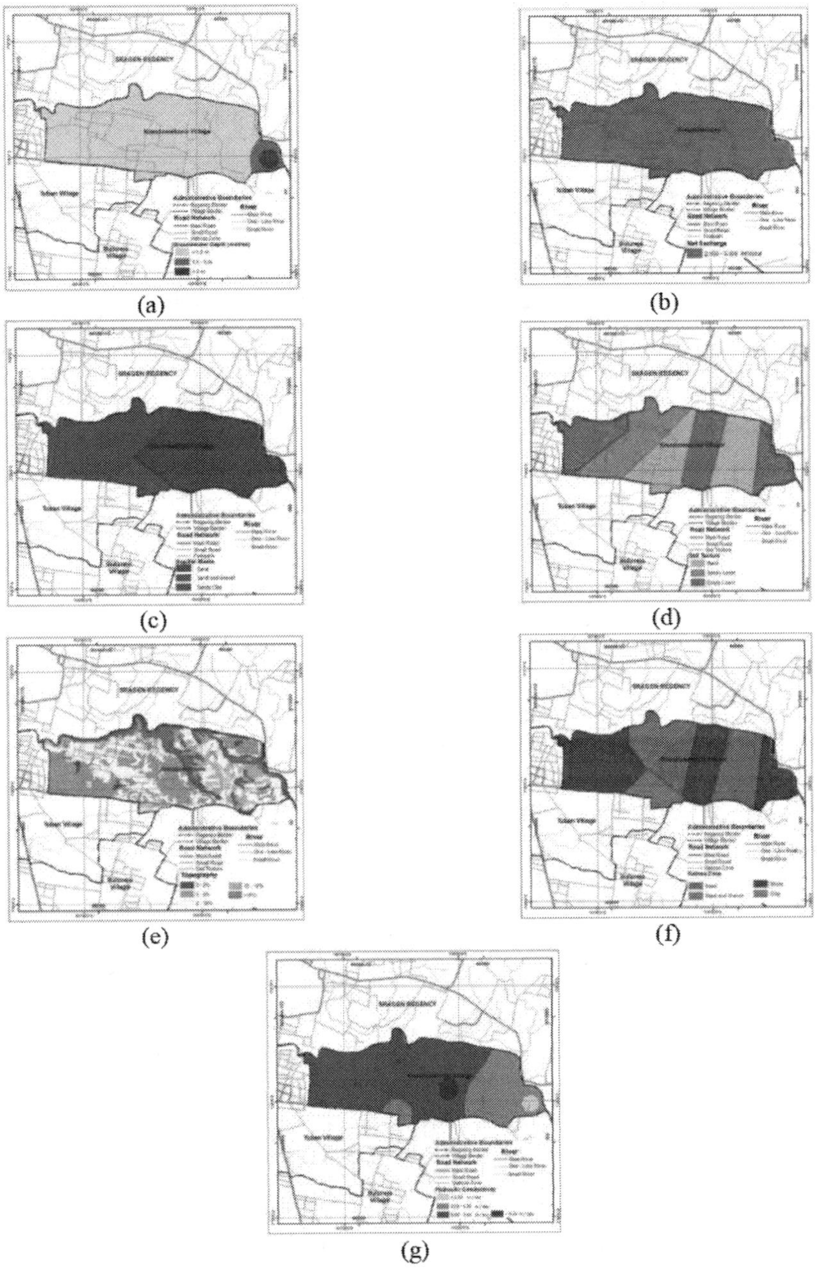

Figure 2. Individual Parameter Map (a) Depth to water (b) Net recharge (c) Aquifer media (d) Soil media e) Topography (f) Vadose zone (g) Hydraulic Conductivity

The results of the individual maps were used to define the DRASTIC vulnerability map. This map can be obtained by calculating the DRASTIC index. The vulnerability index of this method has the lowest value of 119 and the highest value of 159. Based on this value, there are two levels of

IOP Conf. Series: Earth and Environmental Science **989** (2022) 012002 doi:10.1088/1755-1315/989/1/012002

vulnerability, such as the medium level of vulnerability (99.64% of the research area) and the low level of vulnerability (0.36 % of the research area), as presented in Figure 3. Low susceptibility index values are produced in areas that have rock lithology with low water permeability values. Parameters that have a major influence on the results of the vulnerability value of this method are the parameters of the depth of the groundwater table, aquifer rock lithology, and net recharge.

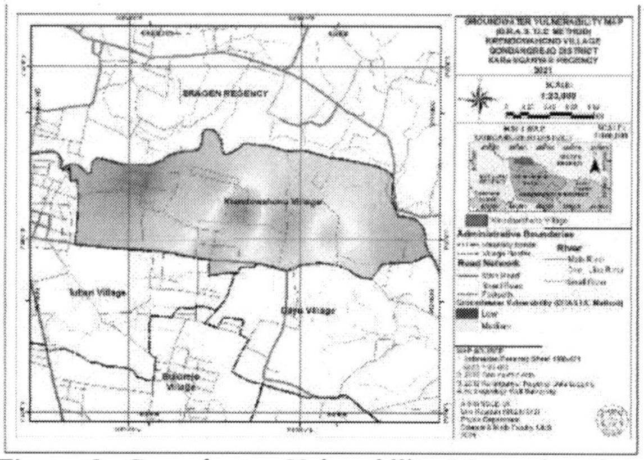

Figure 3. Groundwater Vulnerability Map using the RASTIC Method in Krendowahono Village

3.2. GOD Method Vulnerability Map

The assessment of the level of groundwater vulnerability using the GOD method depends on the parameters of the aquifer type, the type of lithological at the upper layer of the aquifer, and the value of the depth of the groundwater table. Each parameter has a rating value that is determined according to its conditions. The individual parameter maps of the GOD method are presented in Figure 4.

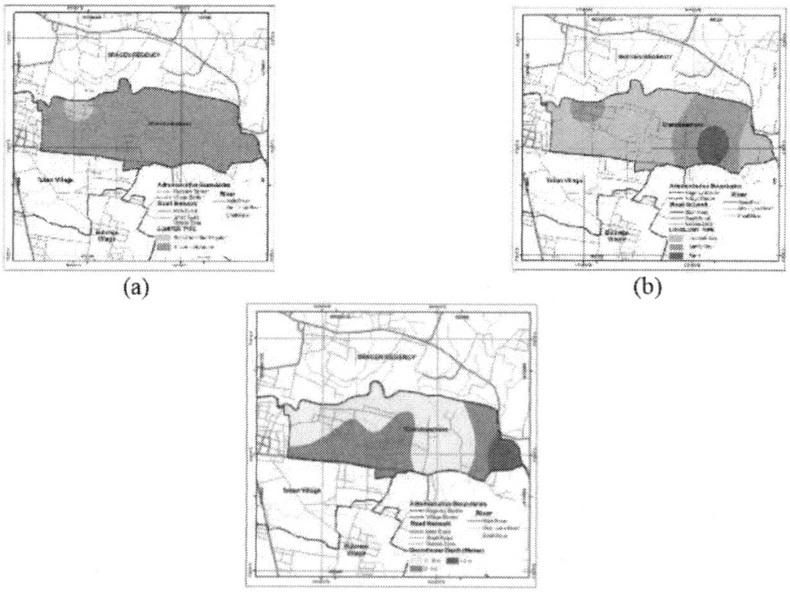

(a) (b)

(c)

Figure 4. Individual Parameter Map (a) Aquifer type (b) Overlying lithology (c) Depth to water table

The individual map results are used to define the GOD vulnerability map. In the results of the GOD method, there are three classes of vulnerability, such as high vulnerability (29.38% of the research area), moderate (65.29% of the research area), and low (5.32% of the research area), presented in Figure 5. The most influential parameter on the value of the GOD vulnerability index is the aquifer type parameter, then followed by the groundwater table depth parameter, and the lithology type at the upper layer of the aquifer.

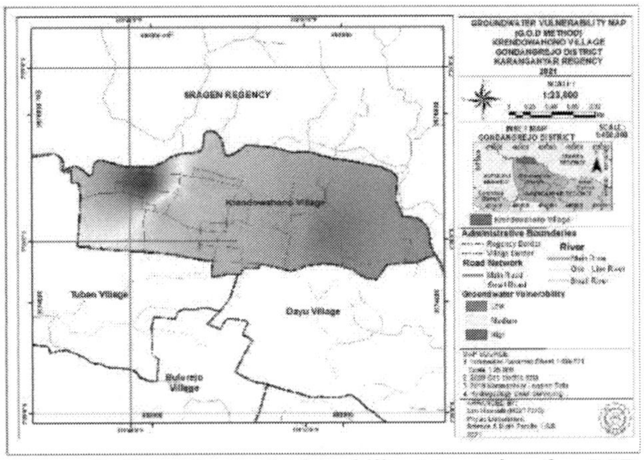

Figure 5. Groundwater Vulnerability Map using the GOD Method in Krendowahono Village

3.3. Comparison of the Vulnerability Results of DRASTIC and GOD Method

Comparison of the vulnerability area of the two methods are shown in Table 1. The results of the vulnerability of the DRASTIC method show two vulnerability classes and the resulting index value is not very significant. This happens due to the research area is carried out with village coverage so that it produces less variable rainfall parameter conditions. At research point locations 1 and 2, the vulnerability assessment using the GOD method resulted in a high level of vulnerability. However, the vulnerability assessment using the DRASTIC method resulted in a moderate level of vulnerability. This happens due to the DRASTIC method considers topographic parameters where the higher the slope value, the lower the vulnerability value. While the GOD method does not consider these parameters.

Table 1. Comparison of the results area of vulnerability class

Vulnerability Level	Result of the DRASTIC Method		Result of the GOD Method	
High	0 Ha	0 %	116.08 Ha	29.38 %
Moderate	393.68 Ha	99.64 %	257.99 Ha	65.30 %
Low	1.42 Ha	0.36 %	21.03 Ha	5.32 %

3.4. Validation with Water Quality Test

Validation of the results of the groundwater vulnerability assessment was carried out to determine the accuracy and suitability of the DRASTIC and GOD methods with field conditions. Validation of the results is done by comparing the quality standards that have been set with the results of the water sample content test. The parameters used in testing water samples are the amount of solute (TDS), pH, salinity, NH_3, and manganese (Mn). Groundwater vulnerability validation was carried out using Spearman's rho correlation analysis. The results of the groundwater sample test are shown in Table 2, and the correlation of the water sample test with vulnerability assessment result was carried out by correlation analysis as presented in Table 3.

Table 2. Results of the test of water samples from resident wells

Parameter	Unit	Point 1	Point 2	Point 4	Point 5	Point 6	Point 7	Point 8
Total Dissolved Solid (TDS)	mg/L	840	612	126	512	349	462	370
pH	-	7	7.2	7.4	6.9	7.3	6.9	6.8
Salinity	$^0/_{00}$	0.844	0.608	0.119	0.505	0.338	0.442	0.36
NH_3	mg/L	0.0112	0.0146	<0.0062	<0.0062	<0.0062	<0.0062	<0.0062
Manganese (Mn)*	mg/L	1.308	<0.0066	0.1285	<0.0066	<0.0066	0.0118	<0.0066

Table 3. Correlation coefficient value

Parameter	Coefficient Values GOD method	Coefficient Values DRASTIC method
Total Dissolved Solid (TDS)	0.714	-0.214
pH	0.330	0.509
Salinity	0.714	-0.214
NH_3	0.857	0.107
Manganese (Mn)*	0.446	0.143

Based on Table 3, the highest correlation coefficient value is the correlation value between the results of the GOD method and the water sample test with the parameters of NH_3, Total Dissolved Solid (TDS), and salinity. The correlation coefficient value in the DRASTIC method occupies the highest value with the pH parameter. Table 3 shows that the GOD method shows better results for regional studies, as soil zoning. This is evidenced by the strong correlation between the results of the GOD susceptibility and the results of the water sample test.

4. Conclusion

The results of the DRASTIC method show that Krendowahono Village has two classes of groundwater vulnerability, such as moderate and low vulnerability. While the results of the GOD method, show three classes of vulnerabilities, such as high, moderate, and low vulnerabilities. Validation of the results of the groundwater vulnerability assessment was carried out to determine the accuracy and suitability of the DRASTIC and GOD methods with field conditions. The highest correlation coefficient value is the correlation value between the results of the GOD method and the water sample test with the parameters of NH_3, Total Dissolved Solid (TDS), and salinity.

Acknowledgment

The author thanked the Geophysical research group of Sebelas Maret University for all forms of support provided so that this research can run smoothly. Thanks are also given to Mr. Sorja Koesuma as the supervisor who has helped the process of writing this proceeding.

IOP Conf. Series: Earth and Environmental Science **989** (2022) 012002 doi:10.1088/1755-1315/989/1/012002

References

[1] Endro 2021 Maret 25 Air Bercampur Gas di Gondangrejo Karanganyar Tidak Layak Konsumsi taken from https://jateng.inews.id/berita/kata-ahli-air-bercampur-gas-di-gondangrejo-karanganyar -tidak-layak-konsumsi (accessed on 27 March 2021)

[2] Mfonka Z, Ndam Ngoupayou J R, Ndjigui P D, Kpoumie A, Zammouri M, Ngouh A N, Mouncherou O F, Rakotondrabe F, and Rasolomanana E H 2018 A GIS-based DRASTIC and GOD models for assessing alterites aquifer of three experimental watersheds in Foumban (Western-Cameroon) *Groundwater for Sustainable Development* vol 7 pp 250–264

[3] Djoudi S, Boulabiez F, Pistre S, and Houha B 2019 Assessing Groundwater Vulnerability to Contamination in A Semi-Arid Environment Using DRASTIC And GOD Models Case Of F'kirina Plain North Of Algeria *IOSR Journal of Environmental Science* vol 13(7) pp 39–44

[4] Sukardi, and Budhitrisna T 1992 Peta Geologi Skala 1:100.000 Lembar Salatiga, Jawa Tengah Bandung : Pusat Penelitian dan Pengembangan Geologi

[5] Hamza S M, Ahsan A, Imteaz M A, Rahman A, Mohammad T A, and Ghazali A H 2015 Accomplishment and subjectivity of GIS-based DRASTIC groundwater vulnerability assessment method: a review *Environmental Earth Sciences* vol 73(7) pp 3063–3076

[6] Wachniew P, Zurek A J, Stumpp C, Gemitzi A, Gargini A, Filippini M, and Witczak S 2016 Toward operational methods for the assessment of intrinsic groundwater vulnerability: A review *Critical Reviews in Environmental Science and Technology* vol 46(9) pp 827-884

[7] Hamamin D F, Qadir R A, Ali S S, and Bosch A P 2018 Hazard and risk intensity maps for water-bearing units: a case study *International Journal of Environmental Science and Technology* vol 15(1) pp 173–184

[8] Sugianti K, Mulyadi D, and Maria R 2016 Analisis Kerentanan Pencemaran Airtanah dengan Pendekatan Metode DRASTIC di Bandung Selatan *Jurnal Lingkungan Dan Bencana Geologi* vol 7(1) pp 19–33

[9] Ahirwar S, and Shukla J P 2018 Assessment of Groundwater Vulnerability in Upper Betwa River Watershed using GIS based DRASTIC Model *Journal of the Geological Society of India* vol 91(3) pp 334–340

[10] Mohammad A H 2017 Assessing the groundwater vulnerability in the upper aquifers of Zarqa River Basin, Jordan using DRASTIC, SINTACS and GOD methods *International Journal of Water Resources and Environmental Engineering* vol 9(2) pp 44–53

[11] Putranto T T, Widiarso D A, and Yuslihanu F 2016 Studi Kerentanan Air Tanah Terhadap Kontaminan Menggunakan Metode Drastic di Kota Pekalongan *Teknik* vol 37(1) p 26

[12] Hastuti D, Yulianto T, and Putranto T T 2016 Analisis Kerentanan Airtanah Terhadap Pencemaran Di Dataran Alluvial Kota Semarang Menggunakan Metode God Dengan Memanfaatkan Data Resistivitas Dan Data Hidrogeologi *Youngster Physics Journal* vol 5(4) pp 277–290

[13] Gogu R C and Dassargues A 2000 Current trends and future challenges in groundwater vulnerability assessment using overlay and index methods. *Environmental geology* vol 39(6) pp 549-559

[14] Rukmana B T S, Bargawa W S and Cahyadi T A 2020 Assessment of Groundwater Vulnerability Using GOD Method In *IOP Conference Series: Earth and Environmental Science* vol 477(1) 012020 IOP Publishing

[15] Putranto T T 2019 Studi Kerentanan Airtanah Terhadap Pencemaran Dengan Menggunakan Metode Drastic Pada Cekungan Airtanah (Cat) Karanganyar-Boyolali, Provinsi Jawa Tengah. *Jurnal Ilmu Lingkungan* vol 17(1) p 159

[16] Siswoyo H 2018 Identifikasi Tingkat Kerentanan Akuifer terhadap Pencemaran di Kecamatan Sumobito Kabupaten Jombang dengan Menggunakan Metode GOD *Jurnal Sains Dan Edukasi Sains* vol 1(2) pp 1–6

[17] Filik Iscen C, Emiroglu Ö, Ilha S, Arslan N, Yilmaz V and Ahiska S 2008 Application of multivariate statistical techniques in the assessment of surface water quality in Uluabat Lake

IOP Conf. Series: Earth and Environmental Science **989** (2022) 012002 doi:10.1088/1755-1315/989/1/012002

Turkey *Environmental Monitoring and Assessment* vol 144(1–3) pp 269–276

[18] Maliva R G 2020 *Anthropogenic Aquifer Recharge* (Issue 5)

[19] Koesuma, S., Sutarno, D., & Ramelan, A. H. (2021, February). Groundwater level in the southern part of surakarta city depicted from vertical electrical sounding. In *Journal of Physics: Conference Series* (Vol. 1825, No. 1, p. 012008). IOP Publishing.

[20] Koesuma, S. 2020. *Metode Geolistrik. Aplikasi Pada Struktur Bawah Permukaan*, Program Studi Fisika FMIPA Universitas Sebelas Maret. ISBN: 978-602-99344-4-1.

[21] Khusna, I. A., & Koesuma, S. 2021. Determination of Groundwater Potential in Pulutan Wetan Village Wuryantoro District Wonogiri Using Vertical Electrical Sounding Method. In *Journal of Physics: Conference Series* (Vol. 1825, No. 1, p. 012011). IOP Publishing.

[22] Marfuatik, L., Koesuma, S., & Legowo, B. (2018, March). Identification of aquifer potential in Karanganyar city by using vertical electrical sounding method. In *Journal of Physics: Conference Series* (Vol. 983, No. 1, p. 012043). IOP Publishing.

Identification of aquifer zone in the southern part of Sukoharjo Regency using VES method as a means of irrigation of agricultural land

Rahma Maulana

Department of Physics, Faculty of Mathematics and Natural Sciences, Sebelas Maret University
Jl. Ir. Sutami 36A Kentingan Jebres Surakarta 57126. INDONESIA

*Email: rahmamaulana12@student.uns.ac.id

Abstract. Sukoharjo is one of the regencies in Indonesia that is famous for its agricultural land. Sukoharjo itself has 12 sub districts spread evenly throughout. In recent years, there have been many conflicts of irrigation water shortages for agricultural land. Therefore, research was conducted on the aquifer zone of the southern district of Sukoharjo. The research was conducted by Vertical Electrical Sounding method to determine the structure of soil in order to improve the efficiency of deep groundwater drilling for irrigation. The study used Schlumberger's configuration with a target zone of deep aquifers for agricultural irrigation. The length of the track used is 275-300 meters. As a result of the study, it is suspected that the aquifer zone in southern Sukoharjo Regency has a depth of about 11.54-83.93 meters below the subsurface. The soil structure in south of Sukoharjo is dominated by sandy clay layers, limestone layers, and sedimentary layers.

1. Introduction

Sukoharjo regency is one of the districts in Indonesia that is famous for its agricultural land. Sukoharjo regency itself has 12 sub-districts that are spread evenly throughout the region. Sukoharjo Regency has an area of 466.66km^2 with a population of 907,587 people in 2020. In terms of meeting the needs of clean water, the people of Sukoharjo mostly utilize shallow groundwater resources. Shallow groundwater is above the first layer that is watertight usually not too deep underground surface. Groundwater (GW) occurs due to the absorption of water from the soil [2].

In recent years, there has been a shortage of irrigation water for agricultural land. This is probably due to the long dry season in the south of Sukoharjo. Some farmers use shallow groundwater as a source of irrigation for their rice fields, but that leaves the population short of water resources for daily needs. Based on these problems, the author conducted a study to find out the depth of groundwater in southern Sukoharjo Regency as a solution to irrigation facilities for agricultural land.

2. Methods

Geoelectricity is one of the geophysical methods used in the exploration of the subsurface. The working principle of the geoelectric method is to study the flow of electricity within the earth and how to detect it on the earth's surface. The first geoelectric method was carried out by Conrad Schlumberger in 1912. The geoelectric method is one of the branches of geophysical science that studies the earth and its environment based on the electrical properties of rocks. These properties are resistivity of type, conductivity, dielectric constant, ability to give rise to one's own electric potential, electric current injected into the earth through two current electrodes and the resulting potential distribution is measured by potential electrodes. The advantage of the geoelectric method is that it does not damage the environment, and is also able to detect up to a depth of several meters according to the length of the trajectory on data retrieval in the field.

The geoelectric method is performed by injecting an electric current into the earth through two current electrodes, then measuring the voltage value from within the earth through two potential different electrodes. The geoelectric method is intended to obtain an overview of the subsurface layers of soil and the possible inclusion of water and minerals at certain depths.[8].

Ohm's law states that the magnitude of the electric current I flowing through a deliverer is always directly proportional to the difference in potential V applied to it [11].

$$J = \sigma E \quad (1)$$

From equation 1 it is lowered to:

$$V = I \, . R \quad (2)$$

$$\rho \approx \frac{\pi s^2}{2b}\left(\frac{\Delta V}{I}\right) \quad (3)$$

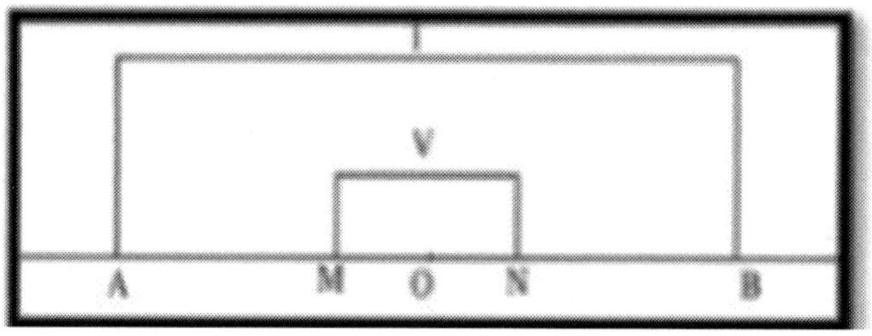

Figure 1. Schlumberger Configuration

The Schlumberger configuration is a four-electrode configuration in which there is a pair of current electrodes namely C1-C2 or A-B and a pair of P1-P2 or M-N potential electrodes. In measurements this configuration is usually often changed at the distance between the current electrodes, and sometimes the potential electrodes remain. In this configuration, it can be used for resistivity mapping and sounding, this configuration is excellent for VES (Vertical Electrical Sounding) and not suitable for CST (constant separation traversing). Has good orientation sensitivity, good lateral sensitivity in the determination of in homogeneity [9].

3. Results and Discussions

Resistivity measurements are carried out with 6 different research location points with the 1-dimensional method (Figure 2). The 1-dimensional method uses the Schlumberger configuration, the MN potential electrode is always placed between two AB current electrodes (Figure 1). This research aims to find out the depth of groundwater in southern Sukoharjo Regency using the Vertical Electrical Sounding (VES) method. Resistivity measurements use OYO resistivitymeter. The length of the stretch of measurement tool is 250- 300 meters long, with a target depth reached as far as 120 meters from ground level. This research was conducted at 6 different measurement points in the South of Sukoharjo Regency namely Kedu, Watu, Tawa, Kara, Kanyar, and Bulu, Sukoharjo Regency, Central Java.

IOP Conf. Series: Earth and Environmental Science **989** (2022) 012003 doi:10.1088/1755-1315/989/1/012003

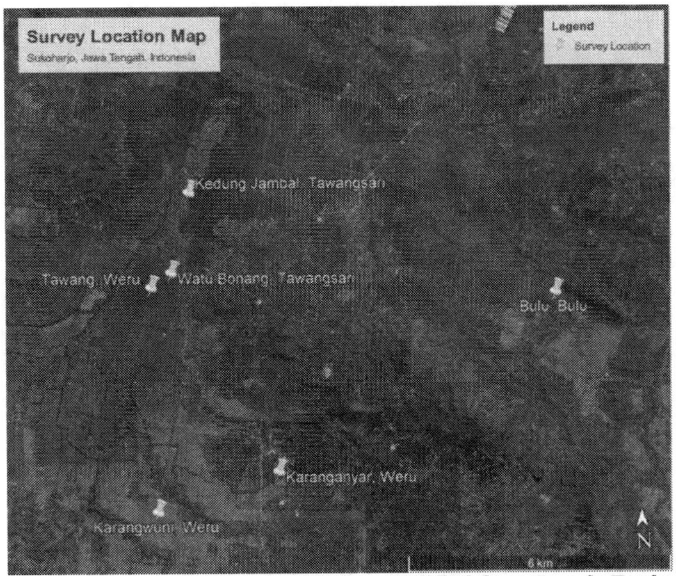

Figure 2. VES Survey Location, modified from Google Earth

The value of resistivity can actually be obtained from the results of pseudo-resistivity inversion using curve matching method by comparing measurement data and model data to get smallest errors. The result of inversion can be used as a reference to determine the aquifer zone that exists below ground level. The log data obtained is:

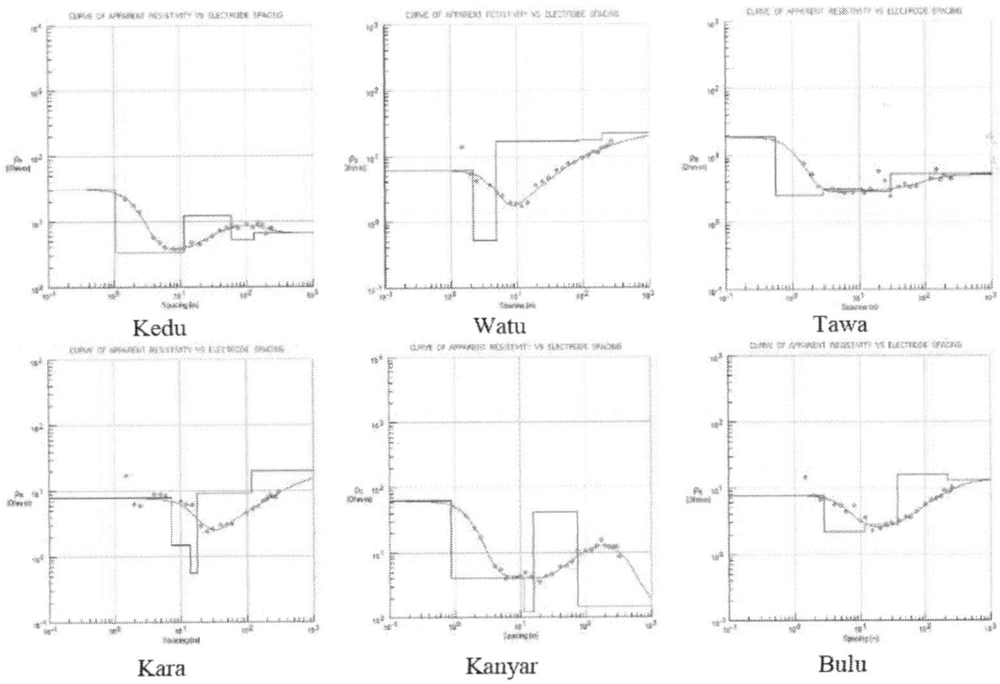

Figure 3. Log Graph of True Resistivity versus Electrode Spacing

The results of pseudo-resistivity inversion are then matched with the rock resistivity table at the measurement site so that it can be indicated as an aquifer zone. Literature on the value of resistivity types can be seen in table 1 [6].

Table 1. Rock Resistivity

Material	Resistivity (Ωm)
Air	~
Pyrite	3 × 10−1
Galana	2 × 10−3
Kwarsa	4 × 1010 to 2 × 1014
Calcite	1 × 1012 to 1 × 1013
Salt Rocks	30 to 1 × 1013
Mica	9 × 1012 to 1 × 1014
Granite	102 to 1 × 106
Gabro	1 × 103 to 1 × 106
Basalt	10 to 1 × 107
Limestone Rocks	50 to 1 × 107
Sandstone	1 to 1 × 108
Shale Rocks	20 to 1 × 103
Dolomite	102 to 104
Sand	1 to 103
Clay	1 to 102
Groundwater	0.5 to 3 × 102
Sea water	0.2

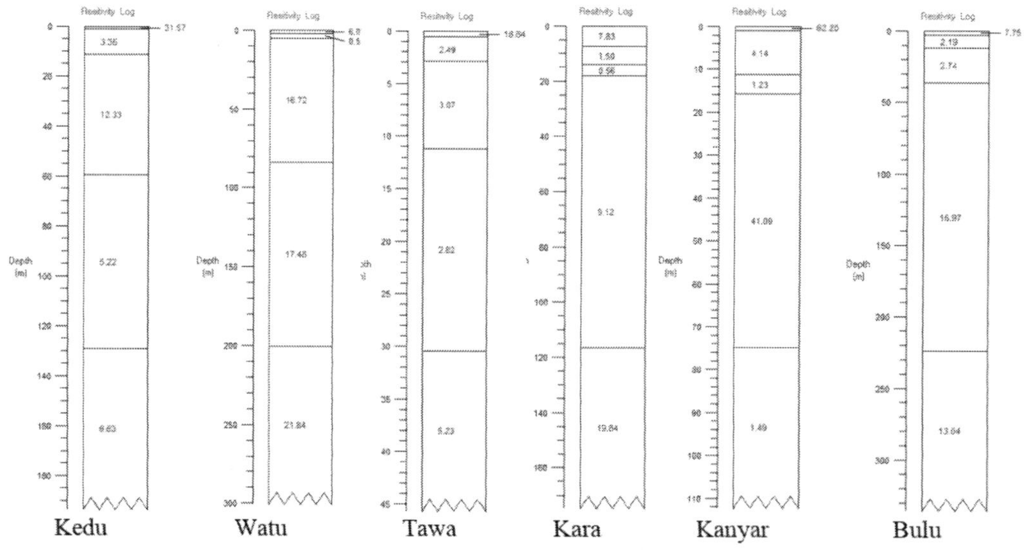

Figure 4. Log Resistivity inversion versus Depth

Based on the results of the inversion of pseudo-resistivity obtained log data in Figure 4 that can be interpreted with table 1 value resistivity of the rock layer. The interpretation results can be used as a reference for determining the aquifer zone below ground level. When measuring also conducted interviews to the community around the measurement site to find out the depth of the water well used. This is done to determine the aquifer zone around the measurement site. Interpretation results show that the rock layer located south of Sukoharjo consists of top soil, sand, gravel sand, sandy clay, clay. The clay layer spreads from Kedu to Bulu shown by Figure 4.

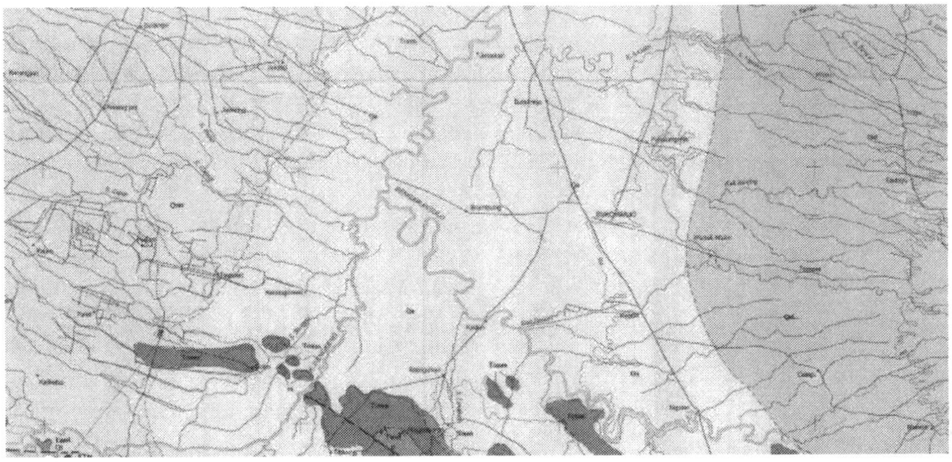

Figure 5. Regional Geology of Sukoharjo Regency, modified from [15]

Regional geological formations at the site of the feed can be seen in Figure 5. The southern part of sukoharjo regency consists of samabipitu formation (Tmss) consisting of sandstone and clay and Mandalika formation (Tomm) consisting of lava dasit-andesite and tuff dasit with diorite hack. Geomorphology measurement area south of Sukoharjo is generally a rice field and residential area of the population. In addition, it is also bordered by Wonogiri Regency and Gunung Kidul Regency which is famous for karst rock formations.

Table 2. Indication of the depth of ground water face from the surface

No.	Sites	Depth of GW
1	Kedu	59.43 meters
2	Watu	83.93 meters
3	Tawa	26.54 meters
4	Kara	22.42 meters
5	Kanyar	74.88 meters
6	Bulu	11.54 meters

The results of measurements obtained based on the table of indications of groundwater level (table 2) show different depths. In Kedu, the ground water level is at a depth of 59.43 meters. In Watu the ground water level is at a depth of 83.93 meters. Then in Tawa the ground water level is at a depth of 26.54 meters. Then in Kara the ground water level is at a depth of 22.42 meters. In Kanyar the ground water level is at a depth of 74.88 meters. While in Bulu the ground water level is at a depth of 11.54 meters. Aquifer zones are scattered due to the influence of geological conditions at each different point.

4. Conclusion

Based on the results of measurements using Vertical Electrical Sounding to identify the aquifer zone in the south of Sukoharjo Regency, it can be concluded that the aquifer zone is located at 11.54 - 83.93 meters below ground level with a rock layer structure in the form of top soil, sand, gravel sand, sandy clay, and clay that is evenly spread across all measurement locations.

Acknowledgment
The author thanked the Geophysical research group of Sebelas Maret University for all forms of support provided so that this research can run smoothly. Thanks are also given to Mr. Sorja Koesuma as the supervisor who has helped the process of writing this proceeding.

References

[1] BPS Statistics of Sukoharjo Regency, Sukoharjo Regency In Figures 2021, Sukoharjo Regency ISSN: 2715-2847: BPS-Statistics of Sukoharjo Regency, 2021.

[2] H. H. A. Matin, S. Rachmawati, L. Kusumaningrum, I. Y. Septiariva and Purwono, "Analysis of Shallow Groundwater Quality in Joho Village,," *International Journal of Engineering Education,* vol. 2, no. 2, pp. 99-102, 2020.

[3] B. Sadjab, As'ari and T. Adey, "Pemetaan Akuifer Air Tanah di Kecamatan Prambanan Kabupaten Sleman Daerah Istimewa Yogyakarta Dengan Metode Geolistrik Tahanan Jenis," *Jurnal MIPA,* vol. 1, no. 1, pp. 37-44, UNSRAT Online.

[4] Sedana, D., As'ari & Tanauma, A., 2015. Pemetaan Akuifer Air Tanah di Jalan Ringroad Kelurahan Malendeng Dengan Menggunakan Metode Geolistrik Tahanan Jenis. *Jurnal Ilmiah Sains,* XV(2), pp. 33-38.

[5] W. Lowrie, Fundamentals of Geophysics Second Edition, Cambridge: Cambridge University Press, 2007.

[6] L. Marfuatik, S. Koesuma, B. Legowo and Darsono, "Identification of aquifer potential in Karanganyar City by using vertical electrical sounding method," *Journal of Physics: Conference Series,* **983**(1), pp. 1-7, 2018.

[7] Febriana, R. K. N., Minarto, E. & Tryono, F. Y., 2017. Identifikasi Sebaran Aliran Air Bawah Tanah (Groundwater) dengan Metode Vertical Electrical Sounding (VES) Konfigurasi Schlumberger di Wilayah Cepu, Blora Jawa Tengah. *Jurnal Sains dan Seni ITS,* VI(2), pp. 29-33.

[8] P. K. Bhattacaharya and H. P. Patra, Direct Current Geolectric Sounding, Amsterdam: Elsevier Publishing Company, 1968.

[9] F. Waskito, T. Yulianto and D. J. Suprapto, "Analisis Rembesan Pada Bendungan Cengklik Menggunakan Metode Resistivitas Konfigurasi Schlumberger Dan Uji Permeabilitas Untuk Menghitung Debit Rembesan," *Youngster Physics Journal,* vol. 5, no. 4, pp. 425-432, 2016.

[10] Brahmantyo, A. & Yulianto, T., 2014. Identifikasi Bidang Gelincir Pemicu Tanah Longsor Dengan Metode Resitivitas 2 Dimensi Di Desa Trangkil Sejahtera Kecamatan Gunungpati Semarang. *Youngster Physics Journal,* III(2), pp. 83-96.

[11] S. Koesuma and Sulastoro, "Identifying of ground water level by using geoelectric method in Karanganyar, Central Java, Indonesia," *Journal of Physics: Conference Series* **776**, pp. 1-6, 2016.

[12] Y. Yuniardi, Hendarmawan, Abdurrokhim, . V. Isnaniawardhani, F. Mohammad, M. K. Alfadli and P. Ridwan, "Pendugaan Akifer Airtanah Dengan Metode Geolistrik Konfigurasi Schlumberger Di Lereng Utara Gunungapi Tangkubanparahu," *Jurnal RISET Geologi dan Pertambangan,* vol. 29, no. 2, pp. 239-253, 2019.

[13] F. W. Maulana, "Potensi Akuifer Daerah Desa Karangmojo Kecamatan Weru Kabupaten Sukoharjo Propinsi Jawa Tengah Berdasarkan Data Geolistrik," *Jurnal Teknologi Technoscientia,* vol. 6, no. 1, pp. 55-64, 2013.

[14] K. Nisa, T. Yulianto and S. Widada, "Aplikasi Metode Geolistrik Tahanan Jenis Untuk Menentukan Zona Intrusi Air Laut Di Kecamatan Genuk Semarang," *Berkala Fisika,* vol. 15, pp. 7-14, Januari 2012.

[15] Surono, B. Toha and I. Sudarno, "Peta Geologi Lembar Surakarta, Giritontro, Jawa," Pusat Penelitian dan Pengembangan Geologi, 1992.

[16] Telford, . M. W., Geldart, L. P., Sheriff, R. E. & Key, 1976. *Applied Geophysics.* New York: Cambridge University Press.

IOP Conf. Series: Earth and Environmental Science **989** (2022) 012004 doi:10.1088/1755-1315/989/1/012004

Lesson learned and enhanced adaptation strategies for isolated indigenous community in West Papua

Lazarus Indouw [1,*], Rudi A Maturbongs[2], and Saraswati Prabawardani[3]

[1] Environmental Science of Graduate Program, Universitas Papua, West Papua, 98314, Indonesia
[2] Faculty of Forestry and Graduate Program, Universitas Papua, West Papua, 98314, Indonesia
[3] Faculty of Agriculture and Graduate Program, Universitas Papua, West Papua, 98314, Indonesia

[*]Email: lazarus.indow@gmail.com

Abstract. Isolated indigenous communities are groups of people who live their lives very apprehensively. Most of the people live in places that are difficult to reach geographically, such as inland, beaches, forests, hills and remote islands (outer, inland, and left behind). Remoteness is the biggest contributing factor to their inability to be part of the development process and coupled with climate-related disasters. Descriptive methods with in-depth interviews and Focus Group Discussion (FGD) are used to explore local knowledge in efforts to adapt to climate change and the introduction of adaptation strategies, especially related to enhance local knowledge (SASI).

1. Introduction

Isolated indigenous community as part of remote community groups, both physically and geographically, as well as socio-culturally and their lives still adhere to local wisdom. Most of these communities live in isolated and difficult to reach areas [1,2]. So that their livelihoods depend on local natural resources with simple technology [3], a subsistence economy and limited access to basic social services such as educational infrastructure, places of worship, health services, economic and housing that cannot be properly equipped [4,5,6,7,8].

The typology of indigenous peoples in Indonesia based on social, cultural and economic aspects is divided into 4 categories, namely hunting and gathering social communities, shifting cultivation social communities, rural social communities and urban social communities [9]. West Papua is included in the type of shifting cultivation social community which is also found in inland forests in Papua, where isolated indigenous communities generally live from shifting cultivation, have an orbital area of life to earn a living, have a more varied understanding of technology, and can interact with other people. the outside world, although still constrained by geographical disparities [10].

To utilize natural resources in an isolated indigenous community, they remain based on environmental management based on local wisdom which is the custom and habit of a group of traditional and hereditary customary law communities, known as Sasi. Sasi is local knowledge to maintain forest conservation, food security, climate change and disasters for indigenous peoples found in river, sea, coastal, and forest areas [11,12,13,14,15].

Along with economic demands, the entry of foreign cultures, the shift from agriculture to non-agriculture (oil palm plantations), the shift of local food to rice, expansion of infrastructure, technology, and immigrants have weakened Sasi in the isolated indigenous community [16,17,18]. Besides that, Sasi is generally held by the traditional elders and is still verbal and secret and also the younger generation has a different perception so that it is rarely passed down to the next generation. This can lead to degradation and the threat of loss of local knowledge.

Efforts to preserve local knowledge of Sasi are urgently needed, especially in increasing the resilience of the isolated indigenous community to climate change and climate-related disasters which have increased in recent times. This study would like to enhance adaptation strategies through the development of local knowledge into long-term sustainable development to increase the prosperity of the isolated indigenous community. We hope this research can contribute to the resilient isolated community by focussing on adaptation strategies through enhancing the Sasi in West Papua Province.

2. Methods

2.1. Concept development

The method used in this study is a descriptive method with direct observation techniques, interviews and Focus Group Discussions to gain respondent's profile, perception and participation of the isolated indigenous community in the following areas (Figure 1) :

1. Yamboi Village, South Manokwari Regency in the lowland area;
2. Sakumi Village, Arfak Mountain Regency in the highland area; and
3. Watitindau Village, Teluk Wondama Regency in a coastal area

from efforts to maintain, preserve and develop adaptation strategies for climate change and climate-related disasters.

This study involved 30 families of each village who were selected by non-proportional stratified random sampling from those living with the implementation of Sasi in the lowland, highland and coastal areas.

Primary data was procured from direct interviews with respondents using a questionnaire. The dependent variables studied were the lesson learned and enhanced of Sasi while the independent variables include family characteristics, perceptions, and participations. Family characteristics measured included the age of the husband, age of the wife, length of education of the husband, length of education of the wife, number of family members, and income per capita of the family.

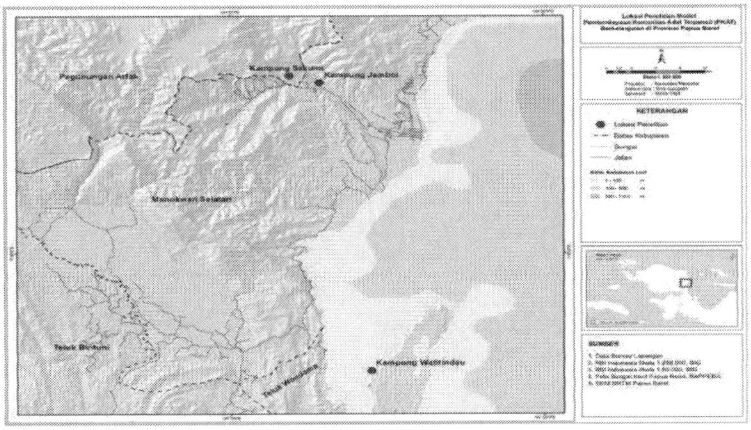

Figure 1. Research site of the isolated indigenous community

The family characteristics, perceptions, and participations data used in the isolated indigenous community in West Papua are shown in Table 1.

Table 1. Respondent's profile, perceptions and participations data.

No	Family Characteristics	Perceptions	Participations
1	Age of husband	Elements of culture and conservation education	Planning
2	Age of wife	Preventing coastal and isolated damaged area	Implementation
3	Education of husband	Knowledge the boundaries of Sasi area	Maintenance
4	Education of wife	Damage due to human factors	Protection and utilization
5	Number of family members	Damage due to natural factors and climate change	Society participation
6	Income per capita of the family	Benefits of forest and coastal area	Involvement of the parties
7		Management responsibilities	
8		Legal enforcement	
9		Isolated indigenous community program	

The Likert scale is used to measure attitudes, opinions, and perceptions of a person or group of people about social phenomena. In this study, the scores used for respondents' responses were from 1 to 5 with the following criteria [19]:

a. Strongly Agree (SA), given a score of 5, 80%-100%.
b. Agree (A), given a score of 4, 60-79.99%.
c. Indecisive (I), scored 40%-59.99%.
d. Disagree (DA), given a score of 20%-39.99%.
e. Strongly Disagree (SDA), given a score of 1, 0%-19.99%.

Index (%) = (Total Score/ Maximum Score) x 100

2.2. Principal Component Analysis

Respondent's profile, perceptions, and participations data will be analysed with the quantitative approach using the Principal Component Analysis (PCA) [20].

3. Results and discussion

3.1. Family Characteristics

Family characteristics data are presented in Table 2 with the average family from the research site up to 5-10 members.

Table 2. Minimum, maximum, and mean values of the family characteristics

Family Characteristics (unit)	(Min-Max; Average)
Age of husband (year)	37.0 – 60.0; 44.0
Age of wife (Year)	35.0 – 55.0; 39.0
Education of husband (year)	0.0 – 12.0; 6.0
Education of wife (year)	0.0 – 12.0; 6.0
Number of family member (people)	5,0 – 10.0; 7.0
Income/capita/month (thousand Rp)	114.3-621.4; 407.1
Income/capita/month (thousand Rp) during climate-related disaster	78.6-460.7;278.6

IOP Conf. Series: Earth and Environmental Science **989** (2022) 012004 doi:10.1088/1755-1315/989/1/012004

The average value of age for both, husband and wife were in middle adulthood with the age range included in the labour force of 35-39 and 40-44 [21]. As for the education of the respondents, in general, they only graduated from elementary school and didn't perform the standard of the 12-year compulsory education program from the Central Government. The average income per capita per month of the family is 407.1 thousand rupiahs and fall into 278.6 thousand rupiahs during climate-related disaster conditions such as strong winds, heavy rains, high tides due, flood and drought which generally occurred a maximum of up 2 weeks. For conditions near the coastal area, the fishermen carry out mitigation and adaptation efforts by planting in their gardens with local food crops for food security through shifting cultivation of agroforestry systems. Meanwhile, lowland and highland areas are generally waiting for further advanced conditions that they can return to their gardens.

3.2. Perceptions

The distribution of perceptions data about a lesson learned of Sasi can be seen in Figure 2 and Index (%) in Figure 3. 90 respondents from family data of the three study areas were spread from traditional leaders, village officials and communities involved in maintaining Sasi as well as traditional isolated indigenous councils in making decisions related to strengthening Sasi enforcement.

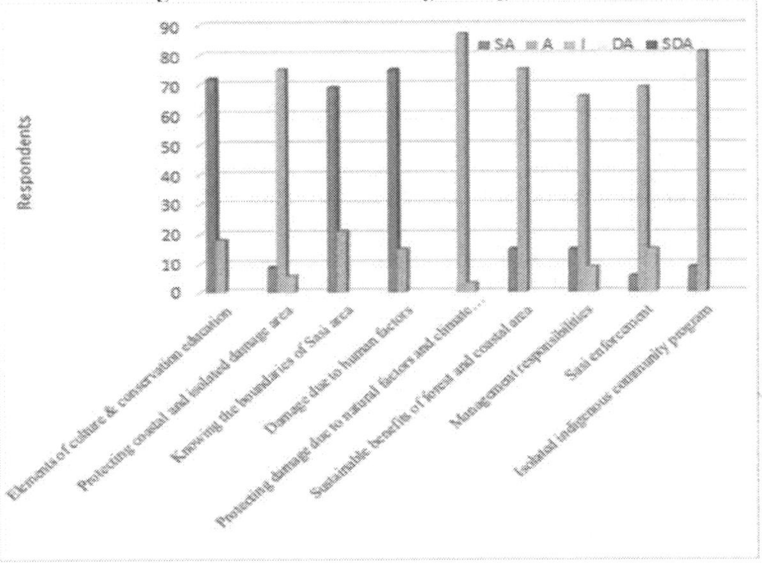

Figure 2. Respondent's perception of Sasi in West Papua.

Figure 2 shows that the strongly agree (SA) of perceptions variables consist of elements of culture and conservation education, knowledge of the boundaries of Sasi area and damage due to human factors. Meanwhile, the other variable comes from the agree (A) perceptions criteria. Then, indecisive (I) criteria are also founded from protecting coastal and isolated damaged area, protecting damage due to natural factors and climate change, management responsibilities and Sasi enforcement. This is missed by Sasi due to the damage area from natural disasters and climate change which have increased in the last decade and there are also Sasi violations that are not honest and reported.

Figure 3. Perception's index of Sasi in West Papua.

Figure 3 represents that the perceptions index of the isolated indigenous community in West Papua with an average index of 86% which is categorized as very high. Therefore, Sasi is still a guideline for an isolated indigenous community in maintaining, protecting, and conserving the natural resources that they can be linked with the vision of Sustainable Development in Papua Region.

3.3 Participations

Data from participations variable are represented in Figure 4 and participations index in Figure 5. Data variables were measured in 7 components namely, planning, implementation, maintenance, protection and utilization, society participations, involvement of the parties and youth generation. The data distribution in Figure 4 expresses the strongly agree (SA) and agree (A) criteria then followed by indecisive (I), disagree (DA) and strongly disagree (SDA) on the involvement of parties and youth generation variables.

Based on the participations index (Figure 5), the involvement of the parties and youth generation has a low score of 39% and 33%, respectively. This value is related to the low involvement of local government organizations (only Public Works and Social Services) in isolated indigenous community empowerment activities that support lessons learned from Sasi. Another problem is the low participation of the younger generation as the successor to the Sasi rule which is only verbal and secret. The young man thought that the Sasi was incompatible with today's technological developments. And also coupled with the pressure of external influences, the economic and land-use change resulted in the degradation of Sasi. Therefore, researchers have promoted the Sasi plus by providing input on climate change mitigation and adaptation efforts as well as disaster risk reduction efforts so that isolated indigenous communities can be resilient in dealing with disasters and the young generation is enthusiastic to continue their ancestral heritage. [12,13,14,15]

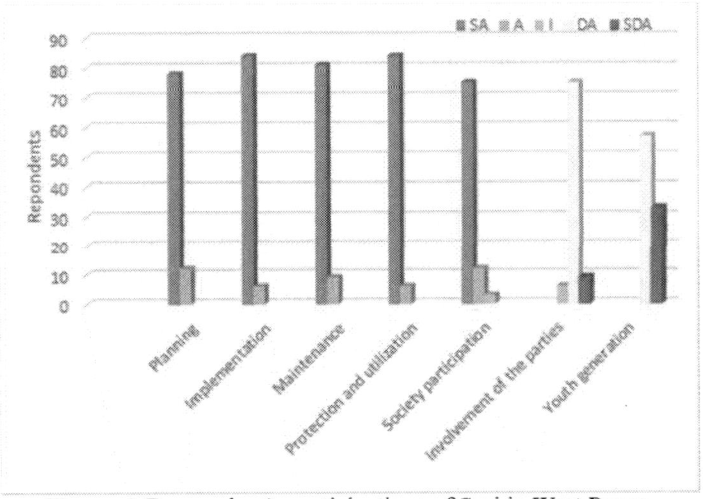

Figure 4. Respondent's participations of Sasi in West Papua.

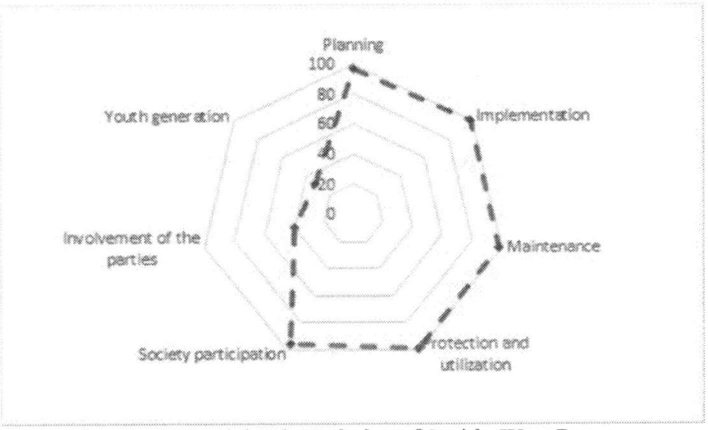

Figure 5. Participation's index of Sasi in West Papua.

3.4. Principal Component Analysis

Furthermore, the family characteristic, perceptions and participations variable are examined using the Principal Component Analysis (PCA). Based on these data, the cumulative value is obtained with a good distribution of data obtained on PC 10 with the following equation:

PC 10 = -0.322 Age of husband – 0.002 Age of wife – 0.062 Education of husband + 0.229 Education of wife + 0.136 Number of family + 0.218 Income/capita/month – 0.005 Income/capita/moth during disaster – 0.065 Elements of culture and conservation education – 0.327 Protecting coastal and isolated damage area + 0.116 Knowing the boundaries of Sasi area + 0.074 Damage due to human factors + 0.243 Preventing damage area due to natural disaster and climate change – 0.053 Benefits of forest and coastal area – 0.259 Management responsibility – 0.071 Sasi Enforcement + 0.431 Isolated indigenous community program + 0.009 Planning – 0.297 Implementation + 0.355 Maintenance -0.074 Protection and utilization – 0.229 Society participation - 0.154 Society participation – 0.154 Involvement of the parties + 0.170 Youth generation.

PC 10 has a cumulative value of 0.808 which is categorized as good data [22] and accounts for 3.8% of the variability data. The variability data indicates that the value is < 5-10%, which means that the equation is valid [23].

4. Conclusion

From the characteristic data, it can be seen that the respondent's age is in productive age with an average education of only graduating from elementary school and there is also a decrease in income when a climate-related disaster occurs. Furthermore, it was found that the perception and participation index data on lessons learned of Sasi was still categorized as very high. However, from the variable participation, especially the involvement of the parties and the youth generation, it is necessary to make improvements in efforts to enhance Sasi. So that Sasi can become local knowledge and at the same time be combined with the latest technology so that it can be in line in maintaining the sustainability of natural resources to support Sustainable Development in Papua Region. It is hoped that this Sasi can become an adaptation strategy from the isolated indigenous community to become resilient in climate-related disasters that are currently increasing due to climate change.

Acknowledgements
The authors would like to thank the isolated indigenous community program in West Papua Province that supported financial and technical during the development of this proceeding. Thanks are also given to Mr. Hendri as the supervisor who has helped the process of writing this proceeding.

References

[1] Adimihardja K 2007 Dynamics of local culture (Bandung: CV. Indra Prahasta and Pusat Kajian LBPB)
[2] Government of Indonesia 1991 Social Welfare Development of Remote Indigenous Community (Jakarta) 7 pp.
[3] Aritonang R 2004 Local knowledge of Orang Rimba and implications for hunting and gathering strategies (Jakarta: Ministry of Environment).
[4] Galloway McLean K 2017 Advance guard: climate change impacts, adaptation, mitigation and indigenous peoples: a compendium of case studies (Australia: United Nations University) 129 pp.
[5] Altman J 2018 Welfare reform and Indigeneity in remote Australia: A toxic mix, 22 pp.
[6] Sujarwani, et al. 2018 Community Empowerment of Remote Indigenous Communities by the Lingga Regency Government. Riau islands. *Journal of Anthropology: Socio-Cultural Issues* 20 (**1**) 17-31 Faculty of Social and Political Sciences Andalas University West Sumatra Indonesia
[7] Hayness M, O'Rourke T, Nash D, Baffour B, York S, Chainey C, Betros G and Memmott P 2019 Aboriginal and Torres Strait Islander preferences for healthcare settings: effective use of design images in survey research *Australian Health Review* 44(**2**) 222-22
[8] Danyluk PJ, Burns A, and Scott D 2020 Becoming a teacher in a rural or remote community: The experiences of educational assistants Rural Teacher Education ed Corbett M and Gereluk (Singapore: Springer Nature) pp 183 – 202.
[9] Bappenas 2013 Indigenous Peoples in Indonesia: Towards inclusive social protection (Jakarta: Directorate of Community Welfare and Protection) 144 pp.
[10] Social ministry of the Republic Indonesia 2015 Social Empowerment of Remote Indigenous Communities (Jakarta) 22 pp.
[11] Thornton PK et al 2018 A framework for priority-setting in climate-smart agriculture research *Agricultural System* 167 161-175.
[12] Rogers N, Stephens S and McClintok K 2018 Literature Review of Education and Cultural Identity Programmes for Indigenous Youth (New Zealand: Te Rau Matatini) 20 pp.
[13] Smith P et al 2020 Which practices co-deliver food security, climate change, mitigation, and

adaptation and combat land degradation and desertification *Global Change Biology* 26 (**3**) 1532-1575.
[14] Kakoty S 2018 Ecology, sustainability and traditional wisdom *Cleaner of Production* **172** 3215-3224.
[15] Hasan ASMM 2019 Biogas potential from municipal and agricultural residual biomass for power generation in Hazaribagh, Bangladesh – A strategy to improve the energy system *Renewable Energy Focus* **29** 14-23.
[16] Boli P, Yulianda F, Damar, A, Soedharma D, and Kinseng R 2014 Benefit of Sasi for Conservation of Marine Resources in Raja Ampat Papua *JMHT* XX (2): 131-139.
[17] Persada MPR, Mangunjaya FM and Tobing ISL 2018 Sasi as Natural Conservation Culture Resource in the Maluku Islands *Science and Culture Journal* 41 (59): 6869-6900.
[18] Prasetyo N, Carr A, and Filep S 2019 Indigenous Knowledge in Marine Ecotourism Development: The Case of Sasi Laut, Misool, Indonesia https://doi.org/10.1080/21568316.2019.1604424.
[19] Pranatawijaya VH, Widiatry, Priskila R, Adidyana PB and Putra A 2019 Web-Based Survey Questionnaire Application Development using Likert and Guttman Scales *Journal of Science and Informatics* 5 (2): 128-137.
[20] van Delsen MSN, Wattimena AZ and Saputri SD 2017 Use of Principal Component Analysis Method to Reduce Inflation Factors in Ambon City *Journal of Applied and Mathematical Sciences* 11 (2): 109-118.
[21] BPS 2021 West Papua in 2021's Figures (Manokwari : BPS) 552 pp.
[22] Widarjono A 2010 Applied Multivariate Statistical Analysis (Yogyakarta: UPP STIM YKPN)
[23] Muhammadi, Aminullah, M.E and Soesilo B 2001 Dynamic System Analysis: Environment (Jakarta: UMJ Press)

Agroforestry and climate smart agriculture to improve food security and resilience indigenous people in Teluk Patipi District Fakfak Regency West Papua Province

Arnoldus Tuturop [2,*], Julius D. Nugroho [1], and Avalinus Warawarin [2]

[1] Faculty of Forestry and Graduate Program, Universitas Papua, West Papua, 98314, Indonesia

[2] Environmental Science of Graduate Program, Universitas Papua, West Papua, 98314, Indonesia

[*]Email: arnoldustuturop26@gmail.com

Abstract. Shifting cultivation, as traditional agroforestry system with trial-and-error to provide sustainable land-use type, has been developed based on social, cultural and economic realities of local community. Development of shifting cultivation in Papua is usually to achieve household food security and economic access if their own needs fulfilled. However, the local agroforestry practices (LAPs) were still lack information on species diversity and technology. Enhancing the knowledge and skills of the LAPs are needed to improve modified agroforestry system for food security and climate change adaptation. Data was collected from households of Teluk Patipi District of Fak-Fak Regency using purposive sampling method through farm inventory. Through the modified agroforestry practice (MAPs), the production scale is expected to increase the local community income as well as to prepare food security in extreme climate conditions, the preservation of typical species, environmental conservation, and support low-carbon development strategy in West Papua.

1. Introduction

Shifting cultivation is a dry land farming system and most of them have implemented traditional agroforestry systems, generally in the tropics region including Papua which are carried out based on indigenous community experiences from generation to generation through a trial-and-error processes in cultivating land and in general to meet the needs and food security first and if there are leftovers to be sold as necessities of daily life [1,2]. Papua remote indigenous communities (RIC) in clearing shifting cultivation began with the cutting of useless trees (macaranga (*macaranga spp.*), ficus (*ficus sp.*) dan pulai (*Alstonia scholaris*)) and leaving of commercial trees (merbau (*Intsia bijuga*), and matoa (*Pometia pinnata*)) and fruits (breadfruit, nutmeg, durian, avocado, banana, and langsat (*Lansium domesticum*)) that were useful [3,4]. Then the former trees that are cut down along with the dried branches are burned and the ashes are used as fertilizer for planting seasonal crops such as sweet potatoes, cassava, taro, peanut, vegetables (mustard greens, gedi (*Abelmoschus manihot*), corn, chili, water spinach) that known as the local agroforestry practices (LAPs) [5,6,7,8,9].

Since 1980, RIC in Teluk Patipi District have been cultivating massoia plants in their shifting cultivation gardens with an average of 30-50 trees per ha. They do this cultivation in an effort to increase their income, especially from harvesting massoia bark which has a better price than other commodities with a selling price in the market reaching US$ 2.7 – 4.0/ kg and also conditions that are difficult to find in nature due to the harvesting system is directly cutting down the trees. The efforts made by these RIC are also in the sense of overcoming problems related to the sustainability and preservation of massoia plants in Papua and at the same time maintaining environmental conditions and be able to preserve the environment and enhanced traditional ecological knowledge [10,11,12].

Agroforestry can mitigate the climate change issue through the process of carbon storage in both vegetation and soil that known as organic carbon pools. Several studies have been conducted to account carbon potential in agroforestry system together with the sustainable economic and social development to improve livelihood through simultaneous production of food, nutrition, and health. Dominance of LAP in Papua provide opportunity to develop carbon sequestration, livelihood, biodiversity conservation especially for endemic or specific species, soil fertility enhancement and rural employment [13,14,15,16,17].

Climate change has an impact on increasing hydrometeorological disasters in West Papua from 2010-2019 data with reaching of 42% [18]. This has an effect on the availability of water that is less during El Nino and excess water during La Nina [19,20]. Therefore, rain harvesting techniques are needed in the form of bio-pores infiltration hole in an effort to anticipate climate change and also in climate smart agriculture efforts, especially for seasonal crops in LAP activities [21].

In order to increase capacity building of RIC, climate smart agriculture and modified agroforestry practices (MAP) are urgently developed, especially in increasing the resilience communities to climate change and climate-related disasters which have increased in the last decade. This study would like to enhance mitigation strategies through the development of bio-pores infiltration hole and enhancing of carbon potential into long-term agroforestry sustainable development to increase the prosperity of the RIC in Teluk Patipi District.

2. Methods

2.1. Study area

The research was conducted in Teluk Patipi District, Fakfak Regency, West Papua Province which is geographically located between LS 02°33′25″ - 02°55′35″ and BT 132°16′54″ with elevation ranging from 183 – 452 m a.s.l. The average of annual rainfall for period 2002-2020 approximately 3321 mm, with a range 1781-4641 mm. While, the average temperature per year during the period is 26.3°C with a range of values from 25.5-26.8 °C and the average humidity per year during the period is 85.1% with a range of values from 82.7% - 87.5% [22]. The major soil type in Teluk Patipi District is dominated by Inceptisols, which is characterized by a young soil development level. So that the value of the availability of nutrients is low, which ranges from 1.7 me/100 g in the top 20 cm to les to 0.8 me/100 g below 20 cm [23].

In the Teluk Patipi, there are 11 villages located on the coastal area, namely Sum, Patipipasir, Puar, Us, Adora, Muhri, Degen, Tetar, Offie, Tibatibanana,, and Mawar (Figure 1). However, the area where massoia plantations have been developed since 1980 has only been implemented in 7 villages, namely Sum, Puar, Us, Adora, Muhri, Degen, Tetar with a total area of 21 ha. Planting was initiated by the Tuturop, Bahamba and Kabes family in Us, Puar, Muhri, Degen and Tetar villages, which later became seed gardens. Then, in 2000 and above, 2 other villages were developed, namely Sum and Adora as well as 6 other family that were newly involved (Hindom, Kaninggai, Iha, Karoute, Muri, and Sagas). Henceforth, the management of the massoia garden is given to the first child as inheritance rights.

As for massoia increments at the age of 3 years, it is no longer possible for agricultural crops to grow because the canopy area is thick enough and covers the land below. So that the RIC opens new land again for the process of fulfilling food crops as food security. Besides that, the massoia garden is cared for until the age of 15 years for the massoia bark to be harvested

For the implementation of this study in an effort to develop demonstration plot of bio-pores and MAP through increasing carbon potential and economic were carried out in 7 sample plots (red boxes) on the massoia agroforestry gardens of RIC (Figure 1).

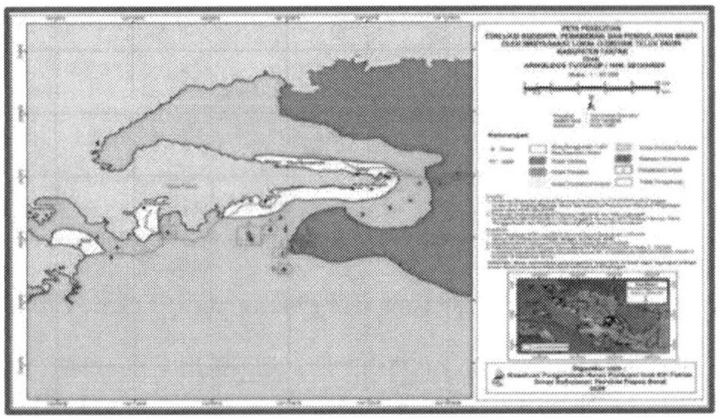

Figure 1. Research site for MAP and bio-pores

2.2. Method of data analysis

This study involved 30 families of the coastal area in Teluk Patipi District who were selected by non-proportional stratified random sampling from those who manage of massoia LAP. Primary data was obtained from direct interviews with respondents using a questionnaire in the series of FGD. The data needed from the community is the selection of type of forestry plants, fruit and seasonal crops based on local knowledge and good price in the market.

Furthermore, an effort to determine carbon potential originating from the carbon pool consisting of biomass, understorey, necromass, litter and soil using the following equation [24]:

$$\Delta C_{t,t+1} = A_t[\sum_{i=1}^{n}(a_{1,t}(C_{1,t+1} - C_{1,t}) + (a_{i,t+1} - a_{i,t})C_{i,t}] \quad (1)$$

where ΔC is the annual change in carbon stock (biomass, understorey, necromass, litter and soil) in the landscape (ton yr^{-1}), A the area (ha), c the change carbon stock, a the change in area for class i, and C the carbon stock per year per area (ton yr^{-1} ha^{-1}).

Estimates of economic to calculate the profits obtained by RIC using the net present value (NPV) formula with the following equation [24]:

$$\text{NPV} = \sum_{n}^{t}(R_t - C_t)e^{-rt} \quad (2)$$

where R_t is the Revenue per hectare in time t, C_t the Cost per hectare in time t, r the Rate of Discount, and t the rotation age in years.

For the calculation of the water balance, the first step is to measure the effective rainfall through the following equation [25]:

$$R_{80} = \frac{m}{n+1} \quad (3)$$

where R80 is the rainfall by 80%, m is the preferred rainfall ranking and n the amount of data.

After that, checking the water needs of seasonal crops (palawija and vegetables) from the selection of species determined by the RIC in the dry and rainy seasons [26]. Meanwhile, the water requirement for secondary crops and vegetables is obtained from 50% of the R80 effective rainfall.

3. Results and discussion

3.1. Participatory Selection of LAP

A series of FGDs was carried out together with RIC in a participatory effort to select agroforestry plant species based on local knowledge and good prices in the local market in an effort to achieve a more sustainable MAP on the development of carbon potential and economic (Table 1). The benefit types that are considered are divided into 3 categories, namely economic consisting of forestry plants, fodder, fruit trees which are included in multi-purpose tree species (MPTs), seasonal plants (palawija and vegetables), and medicinal plants; environmental containing soil fertility, erosion control, growth, and sociability; and management involving breeding ease and manageability.

Table 1. RIC specific choice for MAP in RIC Teluk Patipi District

Benefit type	Criteria	Species	Specific benefit
Economic	Wood	merbau (*Intsia bijuga*), matoa (*Pometia pinnata*).	Quality of wood, marketability, color of wood
	Fodder	gamal (Glirisideae sepium), kaliandra (*Calliandra calothyrsus),* rumput gajah (*Pennisetum purpureum),*	Leaf production, softness, nutrient content, quick drying (for storage)
	Fruit/ MPTs	breadfruit (*Artocarpus altilis*), nutmeg (*Myristica fragrans*), durian (*Durio zibethinus*), avocado (*Persea americana*), banana (*Musa sp.*), langsat (*Lansium domesticum*)	Fruit production, food security, marketability,
	Medicine	ginger (*Zingiber officinale*), turmeric (*Curcuma longa*), greater galangal (*Alpinia galanga*), curcuma (*Curcuma zanthorrhiza*), aromatic ginger (*Kaempferia galanga*), true cadarmom (*Elettaria cardamomum*), true cinnamon (Cinnamomum verum), massoia (Cryptocarya massoia).	Effectiveness, wide application, production of medicinal elements, cooking spices
	Agriculture plant	corn (*Zea mays*), sweet potato (*Ipomoea batatas*), cassava (*Manihot esculenta*), taro (*Colocasia esculenta*), caladium (*Caladium sp.*), mustard greens (*Brassica chinensis*), water spinach (*Ipomoea aquatica*), gedi (*Abelmoschus manihot*), chilli (*Capsicum annum*).	Food security, marketability
	Non forest timber product (NFTP)	massoia (Cryptocarya massoia), honey bee (*Apis mellifera Linneus*).	Honey production, other food production, commercial value of seeds and seedlings, oil production
Environmental	Soil fertility	peanuts (*Arachis hypogaea*),	Amount of litter, root nitrogen-fixing ability, fast biodegradation of leaves
	Erosion control	ketapang (*Terminalia catappa*), breadfruit (*Artocarpus altilis*)	Widely spread roots, water retention ability (wide leaves, many branches), large crown area
	Growth	acacia (*acacia mangium*).	Fast growth
	Sociability	massoia (Cryptocarya massoia)	Lack of negative effects on other plants, presence of positive effects on other plants
Management	Breeding ease	massoia (Cryptocarya massoia), tuber plants.	Ease of using seeds, roots, cuttings
	Manageability	massoia (Cryptocarya massoia).	Resistance to disease and pests, less requirement for weeding and fertilizer

RIC's participatory approach empowers the community's capacity building to make decisions and at the same time they are able to manage it well from an economic, environmental and management perspective to be more sustainable and equitable [27].

IOP Conf. Series: Earth and Environmental Science **989** (2022) 012005 doi:10.1088/1755-1315/989/1/012005

In Table 1, it is shown that the participatory approach of RIC plays a significant role in making decisions based on their local knowledge which will later be useful for building a MAP based on carbon potential and economic for increasing community prosperity in Teluk Patipi District.

3.2. Carbon Potential and Economic Value of MAP

The distribution of carbon potential developed by RIC in Teluk Patipi District can be seen in Figure 2. There are 4 scenarios developed, namely (1) a business as usual (BAU) scenario with the availability of massoia only reaching 50 stems per ha, only a few MPTs and forestry plants, (2) complex 1 scenario which assumes an increase in massoia planting by 50% including MPTs and forestry plants, (3) complex 2 scenario assuming an improve of up to 70% of BAU including MPTs and forestry plants, and (4) complex 3 scenario with an increment of 100% of BAU including MPTs and forestry plants.

In 7th demonstration plots from RIC in Teluk Patipi district, the district towards to apply the complex 3 scenario with a potential carbon value of 80 tC/Ha. This value is almost closed to the three agroforestry systems developed at Balung River Plantation, Sabah, Malaysia of 79.13, 85.40, and 78.28 tC/Ha, respectively [15].

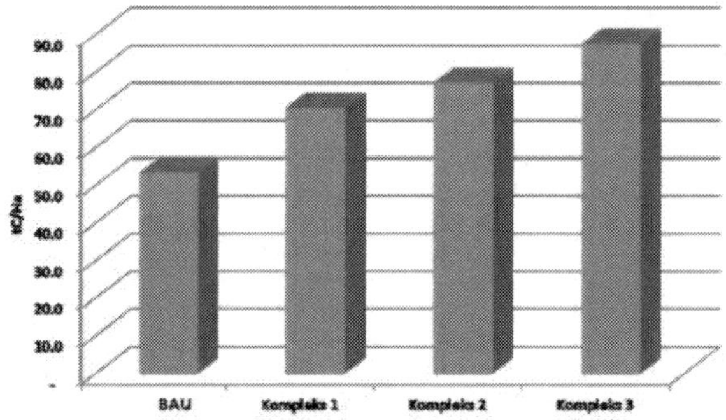

Figure 2. Carbon potential of MAP in Teluk Patipi District

For economic analysis, the profits obtained still rely on the yield of massoia bark followed by harvesting of fruits/ MPTs, medicinal plants and agricultural crops (Table 2). As for medicinal and agricultural plants, it can no longer be planted when the massoia is 3 years old due to the influence of an enlarged canopy and generally plants that are not tolerant of shade.

Table 2. Economic analysis of MAP in Teluk Patipi District

No	Agroforestri System	Initial Cost $/ha	Life Cycle Cost $/ha	NPV Benefit $/ha
1	Maasoia bark + MPTs + medicinal + legume+tubers+vegetables	110	526	10,965
2	Massoia bark + MPTS (1) + medicinal + legume+tubers+vegetables	165	789	16,448
3	Massoia bark + MPTS (2) + medicinal + legume+tubers+vegetables	187	894	18,641
4	Massoia bark + MPTS (3) + medicinal + legume+tubers+vegetables	220	1,052	21,930

Table 2 represents that the NPV of the complex 3 scenario from RIC in Teluk Patipi District of 21,930 $/ha in 15 years of massoia bark harvesting period with an average annual income of 1,595 US$/ha. Assuming the community per family can manage 2 ha, the monthly income equals the regional minimum wage of 244 US$.

3.3 Bio-pores Infiltration Hole

The calculation of the effective rainfall R_{80} from the Period data of 2002-2020 can be shown in Table 3. Rainfall is ranked from 1 to 8 according to colour and then ranking 8 is determined as the effective rainfall. Then, the calculation for effective rainfall on palawija or agricultural crops is done by 50% of R80 to measure the availability of plant water in the dry and rainy seasons. Furthermore, the water availability debit is calculated based on the area of planting seasonal crops reaching 0.5 ha which is compared to the availability of water. So that, we get the need for water shortages in the Teluk Patipi District area and it is necessary to make bio-pores to maintain water balance in the massoia garden (Table 4).

Table 3. R80 of Period 2002-2020

Year	Jan	Feb	Mar	Apr	May	Jun	Jul	Agt	Sept	Oct	Nov	Dec
2002	303	193	395	129	502	638	109	173	133	43	217	295
2003	126	245	186	225	304	322	333	238	397	321	137	257
2004	228	337	197	358	113	361	371	180	631	455	189	166
2005	484	169	197	620	373	378	273	85	164	44	140	282
2006	385	225	288	210	308	575	442	178	445	225	237	171
2007	208.4	132.3	176.3	248.7	457.9	256	338.9	539.1	245.4	247.9	108.5	108.5
2008	200.1	198.3	267.9	154	166.2	192.8	110.8	230.6	130.6	140	151.3	163.7
2009	213.4	257.5	324.1	229.6	243.8	419.7	476.4	493.1	112.4	101.7	182.1	210.7
2010	334.1	62.4	463.3	213	254.8	380.7	264.8	493.1	271.4	208.8	211.4	372.5
2011	246.6	353.8	210.4	303.1	531.4	527.1	324.3	329.3	382.8	210.2	182.1	210.2
2012	253.9	227.9	458.6	135.9	673.4	1163.6	261.7	545.7	355.7	172.6	182.1	210.2
2013	334.1	218.1	150.4	610.3	266.8	297.7	659.2	348.8	477.6	364.4	178.5	166.8
2014	277.7	278.9	211.2	439.4	199.9	362.4	197.2	269.4	104.7	33.2	134.4	228.7
2015	168.7	244.9	197.8	297.3	160	282	58.7	50.7	28.7	86.4	132.3	73.6
2016	230.2	171.7	274.4	367.3	247	284.1	866.4	301.4	259.5	467.6	152.8	339.2
2017	228.7	193.7	342.9	296.9	228.7	393.8	278.4	493.2	384.2	483	126.1	252.6
2018	294.6	138.6	457.1	247.4	282.7	392.9	198.6	244.3	292.2	280.3	205.2	388.4
2019	355.9	209	334.9	191.6	274.8	253.4	277.3	256.1	241	481.1	94.9	119.8
2020	170.6	187.8	333.4	406.3	266.4	590.4	284.2	282.8	77.2	313.2	163.3	130.4

Ranking 1	Ranking 2	Ranking 3	Ranking 4	Ranking 5	Ranking 6	Ranking 7	Ranking 8

Table 4. Water balance and bio-pores needed

Month	R80 (mm)	50% R80 (mm)	Period	Water needs for seasonal crops (mm)				Total (mm)	Water balance (mm)	Areal (m^2)	Debit (m^3)	Biopore
Jan	246.6	123.3										
Feb	225.0	112.5										
Mar	324.1	162.1										
Apr	297.3	148.7										
May	282.7	141.4	Rain	peanuts	tuber	cassava	vegetables					
Jun	392.9	196.5	884.3	202	214	214	60.6	690.6	193.7	5,000	968.5	
July	324.3	162.2										
Agt	282.8	141.4										
Sept	292.2	146.1										
Oct	280.3	140.2										
Nov	163.3	81.7	Dry									
Des	228.7	114.4	785.8	287	305	305	86.1	983.1	-197.3	5,000	-986.5	**18**
Deficit											**-18**	

Table 4 indicates that the water deficit is 18 m^3 which means 18th bio-pores are needed with an area of 1 m^2 and a depth of 1 m. In addition to storing water, this bio-pores can also be used to raise fish so that it can provide the needs of vitamins and minerals for nutritional security and also to increase additional income.

4. Conclusion

The selection of plant species was carried out through a participatory approach from RIC in Teluk Patipi District which really helped them to increase capacity building in managing their local knowledge for economic, environmental and management sustainability. Furthermore, based on the selection of these species, MAP is introduced with the application of the complex 3 scenario with the main concern in the form of increasing the carbon potential that is equal to secondary forest and the economy analysis is equivalent to the regional minimum wage. And then, the water balance analysis was also carried out to determine the water deficit and the bio-pores infiltration hole technology was developed to anticipate climate change and climate-related disaster.

Acknowledgements

The authors would like to thank the Team of Production Forest Management Unit (KPHP Unit XVI) Fakfak Regency that supported data and technical during the development of this proceeding. Thanks are also given to Mr. Hendri as the supervisor who has helped the process of writing this proceeding.

References

[1] Limba RS, Lio A, and Husain YS 2017 Shifting Cultivation System of Indigenous Moronene as Forest Conservation on Local Wisdom Principles in Indonesia *Journal of Sustainable Development* 10 (**4**): 121-129.

[2] Cairns M 2017 Shifting Cultivation Policies: Balancing Environmental and Social Sustainability (USA: CABI) 1057 pp.

[3] Tuturop A, Hendri, Nugroho, JD 2021 Local Knowledge-based of Remote Indigenous Community for Massoia Plant Management: Lesson Learned form Patipi District, Fakfak Regency.

[4] Yuminarti U, Darwanto DH, Jamhari and Subejo 2018 A Comparative Study on the Practice of Shifting Cultivation and Permanent Agriculture to Support Community Food Security (Study on Potato Farming in Arfak Mountains Regency, West Papua Province) Jurnal Ketahanan Nasional 24 (**2**): 215-238.

[5] Pandey, Kumar D, Adhiguru, P, Singh B, and John A 2021 Is traditional agroforestry system of Nagaland bountiful for indigenous and traditional crops species diversity? Indian Journal of Traditional Knowledge 20 (3): 803-809.

[6] Jarrett C, Cummins I, and Logan-Hines E 2018 Adapting Indigenous Agroforestry Systems for Integrative Landscape Management and Sustainable Supply Chain Development in Napo, Ecuador Integrating Landscapes: Agroforestry for Biodiversity Conservation and Food Sovereignty 12: 283-309.

[7] Heredia-R M, Torres B, Cayambe J, Ramos N, Luna M, and Diaz-Ambrona CGH 2020 Sustainability Assessment of Smallholder Agroforestry Indigenous Farming in the Amazon: A Case Study of Ecuadorian Kichwas *Agronomy* 10 (12): 1973; https://doi.org/10.3390/agronomy10121973.

[8] Stanek EC, and Lovell ST(2019. Building multifunctionality into agricultural conservation programs: lessons learned from designing agroforestry systems with central Illinois landowners *Renew Agric Food Syst* pp 1–9.

[9] Vikram NS, Ravi P, Sekar KC, Rawal RS, Bhat ID, Nandi SK, and Dhyani PP 2018 Traditional knowledge and biodiversity conservation: a case study from Byans Valley in Kailash Sacred Landscape, India *Journal of Environmental Planning and Management* 61 (**10**): 1722-1743.

[10] Lalemo LL 2021 Review on the indigenous multipurpose agroforestry tree species in Ethiopia: management, their productive and service roles and constraints Heliyon 7 (9): 1-8.

[11] Nugroho JD., Wanma JF., Susanti CME., Husodo SB., dan Mawikere NL 2019 Decreasing natural stock of massoia [*Cryptocarya massoy* (Oken) Kosterm] in Nabire Regency *Igya Ser Hanjop* 1(**1**):11-18.

[12] Siwalette JD, Hidayat K, Cahyono ED, and Purnomo M 2018 Transformation of Production Organization of Dusung based Nutmeg Plantation (Case Study of Industrial and Farmers' Partnership in Hila Village, Moluccas Province) *Advances in Social Sciences Research Journal*5 (3): 23-36. .

[13] Feliciano D, Ledo A, Hillier J, and Nayak DR 2018 Which agroforestry options give the greatest soil and above ground carbon benefits in different world regions? *Agriculture, Ecosystems & Environment* 254: 117-129.

[14] Goncalves N, Andrade D, Batista A, Cullen Jr L, Souza A, Gomes H, and Uezu A 2021 Potential economic impact of carbon sequestration in coffee agroforestry systems *Agroforestry Systems* 95: 419–430.

[15] Awang Besar N, Suardi H, Phua MH, James D, Mokhtar MB, and Ahmed MF 2020 Carbon Stock and Sequestration Potential of an Agroforestry System in Sabah, Malaysia *Forests* 11: 1-16.

[16] Dhyani SK, Ram A, Newaj R, Handa AK, and Dev I 2019 Agroforestry for Carbon Sequestration in Tropical India *Carbon Management in Tropical and Sub-Tropical Terrestrial Systems* pp 313-331 https://doi.org/10.1007/978-981-13-9628-1_19.

[17] Agevi H, Onwanga R, Kuyah S, and Tsingalia M 2017 Carbon Stocks and Stock Changes in Agroforestry Practices: A Review *Tropical and Subtropical Agroecosystems* 20: 101-109.

[18] National Disaster Management Agency 2020 Disaster data in West Papua (Jakarta: Data Informasi Bencana Indonesia)

[19] Hidayat R, Juniarti MD, and Ma'rufah U 2018 Impact of La Niña and La Niña Modoki on Indonesia rainfall variability *IOP Conf. Ser.: Earth Environ. Sci.* **149** 012046.

[20] Amrianah H 2020 Impacts of El Nino and La Nina Extreme Climate on Socio-Economic and Politics in Indonesia Meraja Journal 3 (1): 35-45.

[21] Verasta T, Maulidan I, Azzahra HA, Sobri ASWSB, and Utami ARI 2021 The Effect of Greenhouse and Biopore on Community Development of Economy and Knowledge of

Citeureup Village During the Pandemic Journal of Innovation and Community Engagement 2 (1): 21-34.

[22] BMKG Fakfak 2020 Meteorology Data in the Period 2002-2020 (Fakfak: Data Analysis)

[23] Musaad I, Tubur HW, Wibowo K, and Santoso B 2017 Pala Fakfak (Bandung: Alfabeta) 13 pp.

[24] Makundi W and Sathaye J 1999 Comprehensive Mitigation Assessment Process (COMAP) – Description and Instruction Manual (USA: Berkeley Lab) 53 pp.

[25] Priyonugroho A 2014 Analysis of irrigation water needs (a case study in the irrigation area of the water keban river in the Empat Lawang Regency) *Jurnal Teknik Sipil dan Lungkungan* 2 (3): 457-470.

[26] Hermantoro and Pusposutarjo 2000 Modeling growth and water use of secondary crops in dry land Buletin Keteknikan Pertanian 14 (2): 139-149.

[27] He J, Ho MH, and Xu J 2015 Participatory Selection of Tree Species for Agroforestry on Sloping Land in North Korea *Mountain Research and Development* 35(**4**): 318-327.

Study of surface and vertical sea temperatures during the process of tropical cyclone formation in the territory of Indonesia (case study 2019-2021)

Anendha Destantyo Nugroho [1,*] and Nur Habib Muzaki [2]

[1] Indonesian Agency for Meteorological, Climatological and Geophysics
[2] College of Meteorology, Climatology and Geophysics

*Email: anendha.nugroho@bmkg.go.id

Abstract. A tropical cyclone is an atmospheric phenomenon that has a system of low air pressure that forms over the oceans in the warm tropics. Tropical cyclones can form in oceans with sea surface temperatures above 26.5°C. In addition to the sea surface temperature which must be more than 26.5°C, the depth of warm sea water temperature strongly supports the formation of tropical cyclones in the territory of Indonesia. In this study, research was conducted on the surface and vertical temperatures of the sea during the formation of tropical cyclones or when cyclone seeds occurred in the case of tropical cyclones that occurred in 2019-2021 in the territory of Indonesia. The data used is processed using Ocean Data View 4.1 to determine the distribution of surface and vertical sea temperatures. Based on the results on this study, the vertical distribution of sea surface temperatures during the formation of tropical cyclones in Indonesia has an average sea temperature higher than 26.5°C which reaches a depth of 64 meters to 70 meters. Then the sea surface temperature during the tropical cyclone formation process averaged 28 °C to 29 °C.

1. Introduction

Tropical cyclone is one of the extreme weather disturbances that begins with the formation of a low pressure center above the ocean which triggers an intensive process of convection and cloud formation. In general, tropical cyclones form at latitudes above 10o in the north and south [7]. Tropical cyclones are low pressure systems with wind speeds of more than 34 knots that occur in the oceans of the tropics [3]. Overseas areas that have warm and humid air mass conditions can cause tropical cyclones [9]. Tropical cyclone formation can occur when it meets several requirements, one of which is a warm sea surface temperature with a sea surface temperature of 26.5°C to a depth of 60 meters [2]. Warm temperatures in the oceans of the tropics are needed as the basis for the formation of tropical cyclones and fuel to drive tropical cyclones [6]. The formation of tropical cyclones has the main requirements, namely a greater coriolis value at latitude 5oC and a sea surface temperature of more than 26 °C [11]. Tropical cyclones have several growth stages, namely the growth stage, immature stage, mature stage and attenuation stage [12]. Tropical Cyclone Lili is one of the tropical cyclones that occurred in the territory of Indonesia on May 9, 2019 in the Banda Sea [4]. Based on the research entitled Tropical Cyclone Lili study and its effect on the weather in NTT, there were no significant changes due to Tropical Cyclone Lili from data on air temperature, humidity, wind speed, and rainfall in the East Nusa Tenggara region [1]. Then another tropical cyclone that has occurred in the territory of Indonesia is the Mangga Tropical Cyclone. The Mangga Tropical Cyclone is a tropical cyclone that occurred on May

21, 2020 in the Indian Ocean southwest of Bengkulu with a maximum wind speed of 35 knots [5]. Furthermore, research on the Mangga Tropical Cyclone shows that the Mangga Tropical Cyclone causes heavy rain accompanied by strong winds and the potential for high waves reaching 4.5 to 6.0 meters in the waters of Lampung and its surroundings [9]. Then based on research on the impact of the Cempaka Tropical Cyclone phenomenon, that the warm sea surface temperature condition reached 28.4oC with an anomaly of sea surface temperature greater than 0.2 °C being one of the factors for the formation of the Cempaka Tropical Cyclone [8]. Based on research on atmospheric conditions during the occurrence of Tropical Cyclone Cempaka and Dahlia, it shows that the sea surface temperature before the occurrence of Tropical Cyclone Cempaka reached 28.9 °C to 30.3 °C with wind speeds reaching 35 to 50 knots [10]. Then during the Tropical Cyclone Cempaka, the sea surface temperature reached 27.5 °C to 29.5 °C. Based on the background that has been explained, that one of the conditions for the formation of tropical cyclones in the oceans of the tropics is a warm sea surface temperature to a certain depth. So that in this research, we will examine the surface temperature at the time of the formation of tropical cyclones in the territory of Indonesia to a certain depth by taking the cases of Tropical Cyclone Lili, Tropical Cyclone Mangga, and Tropical Cyclone Seroja.

2. Methods

2.1. Research Area

This research was conducted in the territory of Indonesia by taking the case of tropical cyclones from 2019 to 2021 at coordinates 6° North Latitude to 11° South Latitude and 95 East Longitude to 141° East Longitude as shown in Figure 1 and the path from the tropical cyclone formation stage can be seen in Figure 2. Then for the time and location of the tropical cyclone taken when the tropical cyclone is in the mature stage as shown in Table 1.

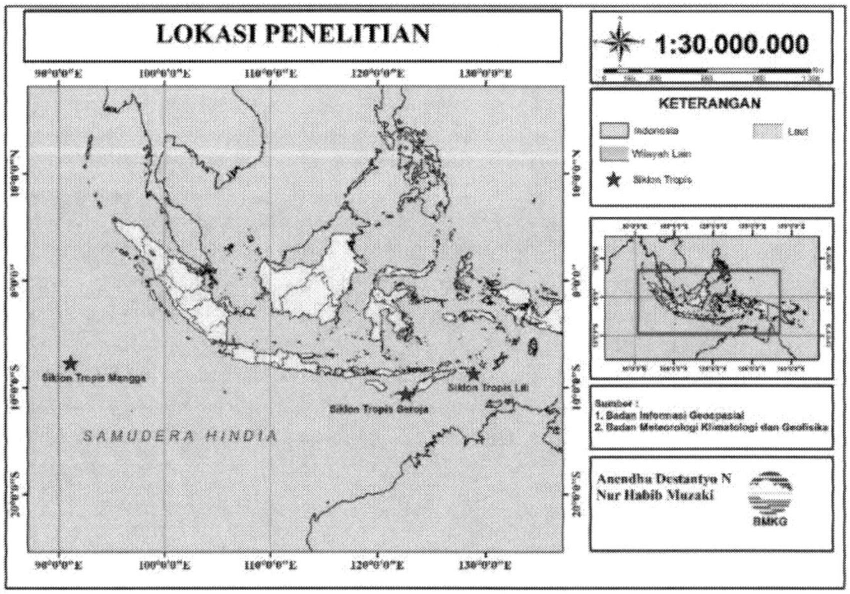

Figure 1. Research Area

IOP Conf. Series: Earth and Environmental Science **989** (2022) 012006 doi:10.1088/1755-1315/989/1/012006

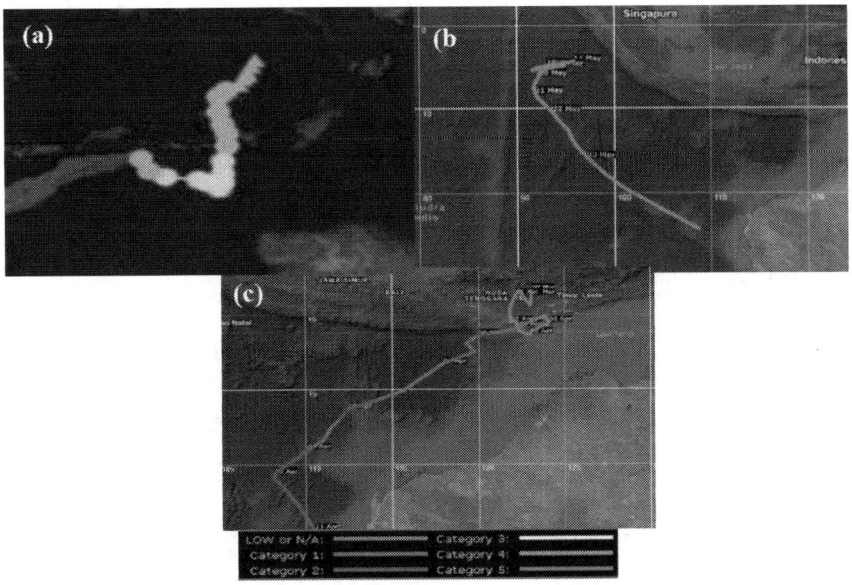

Figure 2. Pathway of Tropical Cyclone Formation (a) Tropical Cyclone Lili, (b) Tropical Cyclone Mangga, and (c) Tropical Cyclone Seroja.

Table 1. Time and location of tropical cyclones at maturity.

Date	Tropical Cyclone	Longtitude	Latitude
8 May 2019	Lili	-8.7	128.9
21 May 2020	Mangga	-7.7	91.1
5 April 2021	Seroja	-10.6	122.6

2.2. Data

The data used in this study is data on sea temperature to a depth of 100 meters during the tropical cyclone formation stage. The data used is obtained from Copernicus Marine which can be downloaded at https://marine.copernicus.eu/ . The data used has a spatial resolution of 0.083° x 0.083° and a temporal resolution of 12 hours. The time and location of the data used in this study can be seen in Table 2.

Table 2. Time and location of tropical cyclones during formation stage.

Tropical Cyclone	Date	Latitude	Longtitude
Lili	6 May 2019 12:00:00 UTC	6.3	129.5
	7 May 2019 12:00:00 UTC	6.4	129.2
	8 May 2019 12:00:00 UTC	7.6	129.1
Mangga	18 May 2020 12:00:00 UTC	5.4	91.7
	19 May 2020 12:00:00 UTC	5.6	92.7
	20 May 2020 12:00:00 UTC	6.9	91.9
Seroja	2 April 2021 12:00:00 UTC	11	122.3
	3 April 2021 12:00:00 UTC	10.7	123.8
	4 April 2021 12:00:00 UTC	10.4	123.2

IOP Conf. Series: Earth and Environmental Science **989** (2022) 012006 doi:10.1088/1755-1315/989/1/012006

2.3. Method

The method used in this research is as follows:

a. Collecting sea temperature data to a depth of 100 meters based on Table 2.

b. Processing ocean temperature data using the Ocean Data View (ODV) application.

c. The sea temperature will be processed horizontally and vertically and then analyzed by looking at the conditions for the formation of tropical cyclones, namely the sea temperature is more than 26.5°C [5].

3. Result and Discussions

3.1. Tropical Cyclone Lili

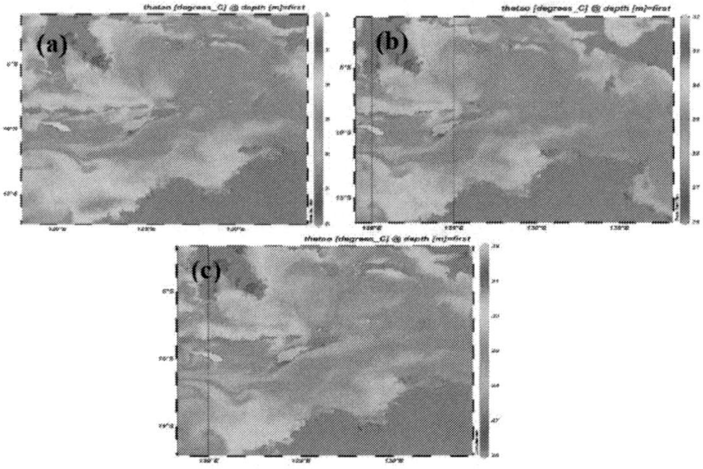

Figure 3. Sea surface temperature (a) 6 May 2019, (b) 7 May 2019, and (c) 8 May 2019.

Based on Figure 3, it can be seen that the sea surface temperature in the area where Tropical Cyclone Lili is formed has a temperature of more than 26.5 °C. This warm temperature makes it a source and one of the factors for the formation of Tropical Cyclone Lili. The temperature on May 6, 2019 reached 29.6 °C in the Tropical Cyclone Lili area at the stage of formation. Then for 7 and 8 May 2019, sea surface temperatures were seen to reach 28.8 °C and 29.1 °C.

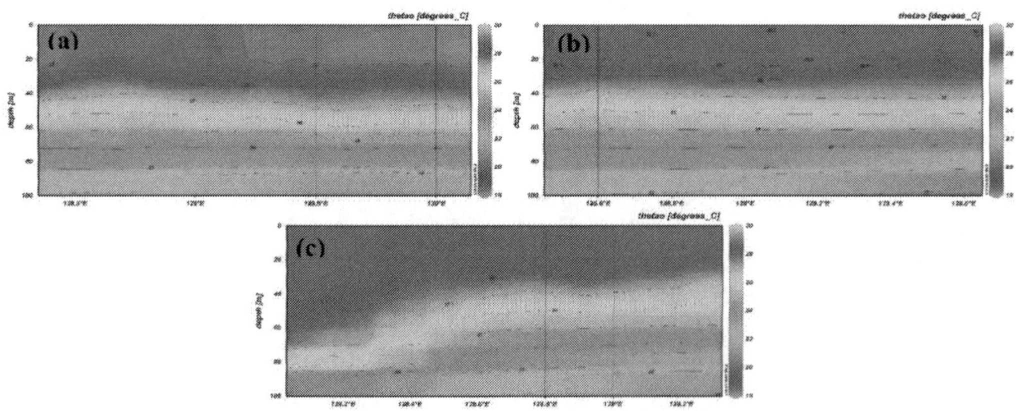

Figure 4. Distribution of sea temperatures to a depth of 100 meters (a) 6 May 2019, (b) 7 May 2019, and (c) 8 May 2019.

IOP Conf. Series: Earth and Environmental Science **989** (2022) 012006 doi:10.1088/1755-1315/989/1/012006

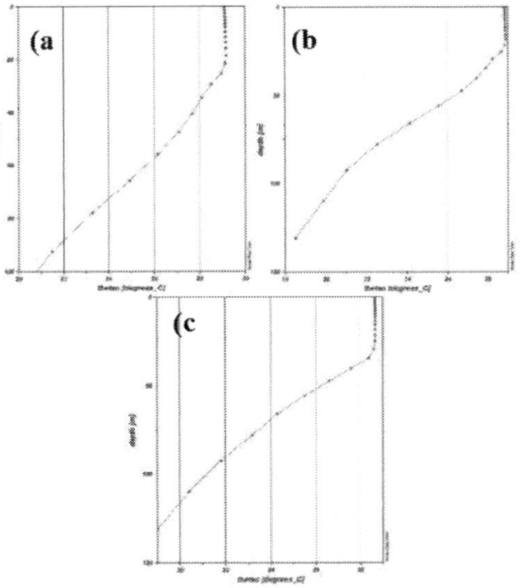

Figure 5. Graph of sea temperature to a depth of 100 meters (a) 6 May 2019, (b) 7 May 2019, and (c) 8 May 2019.

Figure 4 shows the distribution of sea temperatures to a depth of 100 meters in the area where Tropical Cyclone Lili is formed, while Figure 5 shows a graph of temperature to depth according to the point of formation of Tropical Cyclone Lili in Table 2. In Figure 5, it shows that on May 6, 2019, sea temperatures were higher from 26.5 °C to a depth of 54 meters. Then on 7 May 2019 sea temperatures that were more than 26.5 °C reached a depth of 51 meters and for 8 May 2019 the sea temperature reached a depth of 50 meters which had a temperature of more than 26.5 °C. At the stage of the formation of Tropical Cyclone Lili on 6, 7, and 8 May 2019 showed that the sea temperature had a value greater than 26.5 °C to a depth of 50 meters to 54 meters. Then for the sea surface temperature in the area where Tropical Cyclone Lili was formed, it reached 29.6 °C. Warm temperatures that reach a depth of 54 meters during the formation stage of Tropical Cyclone Lilies meet the requirements for the formation of tropical cyclones, namely warm temperatures that reach a depth of 50 meters [5].

3.2. Tropical Cyclone Mangga

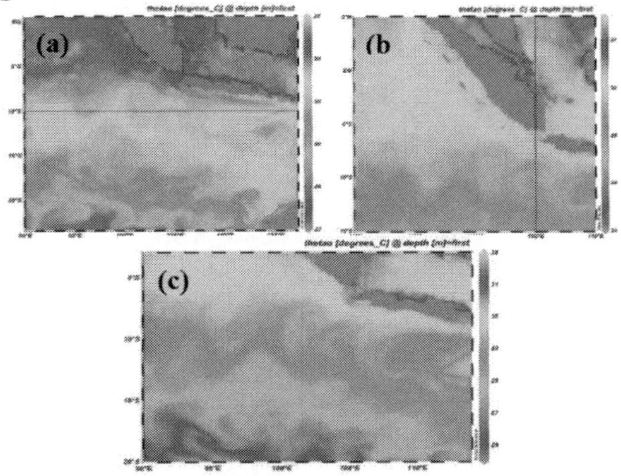

Figure 6. Sea surface temperature (a) 18 May 2020, (b) 19 May 2020, and (c) 20 May 2020.

IOP Conf. Series: Earth and Environmental Science **989** (2022) 012006 doi:10.1088/1755-1315/989/1/012006

Based on Figure 6, it can be seen that the sea surface temperature in the area where the Mangga Tropical Cyclone is formed has a temperature of more than 26.5 °C. This warm temperature makes it a source and one of the factors for the formation of the Mangga Tropical Cyclone. The temperature on 18 May 2020 reached 30.3 °C in the Mangga Tropical Cyclone region at the stage of formation. Then for 19 and 20 May 2020 the sea surface temperature was seen to reach 29.8 °C and 30.1 °C.

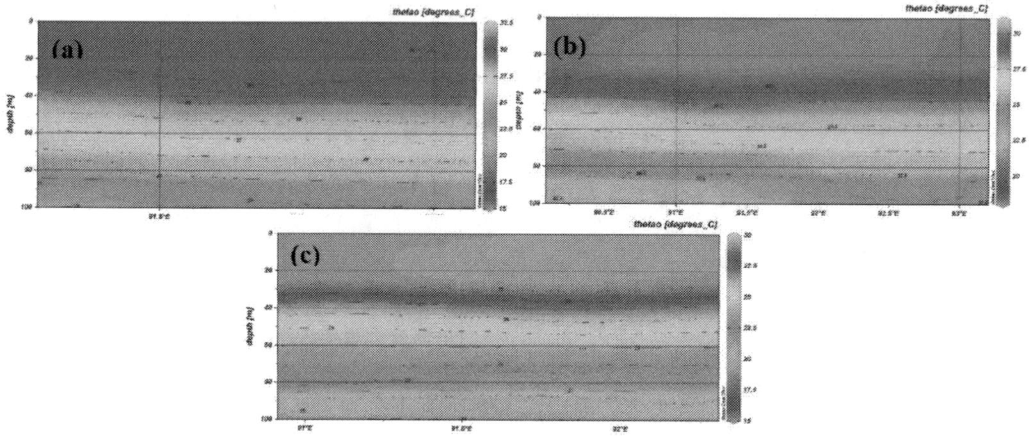

Figure 7. Distribution of sea temperatures to a depth of 100 meters (a) 18 May 2020, (b) 19 May 2020, and (c) 20 May 2020.

Figure 7 shows the distribution of sea temperatures to a depth of 100 meters in the area of the formation of the Mangga Tropical Cyclone, while Figure 8 shows a graph of temperature to depth according to the point of formation of the Mangga Tropical Cyclone in Table 2. In Figure 8, it shows that on 18 May 2020 the sea temperature was higher from 26.5 °C to a depth of 63 meters. Then on 19 May 2020 the sea temperature which was more than 26.5 °C reached a depth of 70 meters and for 19 May 2020 the sea temperature reached a depth of 51 meters which had a temperature of more than 26.5 °C.

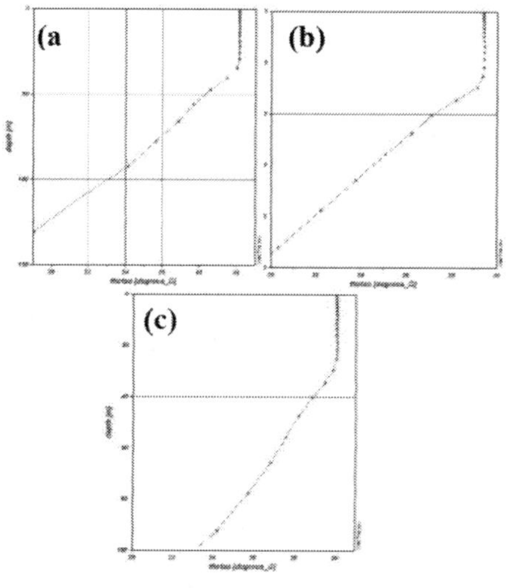

Figure 8. Graph of sea temperature to a depth of 100 meters (a) 18 May 2020, (b) 19 May 2020, and (c) 20 May 2020.

During the formation stage of the Mangga Tropical Cyclone on 18,19, and 20 May 2020, it showed that the sea temperature had a value greater than 26.5 °C to a depth of 51 meters to 70 meters. Then for the sea surface temperature in the area where the Tropical Cyclone Mangga was formed, it reached 30.3 °C. Warm temperatures that reach a depth of 70 meters during the formation stage of this Mangga Tropical Cyclone meet the requirements for the formation of tropical cyclones, namely warm temperatures that reach a depth of 50 meters [5].

3.3. Tropical Cyclone Seroja

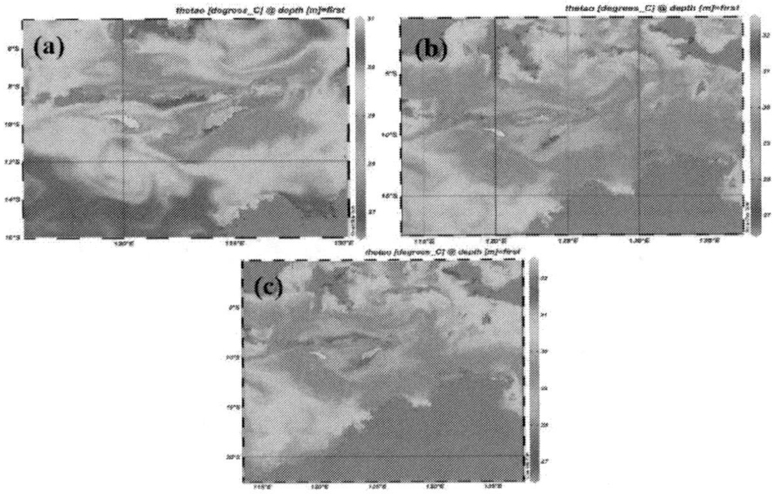

Figure 9. Sea surface temperature (a) 2 April 2021, (b) 3 April 2021, and (c) 4 April 2021.

Based on Figure 9, it can be seen that the sea surface temperature in the area where Tropical Cyclone Seroja was formed has a temperature of more than 26.5 °C. This warm temperature is one of the factors for the formation of Tropical Cyclone Seroja. From April 2, 2021 to April 4, 2021, the Seroja Tropical Cyclone formation area has temperatures that reach 28.1 °C to 28.6 °C.

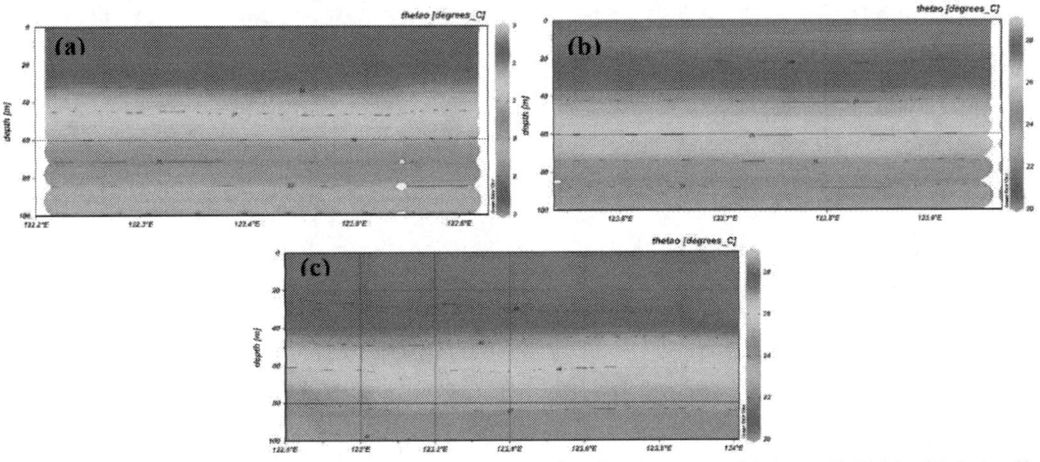

Figure 10. Distribution of sea temperature to a depth of 100 meters (a) 2 April 2021, (b) 3 April 2021, and (c) 4 April 2021.

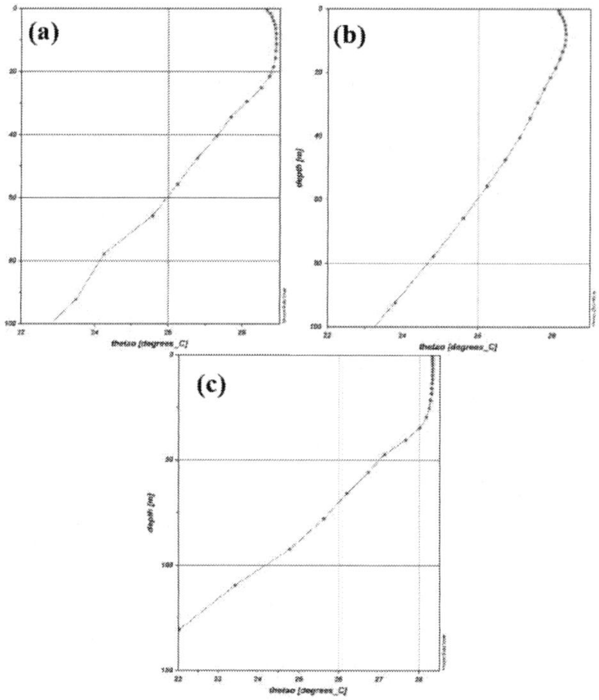

Figure 11. Graph of sea temperature to a depth of 100 meters (a) 2 April 2021, (b) 3 April 2021, and (c) 4 April 2021.

Figure 10 shows the distribution of sea temperatures to a depth of 100 meters in the area of the formation of Tropical Cyclone Seroja, while Figure 11 shows a graph of temperature to depth according to the point of formation of Tropical Cyclone Seroja in Table 2. In Figure 11, it shows that on April 2, 2021 the sea temperature was higher from 26.5 °C to a depth of 53 meters. Then on April 3, 2021, the sea temperature which was more than 26.5 °C reached a depth of 51 meters and for April 4, 2021 the sea temperature reached a depth of 58 meters which has a temperature of more than 26.5 °C. At the stage of the formation of Tropical Cyclone Seroja on April 2,3 and 4, 2021, the sea temperature was greater than 26.5 °C to a depth of 51 meters to 58 meters. Then for the sea surface temperature in the area where Tropical Cyclone Seroja was formed, it reached 28.6 oC. This warm sea surface temperature to a depth of 53 meters is one of the causes of the formation of Tropical Cyclone Seroja. This is in accordance with the conditions for the formation of a Tropical Cyclone which has a temperature of 26.5 °C to a depth of 50 meters [5].

4. Conclussion

Based on the results of the explanation, it can be concluded that from three cases of tropical cyclones in the territory of Indonesia in the years 2019 to 2021, namely Tropical Cyclone Lili, Tropical Cyclone Mangga, and Tropical Cyclone Seroja during the tropical cyclone formation stage by taking cases three days before the formation of tropical cyclones. the sea temperature meets the requirements for the formation of tropical cyclones, namely warm sea temperatures that reach more than 26.5 °C to a depth of 50 meters. Sea temperatures over 26.5 °C reach depths of up to 70 meters in the case of the Mangga Tropical Cyclone. Then for the sea surface temperature at the time of the formation of a tropical cyclone in the territory of Indonesia reached 30.3 °C at the time of the formation of the Mangga Tropical Cyclone.

Acknowledgement
The authors would like to thank the Meteorology, Climatology and Geophysics Agency (BMKG) and the National Oceanic and Atmospheic Administration (NOAA) for providing access to data sources.

References

[1] Azizah, E. 2016. *Kajian Siklon Tropis Lili dan Pengaruhnya Terhadap Cuaca di NTT.* Skripsi. Tangerang Selatan: Program Studi Meteorologi, Sekolah Tinggi Meteorologi Klimatologi dan Geofisika.

[2] BMKG. 2010. *Keputusan No.009 Tentang Prosedur Standart Operasional Pelaksanaan Peringatan Dini, Pelaporan, dan Diseminasi Informasi Cuaca Eksrem.* BMKG. Jakarta.

[3] BMKG. 2019. *Indikator Kondisi Cuaca Maritim dan Parameter Oseanografi Sebelum Kejadian Siklon Tropis Lili.* BMKG. Jakarta.

[4] BMKG. 2020. *Siklon Tropis Mangga Tumbuh di Samudera Hindia Sebelah Barat Daya Bengkulu.* BMKG. Jakarta.

[5] Gray, W.M. 1975. *Tropical Cyclone Genesis. Colorado.* Departement of Atmospheric Science, Colorado State University.

[6] Haryani, S. N. and Any Zubaidah. 2012. *Dinamika Siklon Tropis di Asia Tenggara Menggunakan Pengindraan Jauh.* WIDYA, Vol 29 No. 34: 54-58.

[7] Hastuti, D., Firdianto U. P., and Pribadi Y. H. 2018. *Penentuan Variabilitas Awan Menggunakan Satelit Himawari-8 di Bandara Tunngul Wulung Sebagai Dampak Fenomena Siklon Tropis Cempaka.* JIIF, Vol 02 No. 02: 117-12.

[8] Ninggar, R. D. and Siregar, D. C. 2020. *Analisis Siklon Tropis Mangga dan Dampaknya Terhadap Kondisi Cuaca di Inonesia.* Jurnal Widya Climago, Vol 2 No. 2: 69-76.

[9] Rahmadini, H. N., Azani, A. A., and Fadlan, A. 2018. Identification of Atmosphere Conditions During Cempaka and Dahlia Cyclone Cycle Based on Weather Parameters Analysis and Satellite Imaging. The Annual Basic Science International Conference, 73-84.

[10] Tjasyono, B. 2006. *Meteorologi Indonesia 1*, Karakteristik dan Sirkulasi Atmosfer, BMG, Jakarta – Indonesia.

[11] Zakir A., Sulistya, W., dan Khotimah, M. 2010. *Perspektif Operasional Cuaca Tropis.* Pusat Penelitian dan Pengembangan Badan Meteorologi Klimatologi dan Geofisika. Jakarta.

Towards low carbon development strategies from forestry sector in West Papua

Hendri[1*], Syarifudin Raharjo[2], Egi Suarga[3], M. Faruk Rosyaridho[3], and Julia Kalmirah[3]

[1] Faculty of Forestry and Graduate Program, Universitas Papua, West Papua, 98314, Indonesia
[2] Faculty of Fishery and Marine Science; and Graduate Program, Universitas Papua, West Papua, 98314, Indonesia
3 World Resources Institute Jakarta, 12170, Indonesia

*Email: h.hendri@unipa.ac.id

Abstract. The forestry sector plays an important role in mitigating and adapting to climate change. In the NDC Indonesia Report (2021), forestry and other land use (FOLU) sectors contribute emission based on BAU projected 2030 of 714 Mt CO_2e and emission reduction for both unconditional (Counter Measure, CM1) of 17.2% and conditional (CM2) of 24.5%. FOLU emissions correlated with changes in the primary forest to secondary (degradation) and primary/secondary forest to other uses for agriculture, transmigration, plantations and others (deforestation). Due to climate change, the conditions of degradation and deforestation can bring disaster, especially triggered by hydrometeorological hazards that have catastrophic impacts such as floods, flash floods, droughts, erosion, and abrasion due to changes in land cover. Therefore, low carbon development strategies from the forestry sector are needed to reduce climate-related disasters in West Papua.

1. Introduction

Located in the heart of the Coral Triangle, the Bird's Head Seascape (BHS) of West Papua includes more than 22.5 million hectares of sea and more than 4,000 small islands with endemic birds, mammals, insects and plants as global centers of diverse tropical marine life located on the Sahul shelf. Seascapes imply a diversity of habitats, including 1711 reef fish, 600+ coral species, 17 dolphin and whale species, whale sharks, and critical habitats for globally endangered marine species, such as turtles and cetaceans [1,2,3,4,5]. Besides that, the Papua (West Papua and Papua) and Papua New Guinea regions are included in the third tropical rain forest in the world after Brazil and the Congo, which have high biodiversity both in the highlands, medium, low to coastal areas (mangrove forests) both mineral soils and peat [6,7,8]. Besides mega-biodiversity, the strength side (S) of the SWOT analysis, it was also found that West Papua already has several important documents, including the Regional Action Plan (RAP) for Reducing Greenhouse Gas (GHGs) Emissions of West Papua Province 2017-2021, Environmental Strategic Environmental Studies (SES) Medium-Term Development Plan (MTDP) West Papua Province 2017-2022, SES of Zoning Plan for Coastal Areas and Small Islands (ZPCASI) West Papua Province, Local Regulation of ZPCASI, Disaster Management Plan (DMP) of West Papua Province, and Regional Regulations for the Implementation of Disaster Management in West Papua Province [9,10,11,12,13,14].

However, in the management of natural resources, several problems become the weak side (W), namely the not yet optimal level of welfare of the local community (Papuan Indigenous People, PIP), which is reflected in the low Human Development Index (HDI) which occupies the lowest 2nd rank at the national level (65.09%) after Papua Province (60.44%). HDI is also influenced by the fact that West Papua Province is the 34th youngest province which is still in the development stage so that it focuses on economic development, which has implications for decreasing the carrying capacity and carrying capacity of its environment. Furthermore, it is affected by environmental damage and increased greenhouse gas emissions (GHG) from agriculture, forestry, industry, energy and transportation, and waste [9,15].

The main problem of GHG emissions from the land-based sector included the forestry sector comes from the conversion of forest land into other uses known as degradation, which change of primary forest to secondary forest. Meanwhile, deforestation is changing of forest to non-forest such as agricultural areas, plantations, regional expansion, mining, settlements in the form of land clearing, and no biomass left at all (clear-cutting) [16,17].

Besides internal factors (strengths and weaknesses), external factors from opportunity factors (O) to maintain environmental sustainability are also available in West Papua Province including the Manokwari Declaration (International Conference on Biodiversity, Ecotourism, and Creative Economy, 2018) regarding Sustainable Development Based on Customary Areas in Papua with several important points related to maintaining 70% of the forest area and protecting and strengthening the role of indigenous peoples as stated in the Specified Local Regulation; MoU between Bappenas and the Provincial Government of West Papua (18 June 2018) regarding Low Carbon Development Planning in West Papua Province; and internationally, Indonesia has also ratified the Paris Agreement related to the Nationally Determined Contribution (NDC) of National Emission Reduction and adaptation to climate change impacts, Sustainable Development Goals (SDGs), and the Sendai Framework which is also applied at the sub-national level by making Regional Action Plans [18,19,20,21].

Other external factors from threat factors found in West Papua Province are natural and non-natural disasters, which are categorized as a high-risk class with a value of 144.05 consisting of threats of earthquakes, tsunamis, floods, landslides, droughts, extreme weather, forest and land fires and extreme waves/abrasion [22]. In addition, the impact of climate change triggers hydrometeorological disasters (42%, 2010-2019) that occur on large islands, coasts, and small islands. In general, this hydrometeorological disaster is triggered by an increase in rainfall that is higher or lower than normal. If the rainfall is higher, then floods, landslides and flash floods are caused by changes in forest and land use in the upstream part of the watershed. Meanwhile, low rainfall impacts the ease of fuel in forests and land experiencing fires due to the effects of drought. West Papua region is directly opposite the Pacific Ocean, tropical storms are often passed by the tails of tropical storms, which along with climate change experience an increase in both frequency and intensity. So that it impacts strong winds and heavy rain on the northern coast of West Papua, which is also followed by high waves and strong currents and affects the fisheries sector and abrasion on the coast [23,24,25].

By taking into account the negative factors of the internal and external SWOT analysis, the Low Carbon Development Indonesia (LCDI) in West Papua is needed for sustainable development by paying attention to mega-biodiversity, green economy and increasing the welfare level of indigenous Papuans. The LCDI also assists West Papua in fulfilling the commitments in the Regional Action Plan for emission reductions at the sub-national level and at the same time fulfilling Indonesia's NDC commitments in the ratification of the Paris Agreement.

2. Methods

2.1. Study area

The research was conducted in West Papua Province, which is geographically located between 4° 30' S - 1° 30' N and 128° 50' - 135° 20' E with elevation ranging from 0 – 2.955 m asl. The average annual rainfall for the period 2002-2020 is approximately 3450 mm, with a range of 400-6500 mm. While the

IOP Conf. Series: Earth and Environmental Science **989** (2022) 012007 doi:10.1088/1755-1315/989/1/012007

average temperature per year during the period is 30.5°C with a range of values from 29.0-32.0°C and the average humidity per year during the period is 81.3%, with a range of values from 75.0% - 87.5% [26]. In general, the physiography of the West Papua Province consists of coastal areas and coastal areas islands, lowlands and swamps, highlands and mountains, and basins and attenuation. The Bird's Head region is massive rugged mountains in the southeast consisting of metamorphic and granitic rocks. In the south and west, which gradually decreases in height, it consists of highlands limestone, alluvial plains, and swamps. This lowland is divided in two by the bay which stretches in an East-West direction and is flanked by swamps and plains of recent (recent) and tertiary alluvium material called Teluk Bintuni (resembling a mouth bird) [27].

Geographically, the boundaries of West Papua Province (Figure 1) are as follows:

• the Northern area: Pacific Ocean
•the Southern area: Banda Sea (Maluku Province)
• the Eastern area: Papua Province
• the Western area: Seram Sea (Maluku Province)

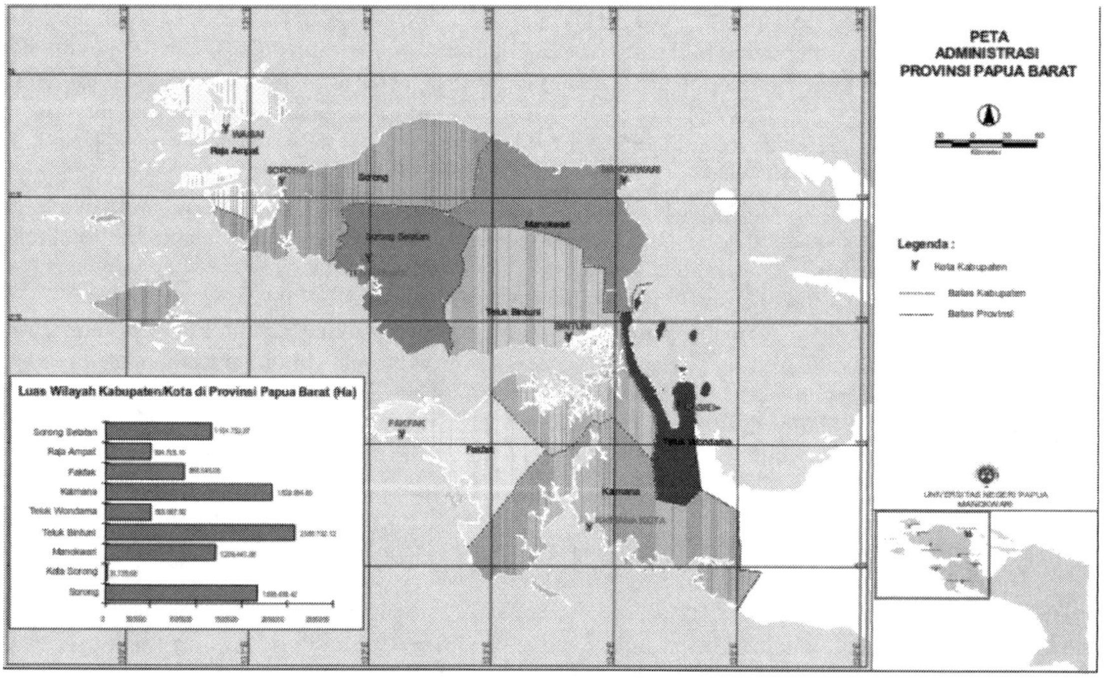

Figure 1. West Papua Province map

2.2. GHGs Emission from Forestry Sector

Emissions from the forestry sector of West Papua in the period 2009-2011, there was a significant change in emissions compared to the period of 2002-2003. Most of the emission sources in the period of 2002-2003 came from the conversion of primary dryland forest to secondary dryland forest/logged over (97.28%). The condition of these changes is much different from the 2009-2011 period, where the emission sources come from several types of changes (Table 1) [19].

Table 1. Emissions from the forestry sector in West Papua Province for the period 2009-2011

LU_CODE	LU-CHG	EM (CO_2.eq)	PERCENTAGE
PSFTSSwF	Primary swamp forest to Secondary swamp forest	468,901.220	27.97
PDFTS/LDF	Primary dryland forest to Secondary/logged dryland forest	421,634.556	25.15
PDFTOL	Primary dryland forest to Open land	315,200.005	18.8
PDFTSh	Primary dryland forest to Shrub	129,233.031	7.71
PDFTSw	Primary dryland forest to Swamp	128,414.621	7.66
S/LDFTOL	Secondary/logged dryland forest to Open land	112,032.475	6.68
PMFTSMF	Primary mangrove forest to Secondary mangrove forest	79,619.622	4.75
S/LDFTB	Secondary/logged dryland forest to Bush	21,216.710	1.27

2.3. Method of data analysis

2.3.1. Concept Development
The preparation of LCDI of West Papua uses the development of a system dynamics model which is a simulation model approach to test, predict, optimize, and analyse policies in the context of sustainable natural resource management. The main data used in the LCDI of West Papua is figured out in Table 2.

Table 2. The main variables of the West Papua LCDI model of forestry

No	Variable	Value	Unit	Estimation Method
1	Availability of forest land	8,411,561 (2010)	Ha	MoEF (2011)
2	Non-forest	945,798 (2010)	Ha	MoEF (2011)
3	Forest Emission	1,018,005 (2010)	tCO_2e	Emission Data x Emission Factor (IPCC, 2006)
4	Total Goss Regional Domestic Product (GRDP)	41,361.7 (2010)	Billion Rp	BPS Papua Barat (2011)
5	Forestry GRDP	1,362.1	Billion Rp	BPS Papua Barat (2011)

2.3.2. Modelling
System dynamics is a complex system, has dynamic and non-linear behaviour, and has feedback, or a closed system (feedback closed system) that focuses on structure and behaviour with interacting feedback. The development of the system dynamics model consists of 2 important stages, namely (1) the conceptualization stage (from story to structure), which produces a Causal Loop Diagram (CLD) and (2) the complexity stage (from structure to behaviour) by building a Stock Flow Diagram (SFD) based on the principle of structural consistency, which is then stimulated to imitate the performance and behaviour of the system being modelled. Furthermore, the formulation of the model is done by testing the consistency of the model is by the objectives or by the behaviour of the real world. After the model is built, then verification is carried out. Next, the system dynamics model is simulated and the results are validated to ensure that the model can represent the actual system (real world) [28].

2.3.3. Simulation and validation of the model

At this stage, the Causal Loop Diagram (CLD) and Stock Flow Diagram (SFD) with Powersim 10 Software are used. CLD consists of causal relationships of variables described in positive and negative relationships. While SFD represents the types of variables in the form of stock (level) & flow (rate). SFD is a more detailed description and pays attention to the effect of time on the relationship between variables so that later each variable can show the accumulated results for a variable level, and the variable which is the level of system activity for each period.

Furthermore, a mathematical model is needed to verify whether conceptual errors in computer programs and implementation do not occur after the inclusion of the formulation model on the variables in the stock flow chart. After verification, validation is carried out to ensure that the behavioural output of the model is accurate and by expectations and realities in the field. The AME (Absolute Mean Error) equation used to validate the model can be seen below:

$$AME = (Si - Ai) / Ai \times 100\% \quad (1)$$

where AME is the absolute mean error, A is the actual value, and S is the simulated value.
The limit of the AME value based on the uncontrolled variable in this study was 30% [29].

3. Results and discussion

3.1. CLD Model from Forestry Sector

The CLD results from the Forestry and Peatlands sub-sector can be seen in Figure 2. The figure shows a causal relationship from one variable to another. Each arrow indicates the effect of one variable on the other because it has a positive (+) or negative (-) loop. Loop + shows that the relationship between these variables is directly proportional. If there is an increase in the value of the variable, it will cause an increase in the value of the variable it affects. On the other hand, a negative loop shows an inversely proportional relationship (not in the same direction). If there is an increase in the value of the variable, it will cause a decrease in the value of the variable it affects.

The number of people or population is correlated with the increase in non-forest areas (deforestation). Likewise, the addition of non-forest areas has implications for the increasing population. So, the relationship between these two variables is called loop reinforcing (positive feedback, R1). Non-forest variables affect the decrease in the availability of forest land. However, on the other hand, the availability of forest land also has an increasing impact on non-forest areas. Thus, this variable relationship is known as loop balancing (negative feedback, B1). Furthermore, the availability of forest land increases the availability of surface water and in turn has a positive impact on the population. The relationship of these variables is included in the balancing loop (negative feedback, B2). Then, the availability of forest land affects the reduction in the rate of forest emissions and the rate of forest emissions contributes to an increase in total emissions, which will hurt the population. So that the variable has a loop balancing relationship (negative feedback, B3).

There are several policy interventions in reducing GHG emissions from the forestry and peatland sub-sectors by increasing forest availability, namely through a forest moratorium, a policy of 70% forest area, afforestation, and forest rehabilitation.

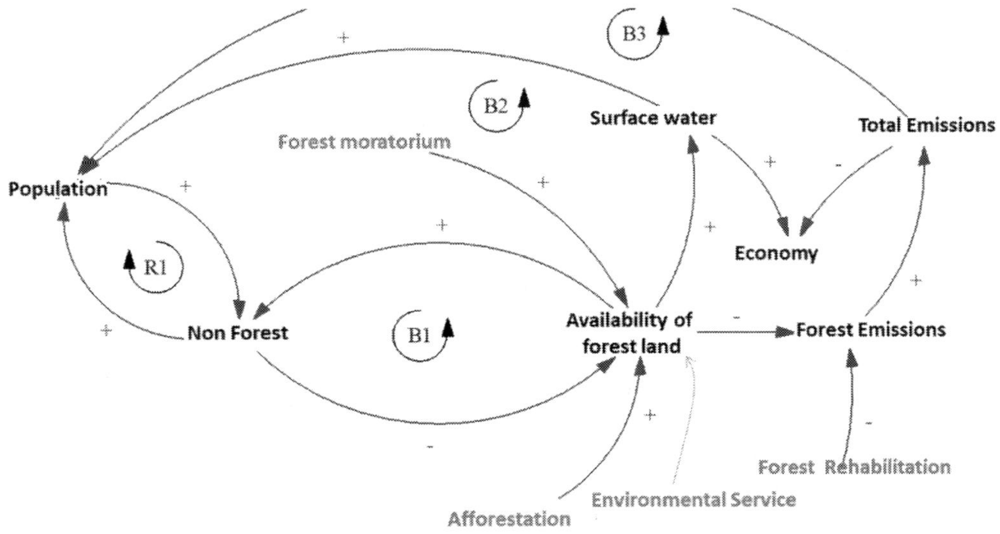

Figure 2. CLD from the forestry sector

3.2. Validation Model

Forest history data for 2010-2020 was used to validate the model (11 years). Validation was carried out for the mineral forest of West Papua in the forest sub-model. The forest area of West Papua fluctuates every year. The years 2010-2017 decreased from year to year (overtime) and subsequently increased and decreased (oscillations), which continued to increase again and tended to slope.

Calculating the error rate from the average value (Ei) obtained a validation value of 0.02%. The validation value means that the system dynamics model built can represent the actual change in forest area over time because the error percentage value is <5% (Figure 3). The model is started with a deviation limit that is still acceptable in the range of 5-10% [29].

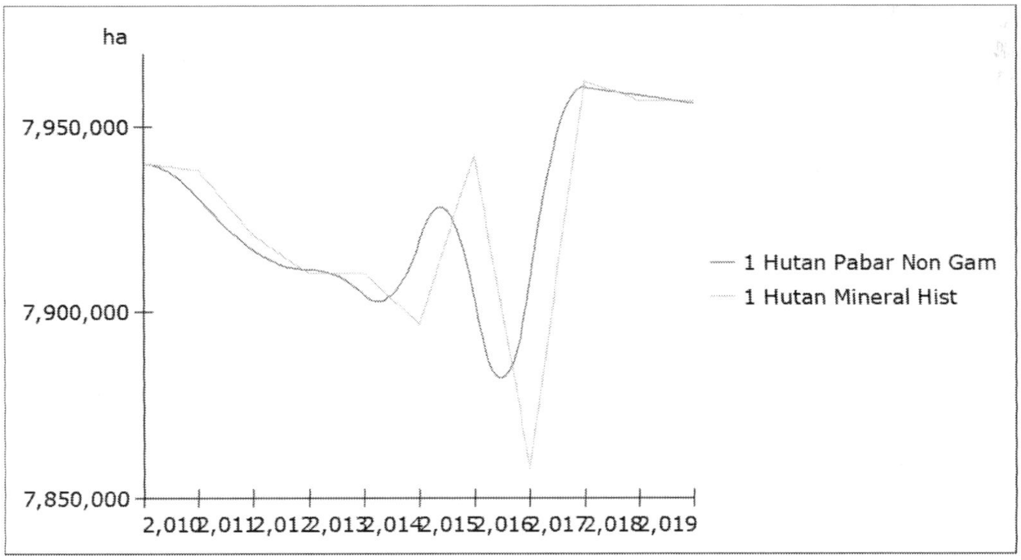

Figure 3. Comparison of historical forest data (green) versus simulated (red)

3.3. LCDI Forestry Scenario of West Papua

There are two scenarios developed in the LCDI of the forestry sector in West Papua, a fair scenario (for a 29% emission reduction) and an ambitious scenario (for a 41% emission reduction) with policy interventions as shown in Table 3.

Table 3. Forestry LCDI scenarios

Policies		Fair Scenario			Ambitious Scenario		
	Baseline	**2030**	**2045**	**2060**	**2030**	**2045**	**2060**
Minimum forest cover area maintained in total forest area (ha)	0	7,850,000	7,850,000	7,850,000	7,900,000	7,900,000	7,900,000
Area of forest and land rehabilitation (ha)	0	100,000	150,000	250,000	150,000	250,000	400,000
Area of additional forest cover (ha)	0	0	0	0	0	0	0

The simulation results show that the policies carried out with a fair scenario impact reducing emissions in 2060 and West Papua becoming a net sink emission reaching an accumulation of 200,000 tons of CO_2 (Figure 4). Likewise, the policies implemented for scenario ambitions have the effect of reducing emissions in 2060 and net sink emissions with a total accumulation of 400,000 tons of CO_2 (Figure 5).

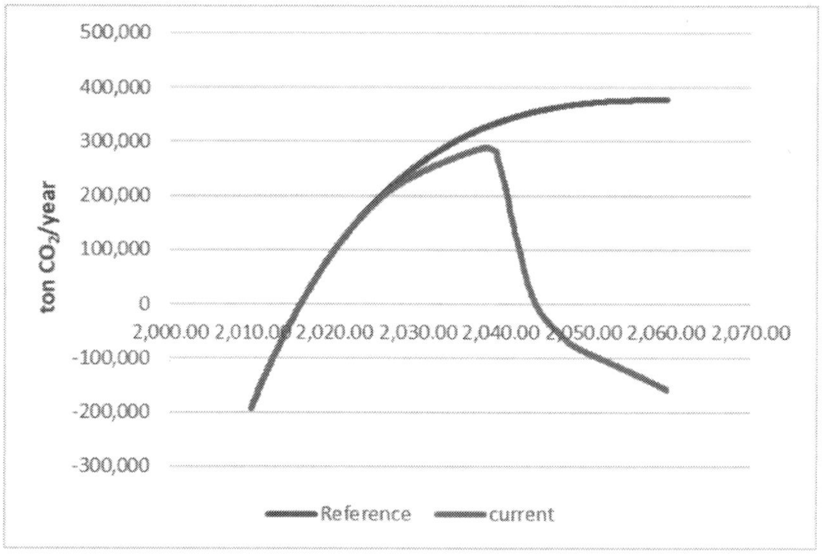

Figure 4. Net sink emission for fair scenario

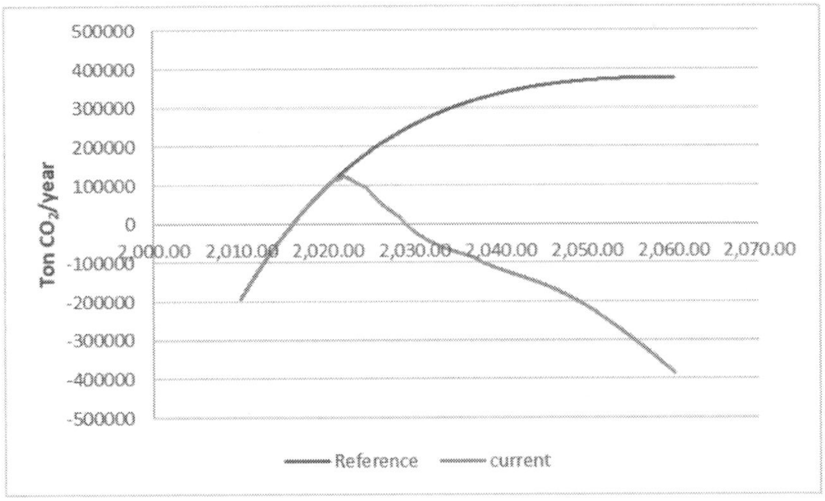

Figure 5. Net sink emission for ambitious scenario

The total GRDP issued for the fair scenario requires a 2.8% increase in the budget for a 29% emission reduction. Meanwhile, the ambitious scenario requires an increase in Total GDRP by 5.6% for a 41% emission reduction.

4. Conclusion

The net sink emission is the LCDI target and Long-Term Strategies for Low Carbon and Climate Resilience (LTS – LCCR) 2060 in West Papua. The calculation of the fair scenario (29% reduction emissions) and ambitious scenarios (41% emission reductions) show values of 200,000 and 400,000 tonnes CO_2, respectively. However, achieving this requires an additional 2.8% Total GRDP for the fair scenario and 5.6% for the ambitious scenario. Thus, the government can take steps to develop work programs related to LCDI in West Papua Province to reduce emissions and at the same time build community resilience to disasters.

Acknowledgements

The author gives the highest appreciation to the LCDI Team from WRI Indonesia, who has assisted in facilitating the preparation of the LCDI document in West Papua Province.

References

[1] Pitcher, TJ, Ainsworth, CH, and Bailey M. 2007. Ecological and economic analyses of marine ecosystems in the Bird's Head Seascape, Papua, Indonesia: I. Fisheries Centre Research Reports 15 (5). University of British Columbia, Canada, 188 pp.

[2] Bailey M, and Pitcher TJ. 2008. . Ecological and economic analyses of marine ecosystems in the Bird's Head Seascape, Papua, Indonesia: II. Fisheries Centre Research Reports 16(1). University of British Columbia, Canada, 192 pp.

[3] Mangubhai S, Erdmann MV, Wilson JR, Huffard CL, Ballamu F, Hidayat NI, Hitipeuw C, Lazuardi ME, Muhajir, Pada D, Purba G, Rotinsulu C, Rumetna L, Sumolang K, and Wen W. 2012. Papuan Bird's Head Seascape: Emerging threats and challenges in the global center of marine biodiversity. Marine Pollution Bulletin, 64 (11): 2279-2295.

[4] Huffard CL, Wilson J, Hitipuew C, Mangubhai S, Erdmann MV, Adnyana W, Barber P, Manuputty J, Mondong M, Purba G, Rhodes K, and Toha H. 2012. Ecosystem-based Management in the Bird's Head Seascape Indonesia: turning science into action. Ecosystem Based Management Program: Conservation International, The Nature Conservancy, and World Wildlife Fund Indonesia. 40 pp.

[5] Purwanto, Brown DA, Matualage D, Rumengan I, Awaludinnoer, Pada D, Hidayat NI, Amkieltiela, Fox HE, Mangubhai S, Hamid L, Lazuardi ME, Mambrasar R, Maulana N, Mulyadi, Tuharea S, Pakiding F, and Ahmadia GN. 2021. The Bird's Head Seascape Marine Protected Area Network-Preventing biodiversity and ecosystem service loss amidst rapid change in Papua, Indonesia. Society for Conservation Biology, Conservation Science and Practice, 2021; 3: e393, DOI: 10.1111/ csp2.393.

[6] Leverage edu. 2020. Largest Forest in the World. https://leverageedu.com/blog/largest-forests-in-the-world/.

[7] Mongobay. 2020. The World's largest rainforests. https://rainforests.mongabay.com/facts/the-worlds-largest-rainforests.html.

[8] World Wildlife Fund 2004 Bird's head and bomberai peninsulas of Indonesian Papua (ESA, Observing Earth).

[9] Bappeda Papua Barat. 2016. Regional Action Plan (RAP) for Reducing Greenhouse Gas (GHGs) Emissions of West Papua Province 2017-2021. Manokwari, 87 pp.

[10] Bappeda Papua Barat. 2017. Medium-Term Development Plan (MTDP) West Papua Province 2017-2022. Manokwari, 938 pp.

[11] Bappeda Papua Barat. 2018. Environmental Strategic Environmental Studies (SES) Medium-Term Development Plan (MTDP) West Papua Province 2017-2022. Manokwari, 35 pp.

[12] Bappeda Papua Barat. 2018. Zoning Plan for Coastal Areas and Small Islands (ZPCASI) West Papua Province. Manokwari, 230 pp.

[13] Bappeda Papua Barat. 2018. Environmental Strategic Studies (SES) Zoning Plan for Coastal Areas and Small Islands (ZPCASI) West Papua Province. Manokwari, 40 pp.

[14] BPBD Papua Barat. 2012. Disaster Management Plan 2012-2016. Manowkari145 pp.

[15] Central Statistics Agency Indonesia 2020 Profil kemiskinan di Indonesia Januari 2020 (Jakarta BPS).

[16] Austin KG, Schwantes A, Gu Y, and Kasibhatla PS. 2019. What causes deforestation in Indonesia. Environmental Research Letter, 14 (2): 1-9.

[17] Shearman P and Bryan J. 2011. A bioregional analysis of the distribution of rainforest cover, deforestation and degradation in Papua New Guinea. Austral Ecology, https://doi.org/10.1111/j.1442-9993.2010.02111.x.

[18] Pemerintah Papua dan Papua Barat. 2018. Manokwari Declaration: Sustainable Development based on Indigenous Area in the Land of Papua. Manokwari, 4 pp.

[19] Tacconi L, and Muttaqin MZ. 2019. Policy Forum: Institutional architecture and activities to reduce emissions from forests in Indonesia. Forest Policy and Economics, 108, https://doi.org/10.1016/j.forpol.2019.101980.

[20] WRI Indonesia. 2018. Connecting the dots: Elements for a jointed-up Implementation of 2030 Agenda and Paris Agreement. https://www.wri.org/research/connecting-dots-elements-joined-implementation-2030-agenda-and-paris-agreement?utm_source=&utm_medium=&utm_campaig=.

[21] Timboe I, and Pharr K. 2021. Nature-based solutions in international policy instruments. Nature-based Solutions and Water Scarcity, Elsevier, 125-147. https://doi.org/10.1016/B978-0-12-819871-1.00015-4.

[22] National Disaster Management Agency. 2020. Indonesia Disaster Risk Index (Jakarta: Data Informasi Bencana Indonesia).

[23] National Disaster Management Agency. 2020. Disaster data in West Papua (Jakarta: Data Informasi Bencana Indonesia).

[24] Hendri, Karuniasa M, Prabawardani S, Syamsudin K, and Pradafitri WS. 2021. Scenario for West Papua contribution for NDC from forestry sector. IOP Conf. Series: Earth and Environmental Science 716 (2021) 012017 IOP Publishing doi:10.1088/1755-1315/716/1/012017.

[25] Hendri, Alianto and Suhaemi. 2016. Coastal Hazard Management due to Extreme Climate Conditions in the Bird's Head Papua. Proceeding: Jogja Earthquake in Reflection 2016 24–26

May 2016.
[26] BMKG Papua Barat. 2020. Climate Data in West Papua. Manokwari.
[27] Bappeda Papua Barat. 2007. West Papua Regional Development and Investment Plan 2007-2026. Manokwari, 21 pp.
[28] Pratiwi Y, Rejo A, Fariani A, and Faizal M. 2021. Source of Carbon Dioxide Emission Equivalent to Plant-based Control Concept with Dynamic System in Prabumulih City, South Sumatera Province. IOP Conf. Ser.: Earth Environ. Sci. 810 012051.
[29] Malahati, M. and T. Masui, The impact of geenhouse gas mitigation policy for land use and forestry sector in Indonesia: Applying the computable general equilibrium model. Journal of Forest Policy and Economics, Vol 109, 102003.

GEOSTAT application design for rapid determination of MATLAB-based earthquake risk level

Takhul Bakhtiar * **and Iman Suardi**

College of Meteorology Climatology and Geophysics, South Tangerang, Banten 15221, Indonesia

*Email: takhul.bakhtiar@stmkg.ac.id

Abstract. In order to support the rapid determination, assessment and research of a major earthquake event that occurs in an area, a fast, simple and easy application is needed to calculate the level of earthquake risk.The GEOSTAT Application is an application created using the MATLAB program. The GEOSTAT Application can be used to calculate the level of earthquake risk using the Gutenberg and Richter equations which are solved by the least squares method and the maximum likelihood method. This application is designed and built using the MATLAB Graphical User Interface Development (GUIDE) which is created and compiled in a pure script. The GEOSTAT Application can help users calculate the level of earthquake risk quickly and easily which is usually calculated manually. In addition, The GEOSTAT Application provides early warning information related to the analysis of the level of earthquake disaster risk.The GEOSTAT Application can be installed on the user's device and has good performance so that it can assist users in calculating the level of earthquake risk quickly and easily.

1. Introduction

Indonesia has a complex tectonic structure because there are many active faults and located on three major tectonic subduction plates in the world. Based on this complex tectonics structure, earthquake activity in Indonesia is very high so that it can cause large and destructive earthquakes. This can threaten human safety and cause casualties. In order to deal with this, one of the mitigation efforts is needed to reduce the loss of life due to the earthquake [5].

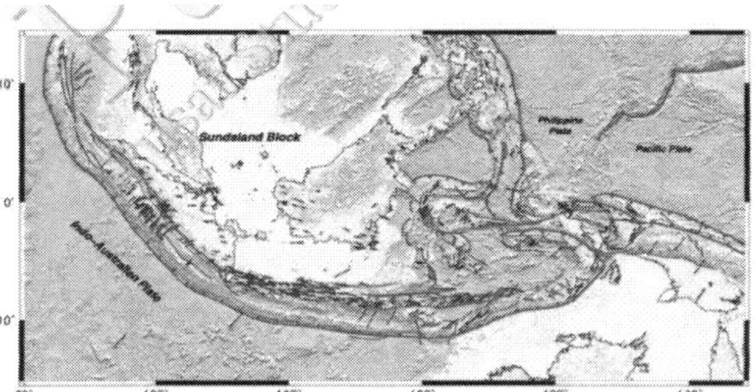

Figure 1. Indonesia Tectonic Map [13]

Large earthquake or destructive earthquakes that have occurred, will cause panic in the people about when the earthquake will happen again. Several academics, researchers and related institutions will investigate, study and examine when the earthquake event will recur. Investigations, studies and research related to the threat of when an earthquake will happen again are aimed at disaster mitigation in the area.

One way to analyze earthquake risk for human safety is by using data from earthquakes that occurred in the past. The earthquake data is used to calculate the frequency of earthquakes in a certain period [8]. The relationship between frequency and earthquakes can be used in order to determine the a-value and b-value parameters. These parameters can be used to calculate the seismicity index, return period and earthquake risk level [9].

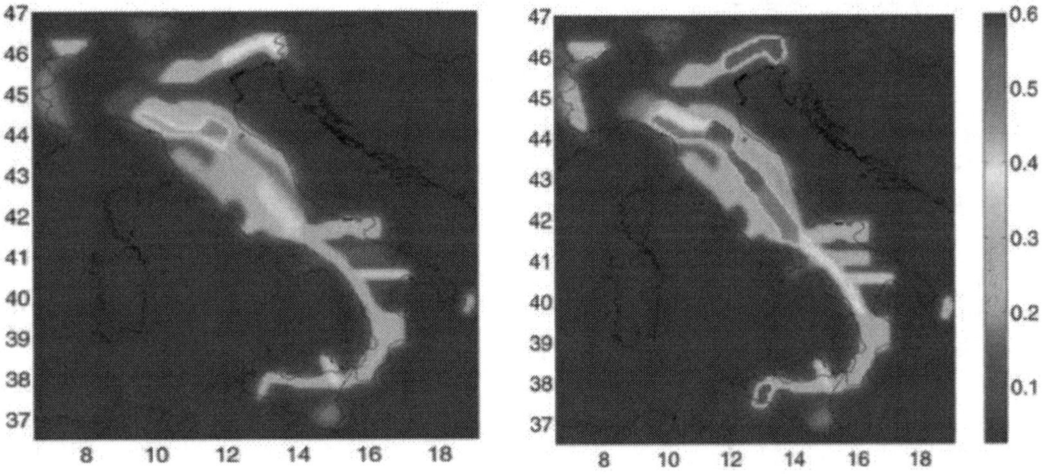

(a) **(b)**

Figure 2. Seismic hazard map 10% probability exceeded in 50 years (a) After the process declustered catalog data (b) Before the process declustered (full non-declusterd) earthquake catalog data [7].

The thing that often done in the process of analyzing earthquake risk is the stage of declustered. Declustered is the process of separating foreshocks, mainshocks and aftershocks from earthquake catalog data [16]. The declustered process can affect the b-value calculation. The calculation of b-value calculated through the declustered process has a higher uncertainty than the calculation without going through the declustered process or full catalog of earthquake data [12]. In addition, the increase in the threat of earthquakes calculated based on the earthquake catalog without the declustered process is higher than through the declustered process [2]. If the calculation is included in the seismic hazard map, the map calculated based on the declustered process will be ignored and cannot be accepted. This is because the results of map that went through declustered process don't based on the worst-case scenario [7].

The complexity and length of the calculation process in determining the level of earthquake risk is also still a problem in the process of analyzing the level of earthquake risk. So we need an application that is simple, fast and easy to be able to analyze earthquake risk. In addition, we need an application that can calculate the level of earthquake risk with the worst scenario for mitigation purposes.The GEOSTAT Application is a solution to answer the complexity of these problems. This application can be used by users quickly, easily and simply. In addition, in the process of analyzing earthquake risk, this application does not have a declustered stage. So for earthquake mitigation purposes this application can calculate the level of earthquake risk based on the worst scenario.

IOP Conf. Series: Earth and Environmental Science **989** (2022) 012008 doi:10.1088/1755-1315/989/1/012008

2. Methods

The GEOSTAT Application design process requires some software such as MATLAB 2016 as the main component in creating, designing and compiling the source code in a pure script. The application design process utilizes the Graphical User Interface Development (GUIDE) facility in MATLAB. In simple terms the design process can be illustrated by the following figure [10].

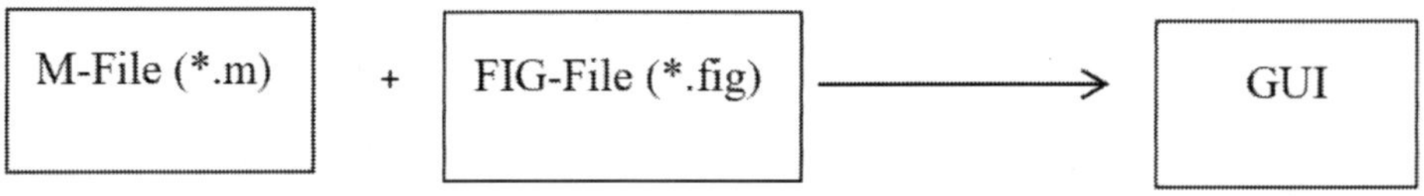

Figure 3. The GUI process [10].

The GEOSTAT Application uses Microsoft Excel as a place to store earthquake data. Canva tools are used to design the background for The GEOSTAT Application. In addition, to test applications, validate systems, determine application performance and application feasibility, several user devices are needed in the form of laptops or computers as test media. The following test media will be used:

Table 1. Application test media

Devices	Specifications	Devices	Specifications
Laptop Lenovo Ideapad 330	- RAM 8 GB - DDR4 - MATLAB version R2016B - HDD 1 TB - AMD A9	Laptop HP	- RAM 4 GB - Intel(R) Coleron CPU N 3060 - Not installed MATLAB - HDD 500 GB
Laptop HP14S	- RAM 8 GB - AMD RYZEN 5 - MATLAB verison R2016B - HDD + SSD	Laptop Lenovo Ideapad 110 15-Ast	- AMD TTH GEN DUALCORE A99400-2.4GHZ TURBO 3.2 GHZ - 4GB DDR4 - AMD RADEON R5 M430 2 GB - 1 TB SATA - MATLAB version R2016B

Research data obtained from several studies. Earthquake data from previous studies will be tested with The GEOSTAT Application to see the level of accuracy this application calculation results by comparing the results of previous studies with the calculation results of The GEOSTAT Application. Several studies that will be tested using The GEOSTAT Application can be seen below:

1. Spatial Variation of b-Values and Their Relationship with the Fault Blocks in the Western Part of the Tibetan Plateau and Its Surrounding Areas [4].
2. Korelasi Tingkat Seismisitas dan Periode Ulang Gempa Bumi di Kepulauan Mentawai dengan Menggunakan Metode Guttenberg Richter [3].
3. Analisa Statistik Seismisitas Sulawesi Utara dan Sekitarnya Periode 1973-2007 [15].

The method that can be used to analyze the relationship between earthquake magnitude and frequency for earthquake mitigation purposes is to use the Gutenberg and Richter equations [9]:

$$Log_{10}N\ (M) = a - bM \tag{1}$$

$N\ (M)$ is the frequency of seismic activity and M is the magnitude, while a and b are constants. The constant a describes earthquake activity in a certain area which is influenced by the research area, the research period and the level of seismic activity in an area. The constant b describes the tectonic parameters of a research area [1].

Then to find out the a-value and b-value parameters, it can be solved using the maximum likelihood method and least square method.

Maximum likelihood formula [14].

$$a = \log n(M) + bM_0 + \log(b\ ln10) \quad (2)$$

and

$$b = \frac{\log e}{\bar{M} - M_0} \quad (3)$$

Least square formula [11]

$$a = \bar{Y} - b\bar{X_\iota} \quad (4)$$

and

$$b = \frac{n\Sigma X_i Y_i - \Sigma X_i \Sigma Y_i}{n\ \Sigma X_i^{\ 2} - (\Sigma X_i^{\ 2})} \quad (5)$$

Parameters a-value and b-value can be used to determine the value of the seismicity index using the following equation [6]:

$$N1(M \geq M_0) = 10^{(a - \log(b \ln 10) - log\Delta t) - bM_0} \quad (6)$$

With :
$N1(M \geq M_0)$ is the seismicity index for $N1(M \geq M_0)$
a and b are constants of the relationship between frequency and magnitude
M is the upper limit of the tested magnitude
$M_0$is the lower limit of the tested magnitude
Δt is the observation time interval

Then after the seismicity index price is obtained, the return period value can be determined using the following equation [6]:

$$\theta(M \geq M_0) = \frac{1}{N1(M \geq M_0)} \quad (7)$$

With :
$\theta(M \geq M_0)$ is earthquake return period
$N1(M \geq M_0)$ is the seismicity index

The earthquake risk value can be determined by using the following formula [6]:

$$P(M \geq M_0, T) = \left(1 - e^{-N1(M \geq M_0)T}\right) \quad (8)$$

$P(M \geq M_0, T)$ is the level of risk of an earthquake with a certain magnitude and time
$N1(M \geq M_0)$ is the seismicity indeks
T is the observation time

3. Result and Discussion

The results obtained from the test installation of The GEOSTAT Application on several user devices can be seen that The GEOSTAT Application has been successfully installed and operated with various specifications. The GEOSTAT Application can be operated on the user's device. available MATLAB such as the **Lenovo Ideapad 110 15-Ast Laptop, HP14S Laptop** and **Lenovo Ideapad 330 Laptop.** In addition, The GEOSTAT Application was also successfully installed and operated without the MATLAB application with a large capacity such as an **HP Laptop**. So that The GEOSTAT Application can be operated with or without MATLAB installed. Here are the results of the installation test on several user devices :

Table 2. Installation test results

Device	Specification	Result	Device	Specification	Result
Laptop Lenovo Ideapad 330	- RAM 8 GB - DDR4 - MATLAB R2016B - HDD 1 TB - AMD A9	Success Installed	Laptop HP	- RAM 4 GB - Intel(R) Coleron CPU N 3060 - Not installed MATLAB - HDD 500 GB	Success Installed
Laptop HP14S	- RAM 8 GB - AMD RYZEN 5 - MATLAB R2016B - HDD + SSD	Success Installed	Laptop Lenovo Ideapad 110 15-Ast	- AMD TTH GENDUALCORE A99400-2.4GHZ TURBO 3.2 GHZ - 4GB DDR4 - AMD RADEON R5 M430 2 GB - 1 TB SATA - MATLAB R2016B	Success Installed

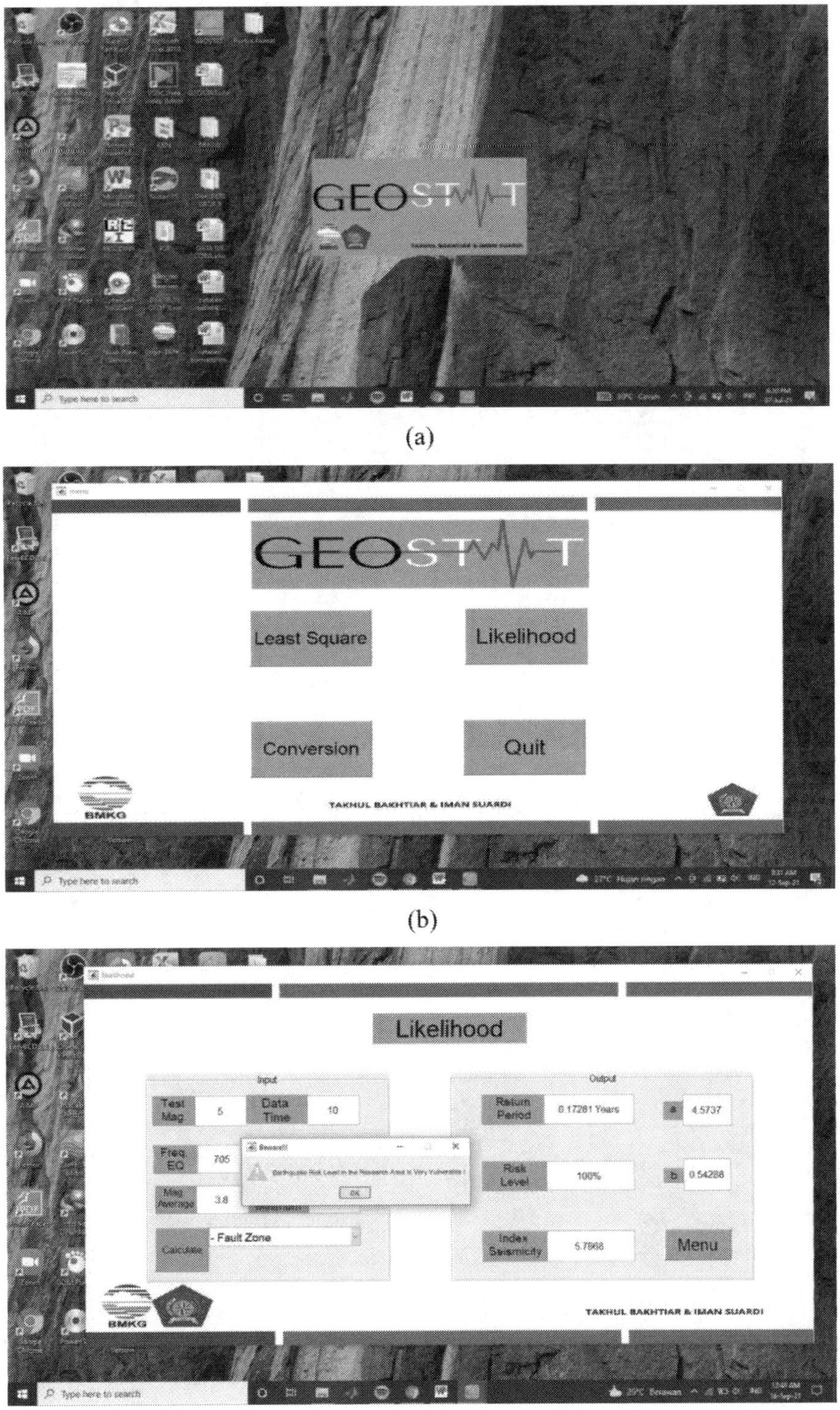

(a) (b) (c)

Figure 4. (a) Splashscreen process (b) GEOSTAT Application Menu (c) GEOSTAT Application calculation results

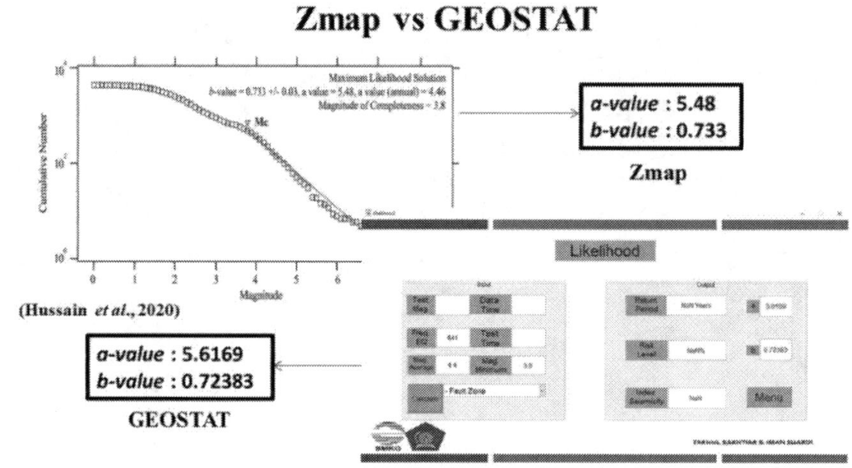

Figure 5. Comparison of Zmap and GEOSTAT calculation results

Table 3. Maximum likelihood validation test results

Parameter	Fidia dkk. (2018)	GEOSTAT
a-value	6.8094	7.0242
b-value	0.8372	0.8686
Seismicity Index (M=5)	2.158	2.4
Return Period (M=5)	0.463	0.4167

Table 4. Least square validation test results

Parameter	Wibowo (2008)	GEOSTAT
a-value	8.2195	8.2195
b-value	1.2219	1.2219
Korelasi	-0.98	-0.98
Seismicity Index (M=5)	1.3085	1.3085
Return Period (M=5)	0.76426	0.76426
Risk Level (M=5)	99.99979	99.99979

Based on the Zmap calculation results, it can be seen that the *a*-value and *b*-value parameters are 5.48 and 0.733, while The GEOSTAT Application calculation results are obtained for *a*-value and *b*-value parameters of 5.6109 and 0.72383. There is no significant difference between two calculation results. In addition, there is no significant difference in calculation results of the maximum likelihood method and least square method between previous researchers and The GEOSTAT Application.

4. Conclusion

The GEOSTAT Application can help simplify, speed up the process of calculating the level of earthquake risk. In addition, The GEOSTAT Application can be installed in various specifications of the user's device. The minimum specification recommendations so that the application performance runs well on the user's device are:

- RAM 4 GB - DDR4
- HDD 1 TB - AMD A9 or Core I3

IOP Conf. Series: Earth and Environmental Science **989** (2022) 012008 doi:10.1088/1755-1315/989/1/012008

References

[1] Bilim F 2019 *Journal of Science and Technology.* **9** 11-15
[2] Boyd O S 2012 *Bulletin of the Seismological Society of America (BSSA).* **102** 909–917
[3] Fidia R, Pujiastuti D and Sabarani A Z 2018 *Jurnal Fisika Universitas Andalas.* **7** 84-89
[4] Hussain H, Shuangxi Z, Usman M and Abid M *Entropy.* **22** 1-13
[5] Ibrahim G, Sendjaja P and Subarjo 2010 *Tektonik dan Mineral Indonesia* (BMKG: Jakarta)
[6] Lusiani E, Anwar S and Nugraha M F 2018 *Jurnal Meteorologi dan Geofisika.* **19** 71-79
[7] Marzocchi W and Taroni M 2014 *Bulletin of the Seismological Society of America (BSSA).* **104** 1838 – 1845
[8] Pawirodikromo W 2012 *Seismologi Teknik dan Rekayasa Kegempaan* (Yogyakarta: Pustaka Pelajar)
[9] Rohadi S, Grandis H and Ratag M A 2008 *Jurnal Meteorologi dan Geofisika.* **9** 101-108
[10] Sugiharto A 2006 *Pemrograman GUI dengan MATLAB* (Yogyakarta: Penerbit Andi)
[11] Supranto J 2008 *Statistika Teori dan Aplikasi Edisi Ketujuh* (Jakarta: Penerbit Erlangga)
[12] Taroni M and Akinci A 2021 *Geophysical Journal Internasional.* **224** 1174-1187
[13] Tim Pusat Studi Gempa Nasional 2017 *Peta Sumber dan Bahaya Gempa Indonesia Tahun 2017* (Bandung: Pusat Litbang Perumahan dan Pemukiman)
[14] Utsu T 1974 *J. Phys. Earth.* **22** 71-85.
[15] Wibowo S N E 2008 Analisa Statistik Seismisitas Sulawesi Utara dan Sekitarnya Periode 1973-2007 *Laporan Kerja Program Pendidikan Diploma III* (Jakarta: Akademi Meteorologi dan Geofisika)
[16] Zhuang J, Ogata Y and Jones D V 2002 *J. Am. Stat. Assoc.* **97** 369–380

IOP Conf. Series: Earth and Environmental Science **989** (2022) 012009 doi:10.1088/1755-1315/989/1/012009

Towards resilience-vulnerability communities to climate change in Doom Island Sorong City West Papua Province

Derek Ampnir [1,*], **Budi Santoso** [2], **and Rudi A. Maturbongs** [2]

[1] Environmental Science of Graduate Program, Universitas Papua, West Papua, 98314, Indonesia
[2] Faculty of Forestry and Graduate Program, Universitas Papua, West Papua, 98314, Indonesia

*Email: derek.ampnir@gmail.com

Abstract. Doom Island is a small island in West Papua from a total of 4.110 small island which is located in the Sorong Archipelago District. This island is directly opposite the Pacific Ocean so it is very vulnerable to climate change induced hazards, including landslides, abrasion, tropical storm, heavy rainfall, water scarcity, and food security. Adaptation strategies especially for vulnerable communities such as baby, children under five years of age, children, pregnant or lactating mothers, persons with disabilities and the elderly is needed to increase their resilience to climate change and the other disasters. Socialization, training, and simulation are introduced to vulnerable communities and followed by in-depth interviews to find out adaptation strategies from local knowledge and harmonized with science.

1. Introduction

In general, Indonesia has 13 disaster threats, namely earthquakes, tsunamis, floods, landslides, volcanic eruptions, extreme waves and abrasion, extreme weather, droughts, forest and land fires, building and residential fires, epidemics and disease outbreaks, technology failures. and social conflict. Meanwhile, West Papua has 11 disaster threats, except for volcanoes and technological failures. Based on InaRISK data [1,2], West Papua is at a moderate Disaster Risk Index (Raja Ampat, Tambrauw, Maybrat, Fakfak and Kaimana) and high Disaster Risk Index (Sorong City, Sorong Regency, South Sorong, Manokwari, South Manokwari, Arfak Mountains, Bintuni Bay and Wondama Bay). The disaster threats of SWOT analysis in West Papua and Indonesia are generally divided into 2 broad categories, namely geology and hydrometeorology [3,4,5,6,7].

However, these disasters have only been studied for large islands and have not paid attention to small islands in West Papua Province which reached 4,110, spread over the Bird's Head of Papua [8]. These small islands are also a strength in SWOT analysis with all the potential, economic and ecological functions [8,9,10].

Weaknesses found in the SWOT analysis for the islands are accessibility, low data and low coordination, especially on the outer islands and also the absence of risk assessment (hazard, vulnerable, exposure and capacity building) [11,12,13]. Therefore, coordination is needed from the National, Sub-National and Small Islands levels [14,15,16].

In terms of opportunities from the SWOT analysis, West Papua is currently also reviewing the Disaster Management Plan which will later become a guide for Disaster Risk Reduction and mainstreaming in development in all regional apparatus organizations and also integration in Regional

IOP Conf. Series: Earth and Environmental Science **989** (2022) 012009 doi:10.1088/1755-1315/989/1/012009

Spatial Plan, Strategic Environmental Studies, Coastal & Small Island Zone Plan and other documents. It is hoped that the document can synergize with policies at the small island level [17,18,19].

In order to increase resilience-vulnerability communities (RVC), vulnerability measurement and assessment are urgently developed, especially in increasing the RVC to climate change and climate-related disasters which have increased in the recently. This study would like to know Vulnerability Index for measuring the RVC in Doom islands, Sorong City, West Papua Province.

2. Methods

2.1. Study area

The research was conducted in Doom Island, Sorong City, West Papua Province which is geographically located 0°53′13″S and 131°14′10″E with elevation ranging from 0 – 30 m a.s.l. The average of annual rainfall for period 2010-2020 approximately 2711 mm, with a range 1824-3941 mm. While, the average temperature per year during the period is 27.1°C with a range of values from 27.0-27.3 °C and the average humidity per year during the period is 85.3% with a range of values from 83.4% - 87.0% [20].

Doom Island is included in the Sorong Islands District and was first served as an administrative center of the Dutch East Indies administration in Papua. Regionally, the island is divided into two villages, namely West Doom and East Doom with a total population of 10,186 which is spread higher in the eastern part by 60%. This is due to East Doom Island is the place for the ship's port and also nearest to the Big Island which can be reached by boat for 10 minutes.

The northern and western parts of Doom Island have seen changes in coastline with an average of 70 m (minimum 55 m and maximum 85 m) and 76 m (minimum 50 m and maximum 104 m). This shoreline change is caused by direct contact with the Pacific Ocean with the potential for higher currents and also the trajectory of the tail of a tropical storm that crosses the northern coast of Papua [21] (Figure 1).

Figure 1. Research site in Doom Island

IOP Conf. Series: Earth and Environmental Science **989** (2022) 012009 doi:10.1088/1755-1315/989/1/012009

2.2. Method of data analysis

This research involved 30 families in Doom Island District who were selected by non-proportional stratified random sampling from those who scattered from the coast and in the center of the island to the northern and western with 4 families each and the eastern and southern coasts of 3 or 4 families each.

Furthermore, an effort to determine the vulnerability index that is divided into exposed population index, social, economic, physical and ecological/environmental vulnerability using the following equation [22]:

Exposed population index = (60% x score) + (40% x score) (1)

Social vulnerability = $(0,6 * \frac{log\frac{population\ density}{0.01}}{\log(\frac{100}{0.01})})$ + (0,1* sex ratio) + (0,1 * poverty ratio) + (0,1 * disability ratio) + (0,1 * age group ratio) (2)

Economic vulnerability = (60% x score) + (40% x score) (3)

Physical vulnerability = (0,4 * house score) + (0,3 * public facility score) + (0,3 * special facility score) (4)

Environmental vulnerability = (0,4 * protected forest score) + (0,4 * primary forest score) + (0,1 * mangrove forest score) + (0,1 * shrub score) (5)

Cliamte change disaster vulnerability = (0.4 * Social vulnerability Score) + (0.25 * economic vulnerability score) + (0.25 * physical vulnerability score) + (0.1 * environmental vulnerability score) (6)

3. Results and discussion

3.1. Respondent Characteristics

Respondent characteristics data are presented in Table 1 with the average family size from the research site of 4 members.

Table 1. Minimum, maximum, and mean values of the respondent characteristics

Family Characteristics (unit)	(Min-Max; Average)
Age of husband (year)	34.0 – 65.0; 50.0
Age of wife (Year)	37.0 – 60.0; 49.0
Education of husband (year)	0.0 – 12.0; 6.0
Education of wife (year)	0.0 – 12.0; 6.0
Number of family member (people)	3,5 – 3.8; 3.7
Income/capita/month (thousand Rp)	550.0 – 1,500.0; 1,125.0
Income/capita/month (thousand Rp) during climate-related disaster	75.0 – 750.0; 375.0

The average age of both husband and wife of family respondents is included in the pre-retirement age group with the age range included in the labour force of 45-49 and 50-54 [23]. Respondent's education can be seen that the average is only graduated from elementary school and does not follow the standard 12-year required education program from the Central Government. The average income per capita per month of the family is 550.1 thousand rupiahs and fall into 75.0 thousand rupiahs during hydrometeorology-related disaster conditions such as strong winds, heavy rains, high tides due, flood and drought which generally developed a maximum of up 2 weeks. During these extreme conditions, fishermen generally look for other jobs or are forced to wait until conditions improve. The land around Doom Island is also limited in terms of yard area, therefore verticulture technique for agroforestry plants are highly recommended for narrow areas with the main focus on local crops such as sweet potatoes and breadfruit for food security.

3.2. Exposed Population Index

Exposed Population Index (EPI) is calculated based on the distribution of weights for population density (60%) and vulnerable groups (40%) which are further categorized into low index classes (score 0.33), medium (score 0.67) and high (score 1). Low index class for population density <500 people/km^2 and vulnerable group <20%. As for the middle and high index classes for a total population density of 500-1000 people/km^2 - > 1000 people/km^2 and vulnerable groups of 20-40% - > 40%. Based on BPS data, the population density of East Doom Village is 119 people/km^2 and 93 people/km^2 for West Doom. While the data for the vulnerable group (12 disabled + 1674 elderly = 1686 people/10,186 people x 100% = 16.6%). So that EPI is included in the index class in the low category.

Disasters related to climate change that occurred on Doom Island from the interview results were floods (rob), landslides, extreme waves and abrasion and extreme weather. From the results of the EPI calculation in Doom Island, it was obtained a value of 0.33 which was categorized in the low index class (Figure 1) [22].

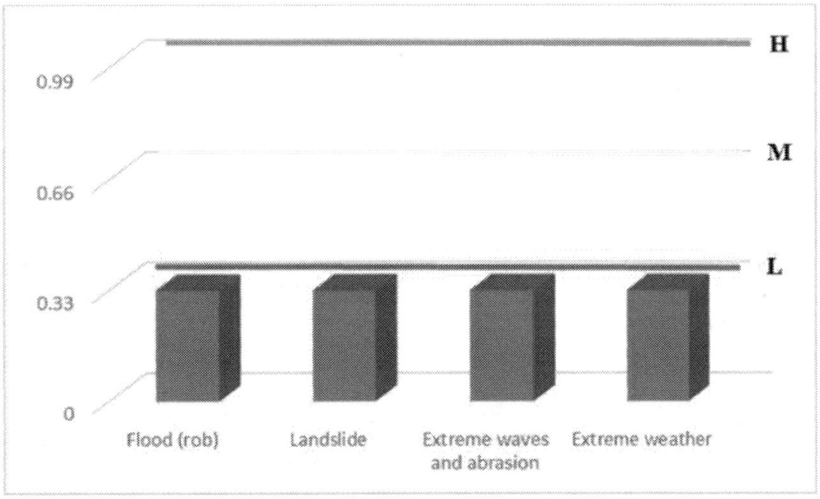

Figure 1. EPI in Doom Island

3.3 Social Vulnerability

Social vulnerability (SV) indicators consist of population density, sex ratio, poverty ratio, ratio of disabled people and age group ratio. The social vulnerability index is obtained from the average weighted population density (60%), vulnerable groups (40%) consisting of sex ratio (10%), poverty ratio (10%), disabled people ratio (10%) and age group (10%).

IOP Conf. Series: Earth and Environmental Science **989** (2022) 012009 doi:10.1088/1755-1315/989/1/012009

BPS data shows that the population density data is 106 people/km^2 with sex ratio 1.07, poverty ratio 0.04, disability ratio 0.097 and elderly group ratio 0.066. So that the value of SV in Doom Island is 0.73 and is categorized in the middle index class [22].

3.4 Economic Vulnerability
The indicators of economic vulnerability (EV) needed are in the form of productive land area in Rupiah (60%) and Gross Regional Domestic Product (GRDP, 40%) As for the low, medium and high classes of the value of the following productive land area is <50 million Rp, 50 - 200 million Rp and > 200 million Rp respectively. Meanwhile, the value of GRDP for the low, medium and high classes is < 100 million IDR, 100-300 million IDR and > 300 million IDR respectively.

The land area affected by climate change based on survey and interview results is < 50 million IDR and <100 million IDR GRDP which is included in the low index class with a value of 0.33. Thus, after being included in the calculation formula, the EV value is 0.33 that is included in the low index class category [22].

3.5 Physical Vulnerability
The indicators needed in formulating physical vulnerability (PV) are the density of houses (permanent, semi-permanent and non-permanent, 40%), the availability of buildings/public facilities (30%) and the availability of critical facilities (30%%). The value of house prices in the low, middle and high classes is < 400 million IDR, 400 - 800 million IDR and > 800 million IDR, respectively. Meanwhile, the value of prices for general and special facilities for low, middle and high classes is respectively < 500 million Rp, 500 - 1 billion Rp and > 1 billion Rp.

Based on the results of surveys and interviews from respondents, it was found that PV affected by climate change in home is included in the medium category and low category for general and special facilities. Therefore, that if included in the formulation, the results of the PV value is in the low – medium category index with a value of 0.47 [22].

3.6 Environmental Vulnerability
The indicator needed for environmental vulnerability is land cover (30% for protection forest, 30% for natural forest, 10% for mangrove forest, 10% for swamp and 10% for shrubs). For protected forest, the affected class values in the low, medium and high categories is < 20 ha, 20 - 50 ha and > 50 ha, respectively. As for natural forest, the affected class values in the low, medium high categories is < 25 ha, 25 - 75 ha and > 75 ha, respectively. As for the area of mangrove forest and shrubs affected for the low, medium and high categories is < 10 ha, 10 - 30 ha, and > 30 ha, respectively. And for swamp forest, the affected area for low, medium and high categories is < 5 ha, 5 - 20 ha and > 20 ha, respectively.

The EV value obtained from the respondents and the survey shows that the areas affected by climate change from protected forests, nature, mangroves, shrubs and swamps is in the low value category and after being included in the formula with the index value of 0.33 [22].

3.7 Climate Change Disaster Vulnerability
Calculation of vulnerability to climate change disasters on Doom Island can be seen in Figure 2. The figure shows that floods (rob), landslides, extreme waves and abrasion, and extreme weather have a disaster vulnerability value of 0.52 which is in the low - medium class category.

Physical damage in the form of houses is the main focus of disasters caused by climate change, especially from abrasion disasters which cause damage to several houses. The mitigation implemented by the local government has not been able to overcome abrasion due to technical errors with erosion-retaining stones derived from cement which are rectangular in shape which are not suitable for abrasion-resisting construction in the northern and western of Doom Island (Figure 3). Therefore, mitigation efforts based on technology that have been developed by the Ministry of Public Works and Housing in standards and guidelines for the construction of building materials and civil engineering are indispensable in efforts to increase RVC [24].

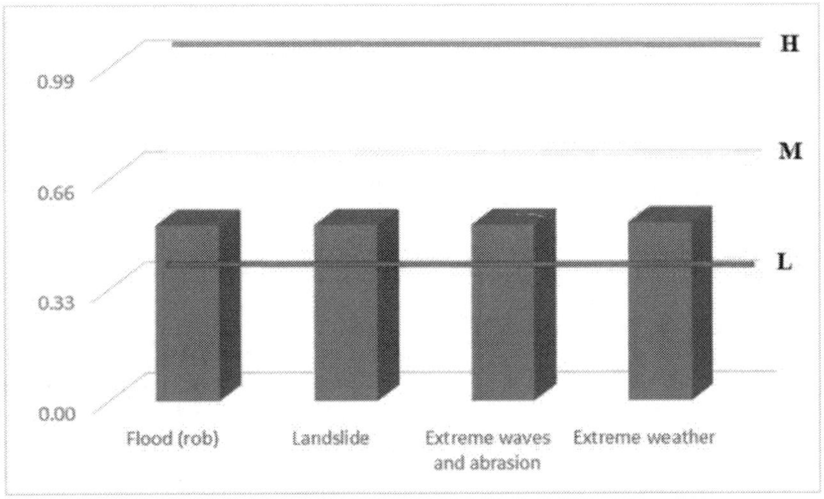

Figure 2. Climate change disaster vulnerability

Figure 3. Mis-construction for abrasion

4. Conclusion

The assessment of vulnerability to climate change disasters was carried out on Doom Island, which is one of the small islands in Sorong City which was originally a historical island in Dutch administration before entering the Big Island. The calculation of the vulnerability value is based on social, economic, physical, and environmental vulnerability. Social vulnerability has a value in the medium - high category, followed by physical vulnerability in the low-medium category, and economic and environmental vulnerability in the low category. In total, the vulnerability to climate change disasters in the Doom Island is categorized in the low-middle class category. Based on these results, what needs to be considered is the construction of abrasion as barrier mitigation, especially in the northern and western areas of Doom Island in the quest towards RCV.

Acknowledgements
The authors would like to thank the Team of Regional Disaster Management Agency of West Papua that supported data and technical during the development of this proceeding. Thanks are also given to Mr. Hendri as the supervisor who has helped the process of writing this proceeding.

References

[1] BNPB 2019 Indonesia Disaster Risk Index http://inarisk.bnpb.go.id/irbi/kabupaten.

[2] BNPB 2020 Disaster Risk Assessment of West Papua (Jakarta) 65 pp.

[3] Putra W, Paripurno ET, and Nugroho NE 2021 Effect of Geological Characteristics on Landslides and Floods in Karangtengah *Engineering and Science Series* (ESS) Vol 1 (**1**): 9-14.

[4] Istiadi Y and Priatna D 2021 Analysis of the determinants and typology of hydrometeorological disaster in Sukajaya Subdistrict, Bogor Regency, West Java, *Indonesia Indonesian Journal of Applied Environmental Studies* 2 (**1**): 41-46.

[5] Kartono K, Purwanto P, and Suripin S 2020 Analysis of Local Rainfall Characteristics as a Mitigation Strategy for Hydrometeorology Disaster in Rain-fed Reservoirs Area *Advances in Science, Technology and Engineering Systems Journal* 5 (**3**): 299-305.

[6] Fauzy A, Muhajir M, Kurniawan MHS, and Ghiffary GG Analytical Research Focus Areas of Disaster: Case Study SCOPUS 2014-2018 *IOP Conference Series: Earth and Environmental Science*, 448 012117.

[7] Lutfi M 2020 Hydrodynamic circulation model in the estuary of Palu river based on numerical calculations *Journal of Engineering Science and Technology* 5(4): 2309 - 2323.

[8] Cohen PJ, Allison EH, Andrew NL, Cinner J, Evans LS, Fabinyi M, Garces LR, Hall SJ, Hicks CC, Hughes TP, vein Jentoft S, Mills DJ, Masu R, Mbaru EK and Ratner BD 2019 Securing a Just Space for Small-Scale Fisheries in the Blue Economy *Front. Mar. Sci.*, 18 April 2019 https://doi.org/10.3389/fmars.2019.00171.

[9] Tegar D, and Gurning ROS 2018 Development of Marine and Coastal Tourism Based on Blue Economy International Journal of Marine Engineering Innovation and Research 2 (2): 128-132.

[10] Sherman KD, Shultz AD, Dahlgren CP, Thomas E, Brooks A, Brumbaugh DR, Gittens L and Murchie KJ 2018 Contemporary and emerging fisheries in The Bahamas—Conservation and management challenges, achievements and future directions Fisheries and Management and Ecology https://doi.org/10.1111/fme.12299.

[11] Bgriguglio, L 2000 Strenghts and weaknesses of small islands The Gozo Observer - Issue 04, December 2000 https://www.um.edu..mt/library/oar/handle/123456789/57605.

[12] Gintings B, Nasution MA, Subhilhar, S, and Harahap RH 2018 Analysis Of Weaknesses Of Coastal Community Economy Empowerment Program (Pemp) And National Program Of Community Empowerment Of Independent Marine And Fisheries (Pnpm-Mkp) On Traditional Fishermen In Indonesia *Munich Personal RePEc Archive* https://mpra.ub.uni-muenchen.de/87762.

[13] Fairnbairn TIJ 2007 Economic Vulnerability and Resilience of Small Island States Island Studies Journal 2 (1): 133-140.

[14] Sindico F 2021 Sub-National Island Jurisdictions and Biodiversity Protection: Moving Towards Greater Resilience and Sustainability *Biological Diversity and International Law* pp 141-155.

[15] Ismail Z 2019 Public Sector Reform and Capacity Building in SIDS (Birmingham: K4D) 17 pp.

[16] Robinson SA, and Gilfillan D 2017 Regional organisations and climate change adaptation in small island developing states *Regional Environmental Change* 17: 989–1004 .

[17] Rao, L and McNaugthon M 2018 A knowledge broker for collaboration and sharing for SIDS: the case of comprehensive disaster management in the Caribbean Information Technology for Development 25 (1) https://doi.org/10.1080/02681102.2018.151036

[18] Roberts JL, Nath S, Paul S and Madhoo YN 2020 Overview, Emerging Issues and a Roadmap of SIDS *Shaping the Future of Small Islands* pp 405-416.
[19] Butcher-Gollach C 2015 Planning, the urban pool and climate change in Small Island Developing States (SIDS): unmitigated disaster or inclusive adaptation? *International Development Planning Review* 37 (**2**): 225-248.
[20] Amrianah H 2020 Impacts of El Nino and La Nina Extreme Climate on Socio-Economic and Politics in Indonesia Meraja Journal 3 (1): 35-45.
[21] Hendri, Alianto & Suhaemi 2016 Coastal Hazard Management due to Extreme Climate Condition in the Bird's Head Papua Jogja Earthquake in reflection 2016 (Jogja: Proceeding) 127-133.
[22] BNPB 2012 Head of BNPB Regulation No. 2 of 2012 concerning General Guidelines for Disaster Risk Assessment (Jakarta: BNPB) 67 pp.
[23] BPS 2021 Sorong Island District in Figures 2021 (Sorong City: BPS) 133 pp.
[24] PUPR 2018 Standards and Guidelines for the Construction of Building Materials and Civil Engineering (Jakarta : Litbang PU) 230 pp.

IOP Conf. Series: Earth and Environmental Science **989** (2022) 012010 doi:10.1088/1755-1315/989/1/012010

Damage, Loss, Economic Pressure, and Time Required For Family Asset Recovery of Garut Flash Flood Survivor

Euis Sunarti[1], Hanna Mariam Erwini[1], Zuhdi Fadhlurrahman[1], Marwiah[1] and Aliya Faizh Fithriyah[2]

[1]Faculty of Human Ecology, Department of Family and Consumer Sciences, IPB University. Kampus IPB Dramaga, Jl. Raya Dramaga, Babakan, Dramaga, Bogor, Jawa Barat 16680

[2]Department of Architecture Landscape, Faculty of Agriculture, IPB University. Kampus IPB Dramaga, Jl. Raya Dramaga, Babakan, Dramaga, Bogor, Jawa Barat 16680

[*]Email: euissunarti@apps.ipb.ac.id

Abstrak. The research was conducted to analyze the effect of damage, losses and economic pressures on the length of recovery of family assets experienced by families victims of flash flood in Garut in 2016. This study used a cross-sectional method and a retrospective study involving 120 examples selected disproportionally stratified random sampling by place of residence namely in shelters and non-shelters (60 families each). The results showed that there were no significant differences between damage and losses and the duration of recovery of assets in families living in shelters and non-shores, while economic pressures were higher for families living in shelters. An important finding of this study is that the average length of asset recovery is 482 months. The regression test results show that the higher the father's education the faster the asset recovery, while the greater the value of damage and the greater the economic pressure, the longer the asset recovery.

1. Introduction

Disaster event have continued to increase in the last decade which resulted in damage to assets and infrastructure and hampered social growth. The United Nations International Strategy for Disaster Reduction [3] notes that between 1998-2017, disasters have caused direct economic losses of US $ 2,908 trillion, of which 77% are climate-related disasters. Furthermore, UNIDSR (2018) also revealed that the highest frequency of climate change-related disasters and with the greatest impact on humans is flooding. The National Disaster Management Agency [11] also noted a similar pattern which in the period 2004-2014, Indonesia had been hit by 11,274 disasters that claimed as many as 193,240 lives and resulted in a total loss of at least IDR 420 trillion. In 2016 the number of disasters in Indonesia reached the highest number of 2,369 events and 92 percent of them were hydro-meteorological disasters which were dominated by floods, landslides and tornadoes [11].

One of the biggest floods that hit Indonesia in 2016 was the Garut flash flood that happened suddenly on September 26. The total estimated damage and losses after the flash flood reached 288 billion rupiah [11]. Disasters generally have a significant impact on the housing sector, which greatly affects the lives of families who are victims. Based on the Regional Agency for Disaster Management of Garut Regency (2016), flash floods affected 2,529 housing units with details of

830 heavily damaged houses, 473 moderately damaged housing units and 1,226 lightly damaged housing units. Responding to these conditions, various assistance was distributed to victims including assistance for shelter namely temporary shelter. Hundreds of families of victims who lost their homes after flash floods were relocated to nine temporary shelters provided by the government. However, the limited capacity of shelters causes many families to be forced to live in their original place of residence (non-shelter) despite all limitations.

Post-disaster ravaged conditions disrupt the survival of the community in various aspects. Therefore, reconstruction and rehabilitation efforts were needed as soon as possible after the disaster. The reconstruction and rehabilitation efforts are one of the government's tasks as stipulated in Law Number 24 of 2007 concerning Disaster Management and Government Regulation Number 21 of 2008 concerning the Implementation of Disaster Management. There are three stages in post-disaster rehabilitation and reconstruction, namely damage and losses assessment, post-disaster needs assessment (human recovery needs assessment) and the preparation of an action plan for rehabilitation and reconstruction. Regarding damage and loss assessment, the Economic Commission for Latin American and Caribbean (ECLAC) in 1972 introduced a calculation method called the ECLAC method. This calculation method was first applied in Indonesia to assess damage and losses after the 2004 Aceh tsunami disaster and the Jogjakarta earthquake in 2006. The ECLAC method analyzes three main aspects, namely damage, loss and economic impact [10] In this study, damage is a direct impact on family assets while losses are indirect impacts which are projections of the productivity constraints of family assets damaged by flash flood.

Although reconstruction and rehabilitation is one of the government's tasks, the community must also take part. Based on the analysis of the Center for Disaster Studies of the Bogor Agricultural University [8], the government generally cannot carry out reconstruction and rehabilitation efforts quickly because it is hampered by regulatory and bureaucratic factors. One of them is related to the mechanism for budgeting for the newly allocated reconstruction and rehabilitation funds in the following year after the incident happened. Therefore, to accelerate recovery efforts, each family should actively seek the recovery of their respective assets. Recovery of economic assets of a family refers to the achievement of family financial status that is equivalent to the family economic situation prior to the disaster [1]. Asset recovery is needed so the family can continue to live well after the disaster and can live independently without relying on assistance.

Recovery ability varies among each household because disasters have different effects on each household itself [7]. The difference in the impact of the disaster is caused by differences in environmental, social and economic conditions. Middle and lower socioeconomic groups generally suffer the most severe losses due to disasters [6]. The impact of the disaster on the poor is because they are more vulnerable in terms of economy, social and environment. Poor people often live in disaster-prone environments, with poor quality home materials, low assets, inadequate education and skills and limited access to social support [4, 6, 14]. This indicates that the asset recovery process will be longer for families with high economic pressure.

According to Conger et al. (2002) economic pressures created by difficult conditions to meet the needs and services that are important for families due to limited resources and income [2]. Often, disasters not only result in the loss of property but also the loss of sources of livelihood for the family that affect the level of family welfare. According to Sunarti (2015), the occurrence of disasters not only disrupts family welfare but also has the potential to impoverish families [15]. Based on the World Bank (2016), every year the disaster drives 26 million people into poverty. Furthermore, family helplessness in a state of post-disaster shortages coupled with the loss of sources of income can encourage debt. This will further aggravate economic pressures experienced by the family while at the same time complicating the family's recovery of assets.

Analysis of the length of recovery of family assets along with the factors that influence them is still very rarely done even though this is very important. Knowledge of the duration of asset recovery and its factors can be used to find out gaps and constraints that still need to be addressed to speed up the recovery process of the assets themselves. Families in shelters and non-shelter are thought to have different time needed in the process of recovering their assets. This is because families in shelters and non-shelters are thought to have different levels of damage and losses as well as economic pressures experienced after the disaster. Therefore, this study was conducted with the aim of: 1) knowing differences in family characteristics, level of damage, losses, economic stresses and the length of recovery of assets experienced by families of victims of flash flood who live in shelters and non-shelters; 2) knowing the influence of family characteristics, the level of damage, loss, and economic pressure on the recovery time of family assets of flash flood victims who live in shelters and non-shelters.

2. Methods

The design of this study was a cross-sectional and retrospective study. This research was carried out at 4 months after the occurrence of flash flood, but the scope of the questions in the questionnaire also included family conditions prior to the occurrence of the disaster. The study involved 120 families of victims of the Garut flash flood who were selected by non-proportional stratified random sampling from those living in shelters and non-shelters (60 families each). Tarogong Kidul Subdistrict was chosen purposively because it is a sub-district with the worst impact on the housing sector with 1,302 damaged houses out of a total of 2,529 houses affected by banjir bandang [12]. The locations of Transito shelters, LPSE, LEC, Islamic Center, Gandasari, and Al-Musadaddiah were purposively selected with consideration that there were victims from the Tarogong Kidul Sub-district. Meanwhile, non-shelters locations were selected purposively in RW 10 (RT 3 and RT 4) and RW 19 (RT 3 and RT 4) because they were one of the locations with the worst impact of flooding because they were right on the edge of the Cimanuk river.

Primary data obtained from interviews directly to respondents using a questionnaire. The dependent variables studied were the duration of asset recovery while the independent variables include family characteristics, disaster damage, losses due to disasters and economic pressures. Family characteristics measured included the age of the husband, age of the wife, length of education of the husband, length of education of the wife, size of family, and income per capita of the family (before the disaster and 4 months after the disaster). Damage and loss data was measured using a method developed by the Economic Commission for Latin America and the Caribbean (ECLAC) known as Damages and Losses Assessment (DaLA) and had been modified by the Center for Disaster Studies IPB. Economic pressure was measured using a questionnaire developed from Sunarti (2005) with 5 criteria, namely welfare status, husband's employment status, wife's employment status, comparison of income and expenses, and comparison of debt and assets. The duration of asset recovery was measured using the formula [16]:

$$X = \frac{\text{Value of damage (IDR)}}{\text{Family saving capacity per month in post disaster}}$$

The ability to save after a disaster was calculated using the assumption of family income minus family expenses each month. The amount of family expenditure was also calculated based on the assumption of the average family needs per month in the Garut region. Family needs were divided into four, namely food expenditure, household expenses, transportation expenses, and unexpected expenses. Food expenditure in the form of food and drink was obtained by assuming the number of family members multiplied by Rp. 200,000 per month. The expenditure of household needs in the form of water and electricity was obtained assuming the number of family members was Rp. 15,000 per month. Transportation expenses in the form of family mobility costs were obtained by assuming the number of family members above school age (except for mothers) multiplied by Rp. 120,000 per month. Expenditures for unexpected needs were obtained assuming the number of family members multiplied by Rp. 20,000.

Data were analyzed using descriptive test, correlation test, independent Ttest, and multiple regression test. There were 15 families who still did not have a source of income for 4 months after the disaster so that the recovery of their assets cannot be calculated. Therefore, only 105 family data were included in the inferential analysis. However, all respondents' data, amounting to 120 families, were still used in descriptive tests.

3. Results and Discussion

Data on family characteristics are presented in Table I. The average family sample in this study is medium-sized families (5-7 members).

Table 1. Minimum, maximum, and mean values of the family characteristics

Family Characteristics (unit)	Shelters (Min-Max; Average)	Non Shelters (Min-Max; Average)	Total (Min-Max; Average)
Age of husband (year)	28.0 – 85.0; 43.3	22.0 – 71.0; 46.0	22.0 – 85.0; 44.5
Age of wife (Year)	23.0 – 68.0; 39.1	21.0 – 70.0; 41.3	21.0 – 70.0; 41.0
Years of husband's education	4.0 – 12.0; 7.55	4.0 – 22.0; 8.1	2.0 – 12.0; 4.5
Years of wife's education	4.0 – 13.0; 7.38	2.0 – 15.0; 7.6	4.0 – 22.0; 7.8
Number of family member (people)	2.0 – 8.0; 4.0	2.0 – 12.0; 4.0	2.0 – 15.0; 7.0
Income/capita/month before disaster (Thousand Rp)	28.5-3300.0; 676.0	42.0 – 3571.0; 763.0	28.5- 3571.0; 692.0
Income/capita/month, 4 month after disaster (Thousand rp)	0.0-1625.0; 354.0	0.0 – 2857.0; 578.0	0.0 – 2857.0; 467.0

Both husbands and wives, on average, were in middle adulthood with an average education of both only graduating from elementary school, or not fulfilling 12 year compulsory education. The average income per capita of the family in 4 months after the disaster (467 thousand rupiah) was lower than the average income per capita before the disaster (692 thousand rupiah). The average income per capita of families living in non-shelters areas was higher than families in shelters, both before the disaster and at 4 months post-disaster.

3.1. Damage

Data on damage experienced by families due to flash flood is presented in Table II. Damage data in this study was measured in six components, namely damage to vehicles, valuables, electronic goods, livestock, buildings, and productive business assets. The total damage suffered by the family ranges from Rp. 100,000 to Rp. 164,050,000 with an average of Rp. 34,784,280. Based on different tests, there was no significant difference between the total damage experienced by families in shelters and non-shores. However, it can be seen that the average value of total damage to families in non-shelters was higher than in shelters. In addition, families in non-shelters did bear higher damage in almost all components of damage, namely damage related to vehicles, valuables, electronic goods, and productive business assets. Only building-related damage had a significantly higher value in families in non-shelters. This was reasonable considering the family living in shelters did come from families who lost their homes due to flash floods. Furthermore, of the six types of damage, building damage was the highest damage with an average of Rp. 18,455,830.

Tabel 2. Minimum, Maximum, Average (in thousand rupiah), and The Results Of Different Tests For Damages In Shelter And Non-Shelter

Component of Damage	Min – Max		Average		*p-value*	total
	shelters	Non-shelters	Shelters	Non-shelters	shelters	Non shelters
Vehicle	0–33,400,000	0–83,500,000	3,374,330	9,014,170	0.007***	6,194,250
Valuables	0–5,500,000	0–11,600,000	237,170	491,00	0.312	364,330
Electronic goods	200,000–9,750,000	0–31,850,000	2,208,330	4,043,330	0.049**	3,125,830
Livestock	0–10,750,000	0–8,050,000	583,500	524,030	0.848	553,760
Building	0–100,000,000	0–100,000,000	24,161,670	12,730,000	0.014**	18,455,830
Productive Asset	0–50,000,000	0–93,000,000	1,725,830	8,624,420	0.017**	5,175,120
Total	100,000–123,750,000	1,800,000–164,050,000	32,352,500	37,216,050	0.468	34,784,280

3.2. Losses

Data regarding losses suffered by families due to flash flood are presented in Table III. Data loss in this study was measured in 7 components, namely losses arising from cleaning activities after floods, losses due to family members becoming sick, husband does not work, wife does not work, decreased land prices, debt arising, and the emergence of dependents. The data in Table III shows that the total losses experienced by families due to flooding ranged from Rp. 150,000 to Rp. 125,714,286 with an average of Rp. 13,008,646. The results of different tests showed that there were no significant differences between the value of total losses in shelters and non-shelters. However, it can be seen that the average value of total losses borne by families in non-shelters areas was higher than families living in shelters. This can be seen from the value of losses in almost every component that is indeed higher in non-shelters families, namely losses due to cleaning, losses due to illness, loss due to husband not working, losses due to decreased land prices and losses due to the emergence of debt. In addition, the data also shows that the loss with the greatest value experienced by the family is loss due to a decrease in land prices, followed by losses due to the husband not working.

Table 3. Minimum, Maximum, Average of Losses (Thousand rupiah) and its Different Tests by residence

Component of Losses	Min – Max		Average		*p-value*	Total average
	Shelter	Non-shelter	Shelter	Non-shelter		
Cleaning the house	150-250	50,100	56.6	150	0.000**	102.9
Sick	0 –1,000	0–3,000	5.35	137.5	0.234	95.5
Do not work (husband)	0-20,200	0–9,000	2,564	2,932	0.514	2,748.6
Do not work (wife)	0–30,000	0–9,000	2,231	1,516	0.333	1,873.5
Decrease of land prices	0–16,200	0–57,600	4,366.2	6,584.4	0.108	5,475.3
debt	0–30,000	0–100,000	1,175	3,855.3	0.191	2,515.2
dependent	0–570	0–7,500	337	27	0.042**	182
Total	150-42,650	50-125,714	10. 78	15.19	0.132	13.008.6

3.3. Economic Pressure

Data on economic pressures experienced by families at 4 months after flash flood are presented in Table IV and Table V. Economic pressure is a form of financial problems experienced by families. Financial problems in this study are seen from the comparison of income per capita of the family with the poverty line, husband's employment status, wife's employment status, comparison of income and expenditure and the ratio of debt to assets. The data in Table IV shows that the average economic pressure experienced by the sample family is 36.83 with a range between 0 and 80%. The results of different tests show that there are significant differences between the economic pressures of families in shelters and non-shelters. Families living in shelters experience higher economic pressure compared to families in non-shelters.

Tabel 4. Minimum, Maximum, Average of Economic Pressure and Result of Comparison Test

Variable	Min – Max		Average		*p-value*	Average
	Shelter	Non-shelters	shelters	Non-shelters		Total
Economic Pressure (index)	0 – 80	20 – 70	43.83	29.83	0.005**	36.83

Before the occurrence of flash flood, the number of families categorized as poor was around 15.8 percent. In Table V it can be seen that after the disaster, the number of poor families doubled to 32.5 percent. The number of poorly categorized families was more in shelters compared to non-shelters. At 4 months after the disaster, there were still 10 percent of husbands who did not work, while wives generally much more did not work (60%). In addition, Table V also shows that as many as 60 percent of families in shelters have greater expenditure compared to post-disaster income.

Tabel 5. Distribution of family (%) by component of economic pressure and temporary residence

Component of Economic Pressure	shelter	Non-shelter	Total
Poverty Status			
Not poor (> Rp. 361.600)	41.7	63.3	52.5
Almost poor (Rp. 241.068- Rp. 361.600)	10.0	20.0	15.0
Poor (< Rp. 241.068)	48.3	16.7	32.5
Husband's job status			
Permanent	30.0	71.7	50.8
Not Permanent	66.7	11.7	39.2
Do not work	3.3	16.7	10.0
Wife's Job Status			
Permanent	13.3	35.0	24.2
Not Permanent	21.7	8.3	15.0
Do not work	65.0	56.7	60.8
Comparison income to expenses			
Expenses < income	40.0	73.3	56.7
Expenses = income	0.0	0.0	0.0
Expenses > income	60.0	26.7	43.3
Comparison debt to asset (rasio)			
No debt	86.7	78.3	82.5
< 50%	13.3	21.7	17.5
≥ 50%	0.0	0.0	0.0

3.4. Duration of Asset Recovery

Data on the length of recovery of family assets are presented in Table VI. The duration of asset recovery was calculated from the amount of damage divided by the amount of income that the family can set aside every month after the disaster. The length of recovery of assets needed by the family ranges from only 4.44 months to 5.107 months with an average of 482 months or the equivalent of 40 years. Based on different tests, there were no significant differences between the length of restoration of assets of families in shelters and non-shelters. However, it can be seen that the average recovery time of assets needed by families in shelters is higher compared to families in non-shelters. At 4 months after the disaster, there were still families that did not have income so that the asset recovery period cannot be calculated.

Table 6. Minimum, Maximum, Average and Long-Term Tests for Recovery of Family Assets in Shelter and Non-Shelter.

Variable	Min – Max		Average		*p-value*	Average total
	shelters	Non-shelters	shelters	Non-shelters		
Duration to recovery asset (month)	4.44 – 5.107	5.73 – 2.620	603.13	368.37	0.148	482.40

3.5. The Effect of Damage, Loss, and Economic Pressure on Family Assets Recovery

The regression test results are presented in Table VII. The data shows that the Adjusted R Square value is 0.290, which means that the model was able to explain 29.0 percent of the factors that affect the duration of asset recovery, while the rest are influenced by other independent variables outside the research. Partially, the duration of husband's education showed a negative influence on the duration of asset recovery. This result showed that the higher the husband's education, the faster the process of asset recovery.

Table 7. Regression Coefficient on the Effect of Family Characteristics, Damage, Loss, and Economic Pressure on the Duration of Family Asset Recovery

Variabel	B	B	Sig
Constanta	333.961		0.344
Age of husband (years)	-0.173	-0.003	0.976
Years of husband's education	-55.877	-0.198	0.022*
Damage (IDR)	0.012	0.539	0.000**
Losses (IDR)	-0.001	-0.022	0.817
Economic pressure (index)	6.108	0.191	0.024*
Adjusted R	0.290		
Sig	0.000***		
F	9.499		

The results also show that economic damage and pressure had a significant positive effect on the length of recovery of family assets. This shows that the higher the value of damage, the longer the time it takes for the family to recover assets. In addition, the higher the economic pressure experienced by the family, the longer the time needed to restore family assets.

3.6. Discussion

The results showed that there were no significant differences between the damage and losses experienced by families living in shelters and non-shelters. However, the average value of damage and loss tends to be higher in families living in non-shelter. Despite higher damage and losses, families in non-shelters experience lower economic pressures and require faster asset recovery times than families in shelters.

The average time needed for an example family to recover post-disaster assets is 482 months or 40 years. The average recovery of assets in shelters was even higher than the total average time of around 603 months or 50 years. The higher recovery time of assets in shelters was also accompanied by economic pressures which were also higher in shelters compared to families in non-shelters. Economic pressure felt by the family regarding the financial conditions experienced by the family. In post-disaster, the income of families in shelters and non-shelters had decreased. This is similar to the findings of the Thorburn (2009) study in the case of the 2006 Tsunami disaster which showed that post-disaster family income declined [17]. Even though both declined, the average income in non-shelters was higher than the average income in shelters. The number of poor families has doubled after the disaster and the number is higher in shelters. In addition, families who have more expenditure than income are more in shelters. This explains the reason for the higher economic pressure in shelters than in non-shelters.

The regression test results show that economic pressure had a positive effect on the duration of asset recovery. The higher the economic pressure experienced by the family, the longer it takes the family to recover its assets. This us in line with the results of Bolin's research (1987) that the family's ability to recover after a disaster was affected by financial conditions. The better the financial ability, the faster the process needed to recover.

In addition to economic pressure, the time needed by the victims' families to recover assets was also influenced by the high value of assets damaged by the disaster. The regression test results indicate that damage had a positive effect on the duration of asset recovery. The higher the damage the family experiences from the disaster, the longer it takes for the family to recover assets. This is in line with the research of Oniki and Dagys (2016) the higher the damage to the family, the longer the time needed to recover [9]. Data shows that the highest damage experienced by families of victims of banjir bandang both those living in shelters and non-shelters was damage to houses. Furthermore, the data also showed that building damage showed significant differences between the two occupancy groups. Families living in shelters experience significantly higher damage to buildings. This is reasonable considering the family that inhabits the shelters was indeed a family whose residence cannot be inhabited anymore due to being washed away by flash floods. The victims who suffered severe damage to houses were generally located very close to the Cimanuk river or even in a zone that should not be inhabited. In addition, the building materials of their homes generally do not meet disaster resistant standards.

Another factor that affects the duration of asset recovery is the length of the husband's education. The regression test results show that the duration of a husband's education has a negative effect on the duration of asset recovery. Families with husbands who have higher levels of education tend to recover assets more quickly. This finding is in line with the results of a study by Frakenberg et al (2013) which found that tsunami victims in Aceh who had higher education tended to have a better life 5 years after the disaster occurred. Better education is associated with higher income because people who have high education will find it easier to obtain higher income jobs [5].

Another interesting thing found in this study is that, although many families in shelters have smaller amounts of income than many even categorized as poor, families in shelters do not feel too anxious because of assistance. In the fourth month after the disaster when this research was conducted, some families in the shelter still received assistance from the government in the form of rice and pocket money for school children. The provision of assistance is thought to have an impact on the enthusiasm of working families in shelters. This is in line with the research of Riswan and Arifika (2012) that women who live in temporary housing feel comfortable waiting and receiving assistance rather than working to increase income. Many families who live in non-shelters even claim to be jealous of the assistance obtained by families in shelters and want to also live in shelters [13].

The results of this study confirm the depth of the impact of the disaster on the lives of the families of the victims. Only in an instant can a disaster sweep away assets owned by the family and it will take a long time for the family to recover the assets. Not only does it take away property, disasters can also take away the source of family livelihood which further worsens family conditions after a disaster. This shows the need for serious efforts to build resilience to disasters given the depth of impact and the length of time needed to recover from disasters. This research will be better if the aspect of assistance is further explored to influence the recovery of assets. In addition, the development of instruments and methods still needs to be developed in order to measure damage and losses better.

4. Conclusions

The results showed that families in non-shelters tended to experience higher damage and losses due to disasters, compared to families in shelters (because generally families living in non-shelters, their homes were better and permanent). However, families in non-shelters experience lower economic pressures and require less time to recover assets (faster) than families in shelters. This is related to job stability and income to families living in non-shelters. However, although the recovery of assets in non-shelters is faster than in shelters, the ability to recover assets in both locations is generally equally low. Therefore, all parties are expected to be more serious in efforts to develop resilience to disasters given the depth of impact and the length of time needed to recover from disasters. Efforts that need to be considered by the government include urban planning so that there are no more families living in high-risk areas of disasters (such as commensurate rivers). The government also needs to review the effectiveness of providing assistance to disaster victims. Do not let assistance intended to stimulate family recovery have a negative impact on family productivity. Further researchers are expected to explore further about the assistance received by victims. In addition, further researchers are expected to further develop methods used to measure damage and losses.

References

[1] Bolin R. 1986. Disaster impact and recovery: a comparison of black and white victims. *International Journal of Mass Emergencies and Disasters*. 35-50

[2] Conger RD, Wallace LE, Sun Y, Simon RL, McLoyd VC, Brody GH. 2002. Economic pressure in African American families: a replication and extension of the family stress model. *Journal of Developmental Psychology*. 38(2): 179-193.

[3] *Economic Losses, Poverty & Disasters 1998-2017*. UNISDR, USA, 2018.

[4] Fothergill, A., & Peek, L. A. 2004. Poverty and disasters in the United States: A review of recent sociological findings. *Natural Hazards.* 32, 89–110.

[5] Frankenberg E, Sikoki B, Sumantri C, Suriastini W, Thomas Deducation. 2013. Vulnerability, and resilience after a natural disaster. *Ecology and Society*. *18*(2): 16.

[6] Hallegatte, S., Vogt-Schilb, A., Bangalore, M., & Rozenberg, J. 2017. *Climate Change and Development Series: Unbreakable; Building the resilience of the poor in the face of natural disasters*. Washington, DC: World Bank.

[7] Lindell M. *Ensiclopedia of natural hazard: Recovery and reconstruction after disaster.* USA. 2013. Texas A&M University

[8] *Model Koordinasi, Sinkronisasi dan Pengendalian Program Pasca Bencana*. [Coordination, Synchronization and Control of Post Disaster Programs]. PSB IPB, Bogor, ID, 2015.

[9] Oniki S. Dagys K. 2016. Recovery from a winter disaster in Tov Provience of Mongolia. *Journal of Arid Environments*. 139: 49-57

[10] Oetari F, Yulianto S, Nugroho A. 2014. Pengembangan sistem informasi penilaian kerusakan dan kerugian bencana menggunakan metode ECLAC. [*Development of information systems for disaster damage and loss assessment using the ECLAC method*]. *Jurnal Teknologi Informasi-Aiti*. 11(2): 101 –202

[11] BNPB. *Rencana Nasional Penanggulangan Bencana 2015-2019*. [National Plan for Disaster Management Year 2015-2019]. Jakarta, ID, 2014.

[12] *Rencana Aksi Rehabilitasi dan Rekonstruksi Korban Banjir Bandang September 2016* . [Action Plan for Rehabilitaton and Reconstruction of September 2016 Flash Flood]. BPBD. ID 2016

[13] Riswan Y, Arifika D. 2012. Perempuan dan bencana: memberdayakan potensi social dan ekonomi perempuan korban banjir lahar dingin merapi. [*Women and disasters: empower the social and economic potential of women victims of Merapi cold lava floods*] *Kawistara*. 2(2): 165-177

[14] SAMHSA. 2017. Greater Impact: How Disasters Affect People of Low Socioeconomic Status. *Disaster Technical Assistance Center Supplemental Research Bulletin*

[15] Sunarti E. Ketahanan Keluarga Indonesia: dari Kebijakan dan Penelitian Menuju Tindakan. [*Indonesian Family Resilience: from Policy and Research Towards Action*]. *Orasi Ilmiah Guru Besar IPB*. 2015.

[16] Sunarti, E, Tati, Atat SN, Noorhaisma R, Lembayung DP. 2005. Pengaruh tekanan ekonomi keluarga, dukungan sosial, kualitas perkawinan, pengasuhan, dan kecerdasan emosi anak terhadap prestasi belajar anak. [*Effect of family economic pressure, social support, marital quality, parenting, and children's emotional intelligence on children's learning achievement*]. *Media Gizi & Keluarga*. 29 (1): 34-40

[17] Thorburn C. 2009. Livelihood recovery in the wake of the tsumani in Aceh. *Buletin of Indonesian Economic Studies*. 45(1): 85-105

IOP Conf. Series: Earth and Environmental Science **989** (2022) 012011 doi:10.1088/1755-1315/989/1/012011

Development of wind monitoring systems with LoRA technology

Purnomo, F.A.*, Yoeseph, N.M., Yulianto, A., Royan, Y.I, and Safi'ie, M,A.

Informatics Engineering Departement, Sebelas Maret University

*Email: fendi_aji@staff.uns.ac.id

Abstract. The purpose of this research is to design and build a wind monitoring system with LoRA (Low Range) technology. This system can observe the value of wind speed, air temperature, and humidity. All of these data must be processed in order to obtain information on the current state of the air. Existing information can be used as a reference for decision making or as a sign of a threat. In this monitoring system, the observation of air conditions uses wireless LoRa (Long Range) technology as a data transmission tool. In this experiment, two LoRa nodes are used and one Lora Gateway is used. There are two sensors used in this case, namely anemometer and BME280. Sensors contained in the LoRa node take data on wind speed, air temperature, and humidity. The LoRa node sends all the data to the LoRa Gateway. LoRa gateway connected by wifi will forward to the server (The Things Network and Firebase). The test results from this study show that the average time for sending LoRa node data to being sent to firebase is 0.25 seconds. LoRA devices with a 3 cm long antenna can communicate at a maximum distance of 400 m.

1. Introduction

Wind is a mass of air that moves due to the influence of air pressure gradients or air pressure differences. The wind moves towards lower air pressure. Wind speed can be used for community needs, but high speeds can cause disasters that can damage buildings such as hurricanes. During 2019, Indonesia recorded 3,814 disaster events dominated by hydrometeorological disasters, such as hurricanes, floods, landslides, and droughts. Hydrometeorological disasters have become a recurring threat throughout the year with a percentage of events reaching 98% [1]. For the benefit of energy conversion, a minimum wind speed of 5 m/s is required to produce small-scale energy [2]. Currently, wind speed monitoring has been developed for weather forecasting [3] and for monitoring the speed of wind power plants [4][5][6]. This study aims to make instruments measuring wind speed, temperature and humidity to see the potential for wind utilization with LoRA technology. LoRa is a proprietary spread spectrum modulation scheme that is derivative of Chirp Spread Spectrum modulation (CSS) [7]. LoRa is different from other technology modules such as GSM, Wifi, Bluetooth and Zigbee modules. LoRa has lower power than GSM / LTE modules and LoRa has a long range of up to 10km further than Wifi, Bluetooth and Zigbee [8]. In Lora-based Internet of Things system, when the sensor on the node detects data, it passes through the gateway and server for data transmission [9]. End-nodes consisting of various sensors and will be placed in an area within the reach of LoRa to monitor that area. Data obtained from these sensors will then be transmitted wirelessly through a LoRa gateway device connected to the cloud server [8].

Due to those characteristics, Lora technology has been used in several fields, such as agriculture[10][11][12] and weather stations[8][13]. Several Flood Early Warning Systems using lora technology had been published [14][15]. This paper will focus on the construction of Lora Sensor Nodes as part of wind speed monitoring. Sensor Node consists of anemometer version YGC-FS-5V-M and sensor BME280 for air temperature and pressure. Experiments were carried out by testing the accuracy of each sensor. In addition, latency tests are also carried out to test the consistency of data transmission against the distance between the gateway and the node.

2. Methods

The architecture of the wind speed monitoring system consists of LoRa nodes connected to several sensors, LoRa gateways, LoRAWAN network servers and databases. LoRA gateway is used to bridge LoRA wireless network to IP network via WiFi and ethernet[16]. This system uses The Things Network (TTN), as a LoRAWAN network server. The Things Network is an open source infrastructure that provides an open set of tools and networks for building IoT applications. The data from the sensors is then stored in a database via HTTP integration with The Things Network. The architecture of the wind speed monitoring system is shown in Figure 1.

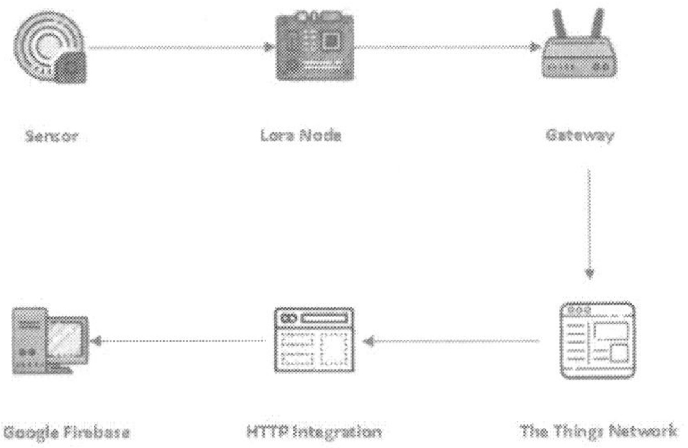

Figure 1. Wind Speed Monitoring System Architecture

LoRa nodes are built with BSFrance Lora32u4 v1.4 as a LoRA wireless module to transmit data from the anemometer sensor and BME280. Anemometer sensors are used to measure wind speed and BME280 sensors to measure air temperature and humidity. Anemometer sensor with type YGC-FS-5V-M. Wind speed is calculated from the output voltage signal. To find the voltage released by the anemometer, use the formula

$$outVoltage = sensorValue * (5.0 / 1023.0). \quad (1)$$

The anemometer speed is obtained by the formula

$$V = 6 * outVoltage. \quad (2)$$

The anemometer and BME280 sensor in this study is shown in Figure 2. This sensor works with a DC voltage of 12-24 Volts, an output voltage of 5 V, a speed range that can be read 0 - 30 m/s and a system error of 3%. The BME280 is a multi-sensor sensor. In this sensor there are 4 data that can be taken, namely air pressure in hPa units, temperature in degrees Celsius units, humidity in % units and proximity in m units. The anemometer sensor calibration results are shown in Figure 3.

IOP Conf. Series: Earth and Environmental Science **989** (2022) 012011 doi:10.1088/1755-1315/989/1/012011

Figure 2. Anemometer and BME280 sensor

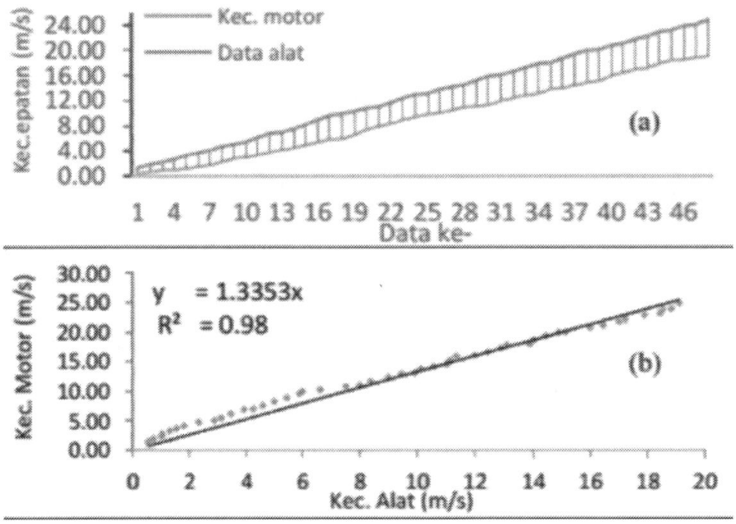

Figure 3. The anemometer sensor calibration results

The block diagram of equipment requirements in the research of Wind Speed, Temperature, and Humidity Sensor Reading Systems Using LoRa Technology is presented in Figure 4.

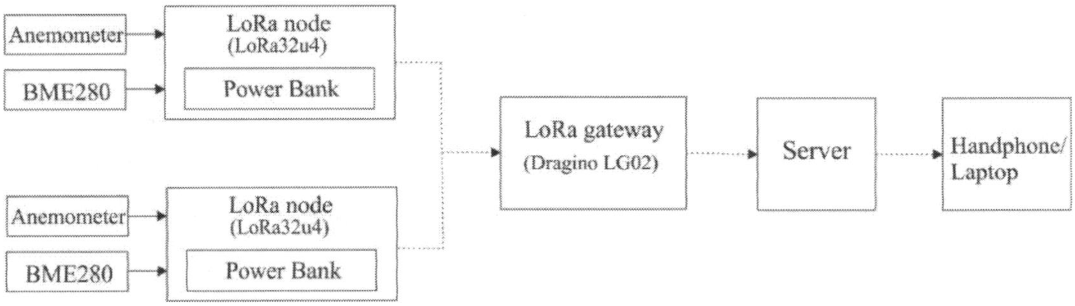

Figure 4. Block diagram of wind speed, temperature and humidity monitoring system

Anemometer Sensor, Anemometer is a wind speed measuring device that is widely used in the field of Meteorology and Geophysics or weather forecasting stations. BME280 sensor, The BME280 is a humidity sensor especially developed for mobile applications and wearables where size and low power consumption are key design parameters. LoRA Node, LoRa Node is a long-range transceiver on an Arduino shield form factor and based on Open source library. The Node allows the user to send data and reach extremely long ranges at low data-rates. LoRA Gateway, LoRaWAN Gateways are the link between the source of data – sensors in the environment – and people Server. The LoRaWAN Network Server is at the core of every LoRaWAN Network that enables connectivity, management, and monitoring of devices, gateways and end-user applications. Smartphone, Devices on end users that are used to monitor value of wind speed, air temperature, and humidity Result and Discussion.

3. Results and Discussion

Testing on the node is focused on testing the time difference between sensor readings and data received by the database. The accuracy of the anemometer sensor in reading the speed is determined by the outVoltage multiplied by the value 6. And the outVoltage is obtained from the multiplication of the sensor Value*(5.0/1023.0). The results of testing the transmission distance and delay in data transmission time are shown in Table 1 and Table 2

Table 1. Test Results of Data Delivery Against Distance

NO	Distance (Meters)	Delivery Status (per 1 minute)
1	200	Sent
2	300	Sent
3	400	Sent
4	500	Not sent
5	> 500	Not sent

Transmission distance testing is done by sending data from the LoRA node to the getway. Distance is measured with the help of the google map application. In testing the maximum distance using a 3 cm long LoRA antenna, it can transmit data as far as 400 m.

Table 2. Time difference between sensor readings and data received by the database

No	*Counter*	Serial Monitor (hour:minute:second)	Firebase (hour:minute:second)	Time Delay (second)
1	0	04:18:20,9	04:18:21,2	0,3
2	1	04:19:23,9	04:19:24,2	0,4
3	2	04:20:26,8	04:20:27,0	0,2
4	3	04:21:29,7	04:21:29,9	0,2
5	4	04:22:32,6	04:22:32,8	0,2
6	5	04:23:35,5	04:23:35.8	0,3
7	6	04:24:38,4	04:24:38,6	0,2
8	7	04:25:41,3	04:25:41,5	0,2
9	8	04:26:44,2	04:26:44,4	0,2
10	9	04:27:47,1	04:27:47,4	0,3
		Average		0,25

Latency testing is done by calculating the time between the data read by the sensor and sent by the node with the data received in the database. This data latency is important to know if the data arrives in the database as soon as the sensor collects the data. The data transmission time is obtained from the sensor reading time from the Serial monitor. The time of receiving the data is obtained from the timestamp of the data obtained in the Firebase console. The data latency is shown in Table 2. In Table 2, the average latency between data transmission and data reception is 0.25 seconds. Nodes for the wind monitoring system can be used to monitor wind speed, temperature and humidity. The transmitted data can be classified as near-zero data latency or so-called near real-time data.

4. Conclusion

The results show that the LoRa Node sends all data to the LoRa Gateway. LoRa gateway connected with wifi will be forwarded to the server (The Things Network and Firebase). The test results show that the average time for sending LoRa node data to be sent to firebase is 0.25 seconds. LoRA devices with a 3 cm long antenna can communicate at a maximum distance of 400 m.

References

[1] BNPB, "Bencana 2019 : Periode 1 Januari 2019 - 31 Desember 2019," *Badan Nasional Penanggulangan Bencana*, 2019. [Online]. Available: https://gis.bnpb.go.id/infografis/bencana/tahun 2019.jpg. [Accessed: 01-Nov-2021].
[2] A. Prasetyo, D. Notosudjono, and H. Soebagja, "Studi Potensi Penerapan Dan Pengembangan Pembangkit Listrik Tenaga Angin Indonesia," 2019.
[3] S. R. Nur Alam, "Jurnal Environmental Science," *J. Environ. Sci.*, vol. 2, no. April, pp. 1–8, 2020.
[4] R. W. Saragih, A. S. Asmita, and A. R. Widayana, "Analisis Kondisi Atmosfer, Indeks Labilitas, dan Citra Satelit Saat Kejadian Puting Beliung di Pontianak Kalimantan Barat (Studi Kasus 17 Juli 2020)," *J. Fis.*, vol. 10, no. 2, pp. 62–71, 2020.
[5] A. J. Ziegelaar, B. C. Travaglione, and M. R. Hodkiewicz, "Sensing system for low cost condition monitoring of remote assets," *IFAC-PapersOnLine*, vol. 53, no. 3, pp. 60–65, 2020.
[6] A. Meyer, "Multi-target normal behaviour models for wind farm condition monitoring," *Appl. Energy*, vol. 300, p. 117342, 2021.
[7] S. Penkov and A. Taneva, "Algorithm development with LoRaWAN for monitoring system," *IOP Conf. Ser. Mater. Sci. Eng.*, vol. 618, no. 1, 2019.
[8] E. Murdyantoro, R. Setiawan, I. Rosyadi, A. W. W. Nugraha, H. Susilawati, and Y. Ramadhani, "Prototype weather station uses LoRa wireless connectivity infrastructure," *J. Phys. Conf. Ser.*, vol. 1367, no. 1, 2019.
[9] J. Wu, X. Tang, and J. Lei, "Design of wireless monitoring system for greenhouse based on LoRa," *J. Phys. Conf. Ser.*, vol. 1650, no. 2, 2020.
[10] F. A. Purnomo, N. M. Yoeseph, S. A. T. Bawono, and R. Hartono, "Development of air temperature and soil moisture monitoring systems with LoRA technology," *J. Phys. Conf. Ser.*, vol. 1825, no. 1, pp. 1–5, 2021.
[11] J. Jin, Y. Ma, Y. Zhang, and Q. Huang, "Design and implementation of an Agricultural IoT based on LoRa," *MATEC Web Conf.*, vol. 189, 2018.
[12] S. Tathod, "Lora Based Smart Agriculture," no. July, pp. 1616–1621, 2018.
[13] I. Putra, R. Rajibussalim, I. Irwandi, and S. Muhammad, "Utilizing LoRa for IoT based mini weather station as STEM learning media to support industrial revolution 4.0," *J. Phys. Conf. Ser.*, vol. 1470, no. 1, pp. 0–6, 2020.
[14] E. Leon, C. Alberoni, M. Wister, and J. Hernández-Nolasco, "Flood Early Warning System by Twitter Using LoRa," *Proceedings*, vol. 2, no. 19, p. 1213, 2018.
[15] M. Ragnoli, G. Barile, A. Leoni, and G. Ferri, "An Autonomous Low-Power LoRa-Based Flood-Monitoring System," 2020.
[16] T. A. Salih and M. S. Noori, "Using LoRa Technology to Monitor and Control Sensors in the Greenhouse," *IOP Conf. Ser. Mater. Sci. Eng.*, vol. 928, no. 3, 2020.

IOP Conf. Series: Earth and Environmental Science **989** (2022) 012012 doi:10.1088/1755-1315/989/1/012012

Disaster risk reduction of mount agung cold lava flood at the Bali cultural centre in Klungkung Regency

Sutarja I.N [1,2,3], Ardana M.D.W [1,2], and Gustave S.P [1,2]

[1] Civil Engineering Study Program, Faculty of Engineering, Udayana University, Bali, Indonesia
[2] Centre for Disaster Studies, Udayana University.
[3] Disaster Risk Reduction University Forum

Email: nsutarja@unud.ac.id

Abstract. The Bali Province Government plans the development of Bali Cultural Centre as a comprehensive cultural information centre to support the development of Culture-oriented Bali Tourism. The Bali Cultural Centre site is situated in an area of post-mining remnants in Klungkung Regency. The sites located at coordinates 8°32'41.35" - 8°34'30.94" South Latitude and 115°24'54.92" - 115°26'6.99" East Longitude. This site is also portion of estuary area of the Unda River which has a high potential for the danger of cold lava flood from the Mount Agung eruption. The structure to manage cold lava floods at the Unda River estuary is being conducted. The construction is divided into two locations which design to satisfactory a 50-year design flood with discharge of 1436.11 m3 per second. The first location is at upstream (north side of Prof. Ida Bagus Mantra Street) using a single rectangular cross section. The second location is at downstream (south area of Prof. Ida Bagus Mantra Street) using a combine double rectangular cross section. These two types of cross sections are designed to accommodate the capacity of water discharge under normal conditions and still able to drain water discharge in the 50-year return period. The presence of this channel allows the cold lava flood flow directly towards the sea-shore and so as to reduce the risk. It is possible that the sediments settle at the base of the channel, therefore, channel shape normalization and periodic dredging for maintenance are essential.

1. Introduction

The Province of Bali as one of the destinations for Cultural Tourism will build a Bali Cultural Centre (BCC) to introduce its various cultures in a centralized manner while simultaneously growing the Balinese economy in a new era. This Provincial Strategic Area will become a Regional Integrated Development Area to ensure the economic growth and development of the Balinese people. The location is in the former Excavation C downstream of the Unda River basin in Klungkung Regency (Figure 1), which is at coordinates 8°32'41.35" - 8°34'30.94" South Latitude and 115°24'54.92" - 115°26'6.99 " East Longitude. In general, this area is very vulnerable to hazards, especially the cold lava flood of the eruption of Mount Agung (Figure 2), because the Unda River originates at the southern foot of Mount Agung. [1] [2].

There are 5 known and recorded eruptions of Agung Volcano since 1800, namely in 1808 where this year an extraordinary amount of ash and pumice was thrown, then the incident in 1821 where currently there is a normal eruption, but the records are not enough, Next was the year 1843 where the eruption of Agung Volcano was preceded by an earthquake with spewed material, namely ash, sand,

and pumice. Then in 1963, where the eruption began on February 18, 1963 and ended on January 27, 1964. The eruption was magnetic and recorded 1,148 people died and 296 people were injured. The last eruption was in 2017 where magmatic eruption started on November 25, 2017

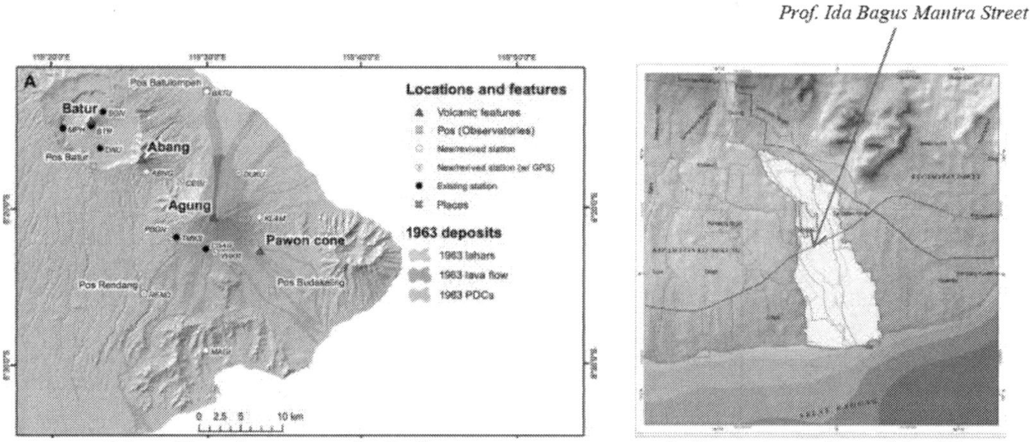

a. Map of lava flows, pyroclastic flows, Unda and land deposits at the time 1963 of the eruption

b. Downstream Part of the River BCC Location

Figure 1. Map of Mount Agung's Cold Lava Flow and Ex-Excavation C in Klungkung Regency [5].

Figure 2. Cold Lava Flood in the Lower Unda River

Based on the Disaster Risk Assessment Document [3], the planned area of the Bali Cultural Centre has a high risk of cold lava flooding from the eruption of Mount Agung, so that mitigation measures are needed to ensure the safety of people who will visit [4] . One of the ways to reduce the risk of cold lava floods is to build infrastructure in the form of straight channels that can channel cold lava floods directly into the sea [5].

Mitigation is carried out in two activities, namely the first survey and planning, the second is the implementation of development. Based on the results of the survey and the appropriate building planning is an open and straight channel from upstream to downstream in the construction area of the Bali Cultural Centre. This building is currently under construction.

IOP Conf. Series: Earth and Environmental Science **989** (2022) 012012 doi:10.1088/1755-1315/989/1/012012

2. Methods

2.1 Design

2.1.1 Area Rainfall Rainfall data obtained from the rainfall stations is in the form of rainfall data at a certain point, while for analysis purposes, what is needed is the area rainfall. Hence, to get the area rainfall data the research used the Thiessen Polygon Method. The rainfall data used in this research was taken from the three stations around the Tukad Unda area which are Besakih Station, Duda Station and Klungkung Station. These three rainfall stations are operated by the Bali Meteorology and Geophysics Agency.

However, before the rainfall data can be used, it must first pass a test for the consistency of the data. The method used is the RAPS (Rescaled Adjusted Partial Sums) method. Consistency test is carried out on annual rainfall data with the aim of detecting deviations in rainfall data, so that it can be concluded whether the data is suitable for use in hydrological analysis or not.

2.1.2 Frequency Analysis The purpose of this rainfall frequency analysis is to obtain rainfall with several return periods. In this analysis, several methods are used to estimate rainfall with a certain return period.

The method used later must be determined by looking at the characteristics of the local rainfall distribution. The return period to be calculated in each method is for a return period of 2, 5, 10, 25, 50, and 100 years.

From the calculation of the distributions above, different results will be obtained, therefore it is necessary to do a test to determine the best result, which is the one with the smallest deviation. There are two methods of checking conformity that are commonly used, namely the Chi-Square Test (X^2 test) and the Smirnov-Kolmogorof method. The calculation results of the two methods are then compared and the one with the smallest deviation is selected.

2.1.3 Design and Hourly Rainfall The design rain analysis method is chosen depending on the suitability of the statistical parameters of the data in question, or it is chosen based on other technical considerations, such as the type of data availability, the length of the data series, the number of parameters of the availability of hydrological data. Therefore, Since there is no rainfall data investigated at each hourly rainfall station, an assumption is taken that the daily rainfall is concentrated for 5 hours every day. The equation used to determine the hourly rain is:

$$I_T = \frac{R24}{t}\left[\frac{t}{T}\right]^{2/3} \tag{1}$$

2.1.4 Flood Hydrograph In most water sheds most of the rainfall will be converted into runoff. This kind of flow can produce high flood peaks. The unit hydrograph theory relates that this kind of flow can produce high flood peaks and the net rain or effective rain, which is part of the total rain that causes surface runoff with the direct runoff hydrograph so that it is a means to calculate the hydrograph due to scattered rain.

In order to construct the flood hydrographs where there are no or very few flood hydrograph observations, it is necessary to look for the characteristics or parameters of the catchment area at first, for example the time to reach the peak of the hydrograph (time to peak magnitude), base width, area, slope, length. the longest channel, runoff coefficient and so on. This research combines hourly rainfall and Nakayasu synthetic unit hydrograph to produce a flood hydrograph which will be used as input to HEC-RAS.

2.1.5 Cross-sectional Capacity Analysis With HEC-RAS To be able to analysis the flood problem, a tool is needed to identify the impact of flooding and look for efforts to overcome it. One of the tools that are currently used to analyse floods by modelling river hydraulics is HEC-RAS.

IOP Conf. Series: Earth and Environmental Science **989** (2022) 012012 doi:10.1088/1755-1315/989/1/012012

HEC – RAS (River Analysis System) is a one-dimensional flow hydraulics model. This program is a program that integrates hydraulics analysis, where program users can interact with the system using the Graphic User Interface (GUI) function. This program can show the calculation of the steady flow surface profile, including unsteady flow, sediment movement and some hydraulic design calculations. In HEC-RAS terminology, a set of data files will be associated with the system stream. Data files can be categorized as data plan data, geometric data, steady flow data, unsteady flow data, sediment data and hydraulic design data.

In addition to showing the flow surface profile, the HEC-RAS program can also be used to perform simulations for steady and unsteady flow models, analysis the size of the reservoir for water regulation needs, and design water infrastructure. HEC-RAS basically consists of 3 (three) components of 1 (one) dimensional hydraulic analysis (one dimensional computation), namely:
1. Simulation of one-dimensional steady flow.
2. Simulation of one-dimensional unsteady flow
3. Calculation of the transport of sediment movement.

2.2 Development
Infrastructure development for cold lava flood control in the Bali Cultural Centre Development Area in Klungkung Regency is in the form of normalizing the Unda River channel. Channels are made of segmental precast reinforced concrete walls reinforced with columns and diaphragm beams.

3. Results and Discussion

3.1 Hydrology Analysis
The rainfall stations namely Besakih, Duda, and Klungkung are found to be consistent by using RAPS. Meanwhile, the area rainfall resulted from Thiessen Polygon is presented in Table 1 as follows.

Table 1. The Thiessen Coefficient

No	Station	Influence Area Fi (km^2)	Thiessen Coefficient Fi/Ftot
1	Besakih	62.37	0.28
2	Duda	46.03	0.21
3	Klungkung	116.10	0.52
	TOTAL	224.50	1.00

Moreover, by applied the frequency analysis, the Log-Pearsson III is found to be the most suitable distribution for the rainfall data distribution based on chi square and Smirnov-kolmogorov test. Hence, the design rainfall is determinet using Log Pearsson III as shown in Table 2.

Table 2. The Rainfall Return Period

Probability	**Return Period**	**Log Pearson III**	
P(x>=X)	**T (years)**	**K_T**	**X_T**
99.99%	1	-4.022	15.476
99.00%	1.01	-2.429	30.071
98.00%	1.02	-2.128	34.087

Probability	Return Period	Log Pearson III	
P(x>=X)	T (years)	K_T	X_T
96.00%	1.042	-1.798	39.115
95.00%	1.053	-1.684	41.022
90.00%	1.111	-1.295	48.223
80.00%	1.25	-0.834	58.449
50.00%	2	0.023	83.551
20.00%	5	0.848	117.806
10.00%	10	1.266	140.228
4.00%	25	1.702	168.183
5.00%	20	1.604	161.484
2.00%	50	1.978	188.729
1.00%	100	2.224	209.043
0.50%	200	2.445	229.27

Based on the probability of rainfall, the coefficient of flow (C) and the pattern of distribution of rainfall intensity, the distribution of net rain per hour can be seen in the following table:

Table 3. The Hourly Rainfall Distribution

Hour	Ratio	cummulative	Hourly Rainfall for Return Period (mm)							
			1.01	2	5	10	20	25	50	100
1	55.03%	55.03%	9.93	27.59	38.90	46.30	53.32	55.53	62.32	69.02
2	14.30%	69.34%	2.58	7.17	10.11	12.03	13.86	14.43	16.20	17.94
3	10.03%	79.37%	1.81	5.03	7.09	8.44	9.72	10.13	11.36	12.59
4	7.99%	87.36%	1.44	4.00	5.65	6.72	7.74	8.06	9.05	10.02
5	6.75%	94.10%	1.22	3.38	4.77	5.68	6.54	6.81	7.64	8.46
6	5.90%	100.00%	1.06	2.96	4.17	4.96	5.71	5.95	6.68	7.40
Daily Rainfall Probability			30.07	83.55	117.81	140.23	161.48	168.18	188.73	209.04
Runoff Coefficient			0.60	0.60	0.60	0.60	0.60	0.60	0.60	0.60
Effective Rainfall			18.04	50.13	70.68	84.14	96.89	100.91	113.24	125.43

Further, the Nakayasu synthetic hydrograph is combined with the hourly rainfall shown in Table 3 to determine the flood hydrograph. The flood hydrograph for each selected return period is presented in Figure 3.

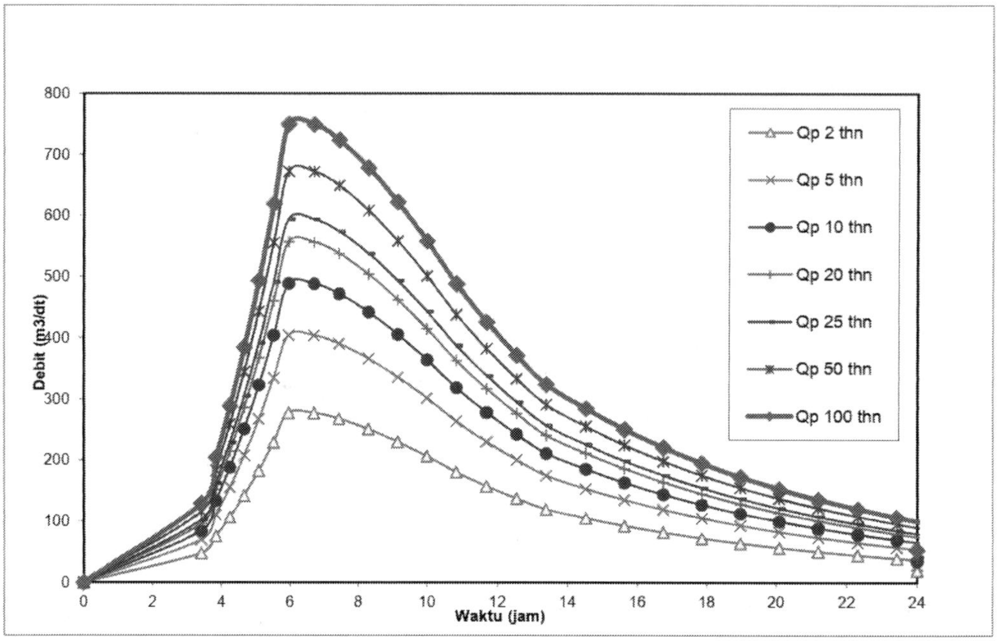

Figure 3. The Flood Hydrograph in Tukad Unda

The Flood analysis in the downstream area of the Unda River was carried out by modelling river hydraulics using the HEC-RAS tool. From the simulation results of the 50-year planned flood discharge, it is found that the design of the river in the lower part of the Unda River as presented in Figure 4 and 5 is able to accommodate the flood design for 50 years return period.

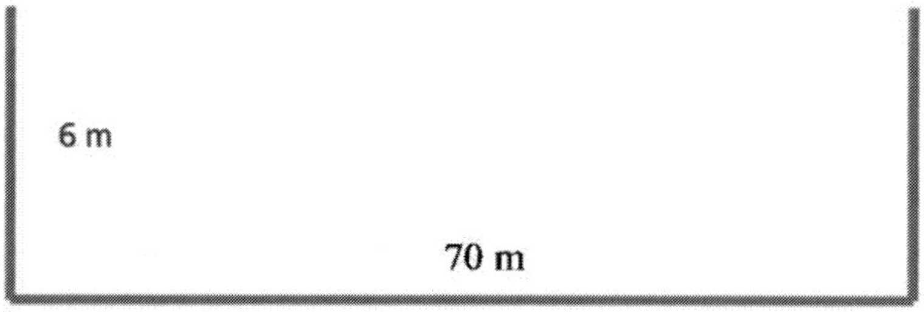

Figure 4. The Shape Of The Cross Section in Upstream Location (Northern Part of Prof. Ida Bagus Mantra Street)

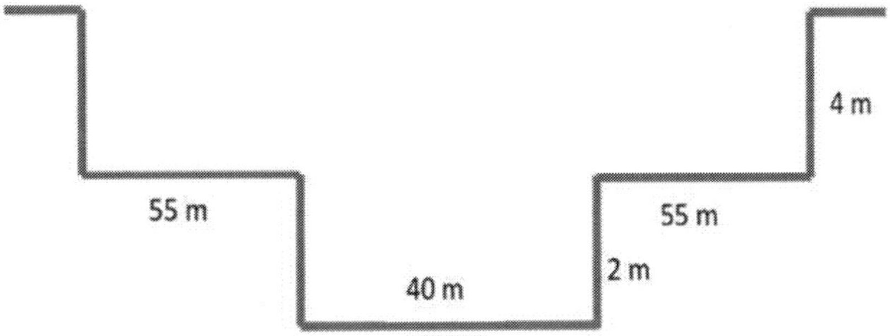

Figure 5. The Shape Of The Cross Section in Downstream Location (Southern Part of Prof. Ida Bagus Mantra Street)

The Geometry data Of Unda River watershed used in HEC-RAS are presented in Figure 6.

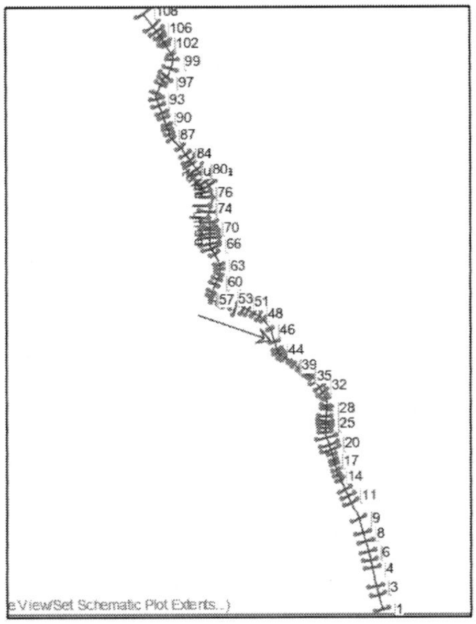

Figure 6. HEC-RAS Geometry Data for Unda River

The flood simulation results of the 50 year return period shows that the design of the river in the lower reaches of the Unda River is able to accommodate the planned discharge as shown in the Figure 7 and 8.

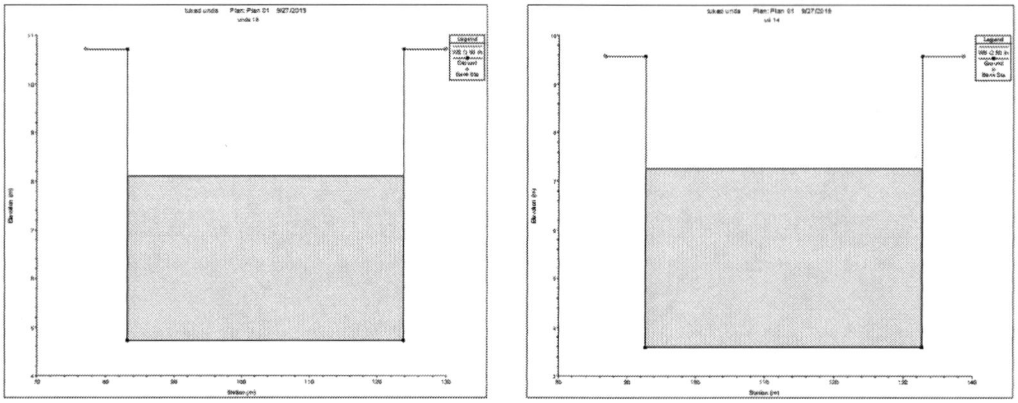

(a) (b)

Figure 7. The Example of Water Level in Cross Section 18 (a) and 14 (b) (Upstream of the Bridge)

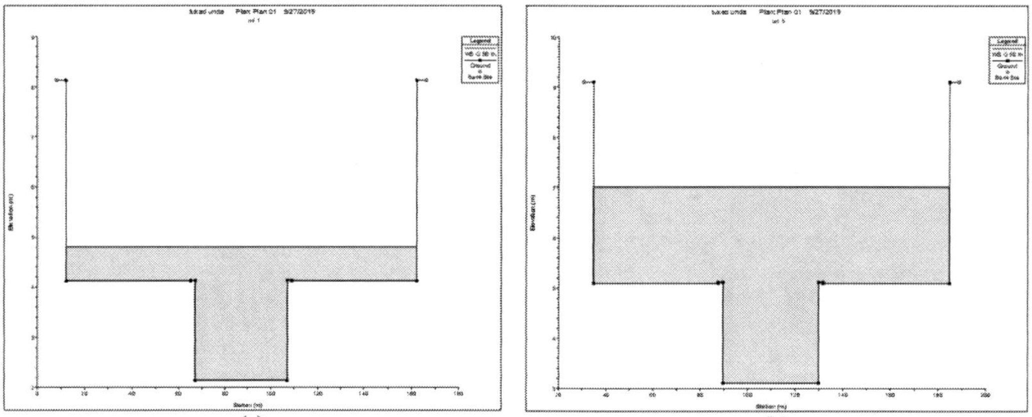

(a) (b)

Figure 8. The Example of Water Level in Cross Section 1 (a) and 15 (b) (Upstream of the Bridge)

3.2 Construction execution

The Ministry of Public Works and Public Housing (PUPR) through the Bali Penida River Basin has started work on the construction of flood control infrastructure downstream of Unda River, in Klungkung Regency, Bali. The river structuring and normalization program can reduce the risk of flooding in the Unda River Water shed (DAS), which is one of the centres of international tourism activities in Bali, namely the Bali Cultural Centre. Construction has started in August 2020 and will be completed in December 2022. Construction for handling cold lava floods at the Unda River estuary is divided into two locations. The first location is at the upstream of Prof. Ida Bagus Mantra Street uses a single rectangular cross section. The second location is at the lower end of Prof. Ida Bagus Mantra Streetuses a double rectangular cross section Figure 9 to be able to drain water discharge under normal conditions and can still drain water discharge in the 50-year return period.

With the construction of this channel, the cold lava flood flow will be directed towards the sea downstream of Unda River. It is possible that there are still sidemen settling on the bottom of the channel, it is necessary to normalize the channel in the form of periodic dredging.

Figure 9. Construction Implementation Downstream (Double Quadrangle)

The wall structure uses two-lane precast reinforced concrete slab panels and is joined by columns and diaphragm beams. Figure 10. The inside is then backfilled with sand. This is a new structural model developed to hold soil.

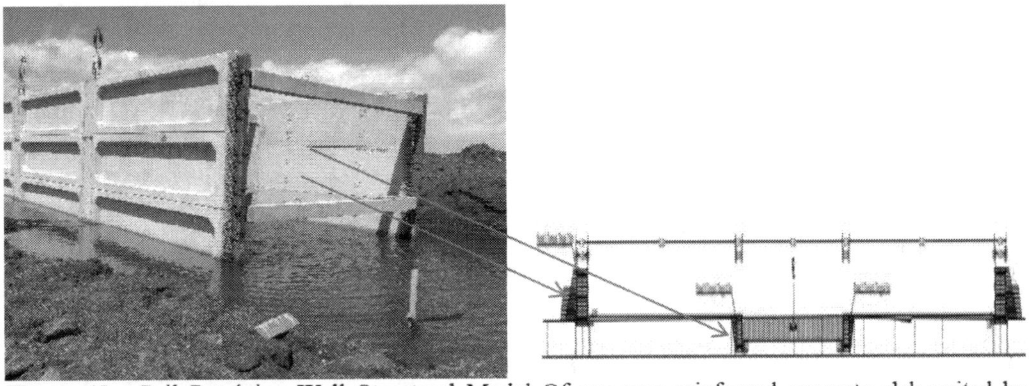

Figure 10. Soil Retaining Wall Structural Model Of one-way reinforced concrete slab united by columns and diaphragm beams.

3.3. Disaster Risk Reduction
Disaster risk can be formulated as, Risk is (hazard x vulnerability) divided by capacity [6]. Reducing the risk of cold lava flooding in the Bali Cultural Centre Development area is carried out in two ways, namely reducing hazard and increasing capacity. Reducing hazard and increasing capacity is carried out by normalizing the Unda River Flow [7].

4. Conclusions and Suggestions

4.1 Conclusions
Reducing the risk of cold lava flooding in the Bali Cultural Centre Development area is carried out in two ways, namely reducing hazard and increasing capacity. Reducing hazard and increasing capacity is carried out by normalizing the Unda River Flow. Normalization of the lower reaches of the Unda River will cause cold lava flows directly to the sea.

4.2 Suggestions

With the construction of this channel, the cold lava flood flow will be directed towards the sea downstream of Tukad Unda. It is possible that there are still sidemen settling on the bottom of the channel, it is necessary to normalize the channel in the form of periodic dredging.

References

[1] As-syakur, A. R., Osawa, T., Miura, F., Nuarsa, I. W., Ekayanti, N. W., Dharma, I. G. B. S., Adnyana, I. W. S., Arthana, I. W., & Tanaka, T. (2016). Maritime continent rainfall variability during the TRMM era: The role of monsoon, topography and El Niño Modoki. Dynamics of Atmospheres and Oceans, 75, 58-77.

[2] Diana H., Wiryajaya I.N., Brian E. P., (2021), Dampak Tidak Langsung Siklon Tropis Seroja Terhadap Cuaca Di Wilayah Bali, Forum Pengurangan Risiko Bencana Provinsi Bali.

[3] Badan Penanggulangan Bencana Daerah Provinsi Bali, 2020, Dokumen Kajian Risiko Bencana Provinsi Bali

[4] Ramadhan P., Sawitri B.U., Asep S., Slamet U.I., and Destia R., 2020, Mitigation Of Disaster Risk Reduction In Pangandaran Regency, Sosiohumaniora - Jurnal Ilmu-ilmu Sosial dan Humaniora, Vol. 22, No. 2, July 2020: 214 – 222, ISSN 1411 - 0903 : eISSN: 2443-2660.

[5] Sutarja I.N., Ardana M.D.W., Rahman A.A., Iman F., Redana I.W., Suarsana I.N., Taufik M., 2021, Laporan Penelitian Risiko Bencana Pada Areal Pengembangan Pusat Kebudayaan Bali Di Kabupaten Klungkung.

[6] Badan Nasional Penanggulangan Bencana, 2020, Indek Risiko Bencana Indonesia (IRBI), Penerbit : Badan Nasional Penanggulangan Bencana, Jakarta

[7] Arif T.W., Aldino F.S., Taubi A.g., 2019, Tinjauan Daerah Terdampak Lahar Dingin; Implikasi Pembuatan Sabodam Terhadap Dampak Sosial dan Ekonomi Pasca Erupsi Merapi 2010 Di Kali Putih, Kabupaten Magelang, Prosiding Seminar Nasional Kebumian Ke 12, Teknik Geologi Fakultas Teknik Universitas Gajah Mada 5-6 September, Hotel Alana Yogyakarta.

IOP Conf. Series: Earth and Environmental Science **989** (2022) 012013 doi:10.1088/1755-1315/989/1/012013

Artificial neural network for landslide vulnerability mapping in Leitimur Peninsula Ambon Island

Ferad Puturuhu[1,*], Projo Danoedoro[2], Junun Sartohadi[2], and Danang Srihadmoko[2]

[1]Faculty of Agriculture, Pattimura University
[2]Faculty of Geography, Gadjah Mada University

[*]Email: feradputuruhu@gmail.com

Abstract. Artificial neural network (ANN) has been widely used in remote sensing to classify the various types of data. Moreover, it provides better accuracy results than statistical methods. In this study, AAN was applied to map landslide vulnerability zone in Leitimur Peninsula Ambon where landslide has frequently occurred in the period 2012 until now resulting in losses to community. The objective of this study was to determine the landslide vulnerability level zones of the study area. The method used was the logistic regression and ANN. The results showed that the largest zone (9957.33ha) or 65.47% of the study area is in low landslide vulnerability level, while the best result of the accuracy test is the ANN analysis with the value of 72.55%.

1. Introduction

The occurence of climate-related disasters such as floods, landslides, whirlwind, tidal waves, and droughts have significantly increased over the last 10 years in Indonesia, and they are responsible for the losses of life and the impacts on the social, economic and environment [1].

In Ambon, landslide is considered as one of the major disaster event which frequently occur during the rainy season especially in the extreme condition (rainfall >100mm/day). It is also associated with the occurence of earthquakes in Ambon island.

Leitimur Peninsula (the study area) is located in the Southern Part of Ambon Island with an area of 152.8 km^2 or 41% of the total area of Ambon Island. Topography of the Leitimur Peninsula is dominated by hilly and mountainous areas with slope steepness ranges from 15% to >45%. The hilly area is mostly occupied by residents with medium density (1315.3 inhabitants / km^2), while the mountainous area is covered by cultivated agricultural activities and mixed cropping [2]. In 2012 and 2013, 43 landslide events are recorded in the Leitimur Peninsula, and they have a significant impact on community due to losses of life and houses damages [3]. So far, the research on the vulnerability to landslide in Ambon is relatively rare, therefore the knowledge of the potentially landslide prone area is limited.

The prediction of occurrence of a potentially landslides are generally represented in the form of landslide vulnerability zone map, and this map is substantial component of hazard management in decreasing the risk of living with landslides. Remote sensing is a method that has been widely used in the field of mapping and land resource management and monitoring [4]. The application of remote sensing method for landslide vulnerability assessment is very broad depending on the objectives to be produced. One of the remote sensing methods used in landslide vulnerability zonation mapping is artificial neural network (ANN). ANN is a computational mechanism that able to acquire, represent,

IOP Conf. Series: Earth and Environmental Science **989** (2022) 012013 doi:10.1088/1755-1315/989/1/012013

and compute a mapping from one multivariate space of information to another given a set of data representing that mapping [5]. ANN has been widely used in remote sensing to classify the various types of remote sensing data and it provides better accuracy results than statistical methods [6]. Artificial neural networks have advantages in combining spectral and non-spectral data, and the the ability to remember data that has been trained (memorization), and to recognize data that has not been trained (generalization). In this study, input data is prepared using the logistic regression statistical method, so, dependent and free variables used in determining the level of landslide vulnerability levels can be separated before being tested by neural network analysis.

By considering the risk of the landslide disaster, the assessment of the landslide vulnerability level needs to be done as an initial step to prevent and reduce the risk of landslides in the Leitimur Peninsula. Therefore, the purpose of the study is to determine the landslide vulnerability level zone in the Leitimur Peninsula of Ambon Island (Figure 1).

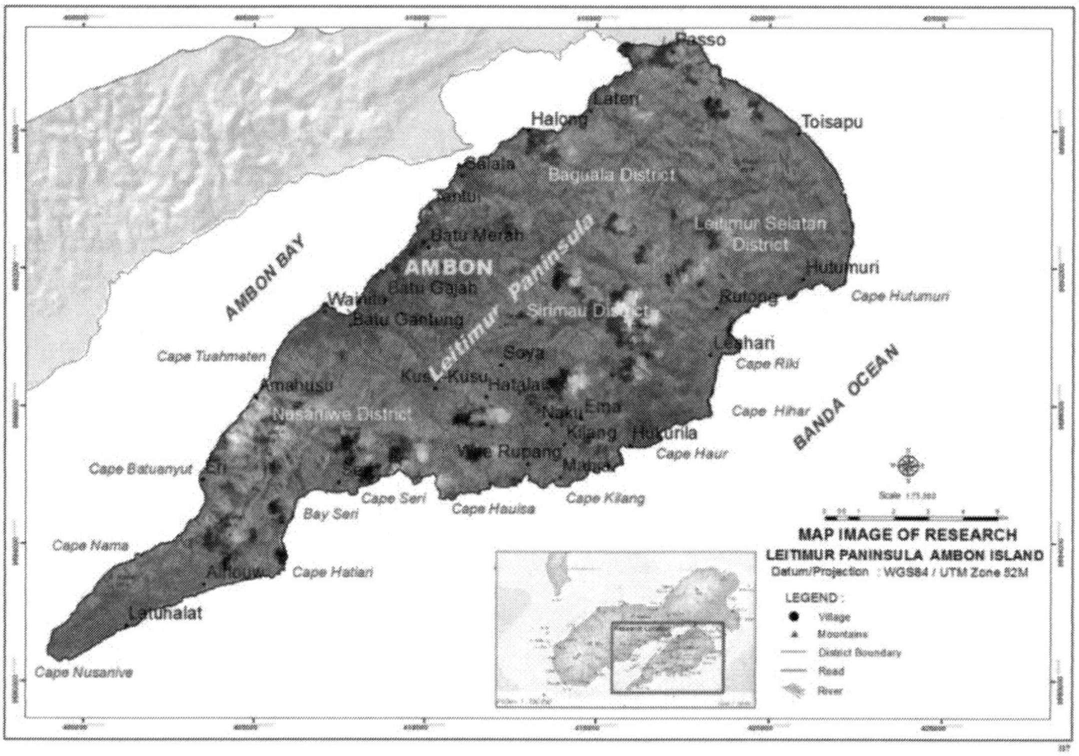

Figure 1. Research Location Map Leitimur Peninsula region, Ambon Island

2. Methods

The method used in this study was artificial neural network and statistical methods. The sampling method was purposive or a non-random sampling technique. The purposive sampling was chosen because this study emphasizes on landslide events in selected land units, so only few land units were taken to be modeled.

IOP Conf. Series: Earth and Environmental Science **989** (2022) 012013 doi:10.1088/1755-1315/989/1/012013

Table 1. The Variable Value of the Landslide Parameter

Varuable	Parameter	Class	Source Data
Geologi	**Lithology** :		Geology Map
	Andesit, Breksi, Liparit, Tuff (Removing Material)(Tpav)	1	
	Granit, Kwarsa (Tidg)	2	
	Peridotit, Serpentinit (JKu)	3	
	Sand Stone (Tjk)	4	
	Limestone (Ql)	5	
	Aluvium (Qa)	6	
Soil	**Types of Soil :**		Data of Soil Unit in Leitimur Peninsula and Field-Checking
	Aluvial (Fluvent), Gleisol (Aquept)	1	
	Kambisol (Tropept), Brunizem (tropept)	2	
	Podsolik (Udult)	3	
	Litosol (Orthent), Rendsina (Rendoll)	4	
Texture	**Texture of Soil:**		Data of Soil Unit in Leitimur Peninsula and Field-Checking
	Smooth	1	
	Rather Smooth	2	
Depth	**Depth of Soil (cm) :**		Data of Soil Unit in Leitimur Peninsula and Field-Checking
	< 50	1	
	50 – 90	2	
	> 90	3	
Landcover	**Landcover:**		Land Unit Map and Supervised Classification
	Forest	1	
	Mixed cropping	2	
	Settlement	3	
	Grasslands and Shurbs	4	
Slope	**Class of Slope:**		Digital Elevation Model (DEM)
	0-3% (Flat)	1	
	3-8% (Sloping/Wavy)	2	
	8-15% (Rather Hilly Sloping/Wavy)	3	
	15-30% (Hilly Sloping)	4	
	30-45% (Rather Steep)	5	
	45-65% (Steep)	6	
	> 65% (Very Steep)	7	
	Morphology :		
	Flat	1	
	Convex	2	
	Sunken	3	
Rainfall	**Total of Rainfall (mm/Yr)**:		BMKG's Data
	1000 – 1500 (Very Low)	1	
	1500 – 2000 (Low)	2	
	2000 – 3000 (Moderate/Medium)	3	
	3000 – 4000 (High)	4	
	4000 – 5000 (Very High)	5	

The variables of lithology, soil, land cover, slope and rainfall were used and grouped into two groups, namely dependent variables which are landslide events, and independent variables which are geology, soil units, soil depth, soil texture, land cover, slope and rainfall, respectively. The secondary data included landsat 8 and quickbird images of 2012, SRTM image of ambon island with 30 x 30 m resolution, climate and seismic data (from Meteorology Climatology and Geophysics Pattimura and Karang Panjang Station Ambon), geology and soil maps of Leitimur Peninsula, and the RBI map (from Geospatial Information Board), respectively. The image was corrected and cut to obtain the study area boundaries in Leitimur Peninsula, Ambon Island. The composite images were created by combining three single channel images in blue, green and red filters [7]. In this study, the combination of 653 bands (SWIR-1, Near-IR, Green) was used to generate false color composites of landsat 8 images. Landsat 8 of Peninsula Leitimur was used for supervised classification and landslide vulnerability analysis.

Maps of geology, soils, multispectral classified land cover, and slope class (derived from DEM-RBI) were digitized using ArcGIS 10 to produce maps in vector format. Then, these maps were overlaid to produce a land unit map which is used as a landslide mapping unit. While, raster format is used in ANN and logistic regression analysis.

The fieldwork was carried out to obtain data on the land condition that will be used to test the accuracy of the method used. Samples of land units was purposively selected and the results were used in logistic regression and artificial neural networks analysis, and in accuracy testing. Based on the results of image interpretation, field work was carried out to find the location of landslides, these data were then used for modelling and validating the occurence of landslides in Leitimur Peninsula Ambon Island.

The landslide vulnerability level was determined using the logistic regression analysis and artificial neural network analysis, a further digital image processing method [8,9].

2.1. Logistic Regression Analysis

Data used in the logistic regression analysis consisted of the dependent (landslide events) and independent variables (lithology, soil, land cover, slope and rainfall), and the variables design was performed in SPSS [10,11]. Because in this study land units were used as the mapping unit (shape vector), so, data of 83 land units was filled based on selected land units. The data was processed using statistical methods, so that the existing data must be numeric. Code numbers for the dependent variable is "0" indicates there is no landslide event and "1" indicates there is landslide event [12,13], while the numbers or numeric codes giving to the independent variables have to be suitable to of each variable of land units. The number of sampling of the dependent variable "0" is 75 means that landslide events are not found in these 75 land units, and the independent variable "1" is 7, means that landslide events are found in these 7 land units.

The feasibility test of model is important before conducting the logistic regression calculation. The test was conducted by using Hosmer and Lemeshow Test. The model is accepted if the model has a significant value > 0.05. Verification of each independent variable should be done with significant test. Variables that do not have a significant difference in affecting the landslide event can then be eliminated or removed. Construction of logistic regression generates variables in the equation that can be used to determine the coefficient of each variable with beta coefficient (ß). This coefficient value represents the relations of each variable in affecting landslide, while the extent of their influence can be seen through significant values (Sig) and the value of the Wald test. The relationship between these variables can be then generated a linear combination of functions of the dependent variable (Y) for vulnerability assessment of landslide (Dai, 2000) as follows:

$$Z = \beta_0+\beta_1L_1+\beta_2ST_2+\beta_3KT_3+\beta_4TT_4+\beta_5PL_5+\beta_6S_6 \quad (1)$$

Where, landslide vulnerability (Z), regression coefficient (β), lithology (L), land unit (ST), soil depth (KT), Land Texture (TT), land cover (PL), soil texture (Tt), slope steepness (S).

The probability values ranged from 0 to 1 are used as databases of each cell on the raster maps. Data is is classified using the natural break method, while the accuracy of landslide vulnerability map is tested using a success rate test. The curve of success rate test shows the comparison between the areas of actual landslides and the model, and the level of confidence of the vulnerability landslide map model.

2.2. Artificial Neural Network

Sample areas (training areas) for the Artificial Neural Network is determined on landsat satellite imagery 8 based on the vulnerability landslide map model. The use of logistic regression is based on the objectivity of the classification results. The position of the training area in ANN is presented in Figure 2.

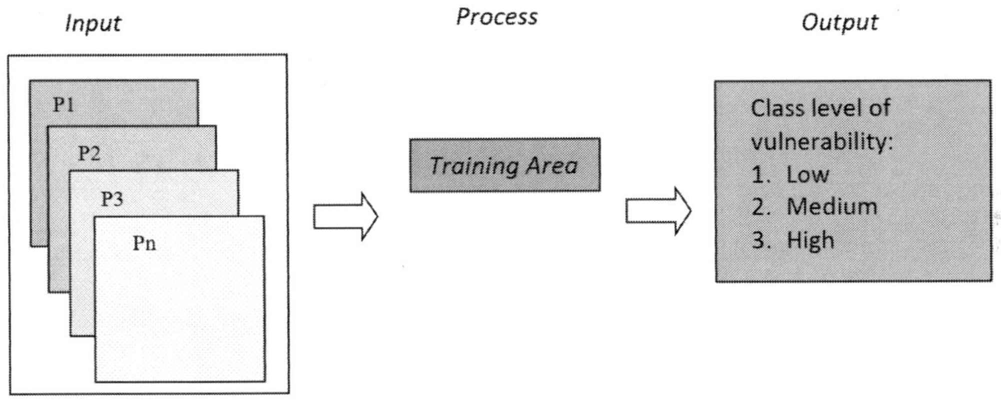

P1,P2, P3,................, Pn = Landslide Parameters 1, 2, 3, ..., n (Spectral Data dan Non-Spectral)

Figure 2. Example of the Position of *The Training Area* in ANN [14]

The algorithm used is the algorithm of the supervised classification non-parametric. The network training with 40 simulation parameters and 6 channels landsat 8, and each one is added with one and two non-spectral channels of landslide parameters (slope and land covers). Simulation process of data entry in an artificial neural network [15] can be seen in Figure 3.

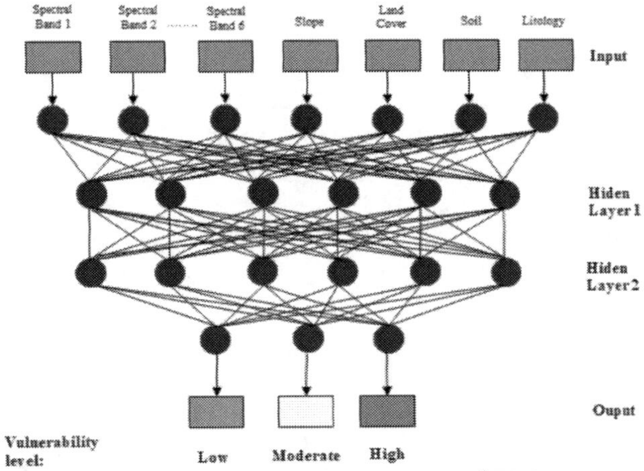

Figure 3. Simulation Process of JST

In this system, artificial neural network training process used a generalization method with three phases, they are the feed-forward of the input pattern of practice, the calculation and the error back propagation [16,17,18] and the adjustment of weights and bias. After the training process is finished, the next step is to carry out the reliability test of the network or the accuracy of the model.

3. Result and Discussion

3.1. The Land Unit of the Research Location
In this research, land unit is used as an approach and as a mapping unit in which land characteristics of the landslide vulnerability levels are examined. Dent and Young [19] define the land units as area of homogeneous ecological characteristics and they are resulted from the interaction of specific land factors. The Unity of the land unit consists of soil, geology, slope and land cover and is presented in a map. The choice of the four factors of the land associated with landslide parameter used to zoning the level of landslide vulnerability. The number of land units in the study is 83 units, and only 7 land units shows the occurence of landslide.

3.2. Parameter of Landslide Events
The parameters of landslide used in the study in Leitimur Peninsula are geology, ground units, texture and depth, slope, and land cover.

3.2.1. Geological Based on the Geological map of Ambon Island and field verification, geologically Leitimur Peninsula is composed by volcanic rocks such as andesite, liparit, breccia, granite, quartz, peridotite and serpentinite. Volcanic rocks are mostly spread in the hilly and mountainous areas with an area of 5173.74 ha, or about 34.02%. In stratigraphic position is above the sedimentary rocks. Sedimentary rocks such as tuff, sandstone, limestone and alluvium in physiographic occupy the plain areas to undulating and hilly, with an area of 10034.26 hectares, or approximately 65.98%, wider than the volcanic rock in Leitimur Peninsula. Size distribution of rock are presented in Table 2.

Table 2. Stone and Wide Development in Leitimur Peninsula

No.	Type of Rock	Distribution	
		Luas (ha)	%
	Volcano' Stones		
1.	Andesite	429.7	2.83
2.	Granite-Kuartz	2,457.84	16.16
3.	Peridotit-Serpentinit	2,286.2	15.03
	Sediment Rock		
4.	Alluvium	1,265.45	8.32
5.	Tuf	5,499.73	36.16
6.	Limestone	1,866.92	12.28
7.	Sandstone	1,402.16	9.22

Geological factors such as physical properties of rocks, position and geological structure factors are considered as factors affect the level of landslide vulnerability. Layered rocks with tilted in the direction of the slope angle will have a high level of vulnerability. Hard Rock and still relatively fresh may also be prone to landslides when it has a lot of cracks / joint in the direction of the angle of slope. Hard Rock with a lot of joints is also more susceptible when formed above the soft rocks with expanding and shrinking properties. Landslide events in the study area are mostly found in tuff rock granite-quartz and serpentinite, because this rock formation is formed by the mixtureable characteristics and easily to separate into rocks and gravels. Serpentine rock often works as a sliding layer as found in Bubawang's landslide event, Amahusu district of Nusanive. Limestone is easily weathered rock showing the occurence of soil creep as found in Silale.

3.2.2. Slopes Topographic influences the vulnerability of landslide event by its slope steepness. Landslide increases with the increasing of the slope steepness [20,21]. 7 slope steepness classes are derived from the DEM RBI with 12.5 m interval contour lines and they are classified according to the USDA. The distribution of slope steepness in the study area is presented in Table 3. This study found that 9 events of landslide occur on slopes of 15-30%.

Table 3. Slope's class of the study area

No.	Slope's Class	Distribution	
		Width (hectare)	Precentage (%)
1.	0-3%	1,285.28	8.45
2.	3-8%	1,447.89	9.52
3.	8-15%	3,234.96	21.27
4.	15-30%	4,243.07	27.90
5.	30-45%	4,377.11	28.78
6.	45-65%	575.62	3.79
7.	>65%	44.07	0.29

3.2.3. Soils Soils distribution in the study area are determined based on the soil map of the Leitimur Peninsula and field study. Soil in the study area are alluvial soil, gleisol, brunizem, litosol, rendzina, kambisol, and podsolic. The effect of soil on the landslide vulnerability event is related to the rate of soil development. Kambisol is the dominant soil found in the study area in particularly on undulating to hilly areas. It covers 9142.03 ha or 60.11% of the study area. The occurence of landslide events is found mostly in Kambisol soil. Soil distribution in the study area is presented in Tabel 4.

Table 4. Soil distribution in the study area

No.	Soil Unit	Distribution	
		Width (Hectare)	% (Precentage)
1.	Aluvial	1,220.51	8.03
2.	Gleisol	187.48	1.23
3.	Podsolik	809.58	5.32
4.	Kambisol	9,142.03	60.11
5.	Brunizem	2,194.15	14.43
6.	Litosol	792.02	5.21
7.	Rendsina	862.23	5.67

3.2.4. Soil Textures There are various properties of soils affecting the occurrence of landslide, they are texture, structure, organic matter, depth, characteristics of soil layers and soil fertility. The parameters of soil used in determining the level of landslide vulnerability in the Leitimur Peninsula is soil texture and soil depth.

The soil texture used in this study is grouped based on the smoothness and the roughness of soil material. The moderately smooth soil texture is the most dominant in the study area covering 14398.42 ha or 94.68% of the study area (Table 5). This class is found in Kambisol, Brunizem, Gleisiol, Rendsina, Alluvial and Litosol soils, while smooth soil texture is found in Podsolic soil. Landslide events are mostly found in the moderately smooth soil texture with slope steepness > 15%.

Table 5. Soil Texture and Width Distribution in Leitimur Peninsula

No.	Soil Texture	Distribution	
		Width (Ha)	%
1.	Smooth	809.58	5.32
2.	moderately Smooth	14,398.42	94.68

3.2.5. Soil Depth The soil depth is one of soil the characteristics that affects the landslide vulnerability. Similar to texture, soil depth is correlated to its soil type. The soil depth of 50-90 cm is largest area with 11224.77 ha or 73.81% of the area and most of the landslide events occur at soil depth of 50-90 cm. This soil depth is found in Kambisol, Rendsina, and Alluvial soils.

Table 6. The Distribution Location and The Width of Soil's Depth in Leitimur Peninsula

No.	Soil Depth (cm)	Distribution	
		Width (ha)	%
1.	< 50	792.02	5.21
2.	50-90	11,224.77	73.81
3.	> 90	3,191.21	20.98

3.2.6. Landcover The land cover that based on the interpretation of Landsat 8 in 2013 and field-checking is reeds, shrubs and moors, forests, mixed farms and settlements. Land cover that is widely spreading in Leitimur Peninsula, is mixed farms with an area of 5358.18 ha, or about 35.23%. Land Cover's deployment is little settlement with an area of 3371.71 ha, or approximately 22.17% (Table 7). Although, not widely spreading yet to settlement is one that affects land cover and its landslide vulnerability and event, especially the settlements are located in areas with slope class of> 15%. While, Land cover that occurred landslide is settlement.

Table 7. Landcover and Distribution Width in Leitimur Peninsula

No.	Landcover	Distribution	
		Width (Hectare)	% (Precentage)
1.	Forest	2,066.42	13.59
2.	Mixed Farms	5,358.18	35.23
3.	Settlement	3,371.71	22.17
4.	Reeds, More and Space Land	4,411.68	29.01

3.3. Logistic regression models for the classification level of landslide vulnerability

3.3.1. Feasibility Assessment Model and Significance Independent Variables The real difference between the classification variables that predicted by landslide classification observed variables known to test the feasibility of the model in LRM, was conducted by Hosmer and Lemeshow Test. A regression model is feasible to be used when there is no real difference between the classification of variables in the prediction of the classification of variables in fact observed by significant existing value. Significant value in the Hosmer and Lemeshow test results are .341 (Sig> 0.05), it means that the regression model tobe continued because of the similarity between the predicted classification with the classification of the real observed or in other words Ho accepted.

Assessment models to be generated must be preceded by assessing the iteration process, with a view to improve the model in order to produce a better model. Results of the LRM seen iterations of "test likelihoold". Based on the results of tests performed on the data of each variable -2 log likelihood values obtained final value is less than the value ie 34 147 -2 log likelihood its initial 53 616. The decline in the value of likelihoold indicate that the results of the resulting models become better and fit for use.

Tests of significance in LRM reconstructed from 6 parameter determining longsorlahan which includes soil type, soil texture, soil depth, geology, land cover, and slope. LRM parameters on rainfall is not included because it has the same classes for the study site Leitimur Peninsula ie including high rainfall 3000-5000 mm / yr, dinumerikan with the number 4, it is proved by trials that included MLR rainfall parameter data, analysis results of iteration history to variables in the equation, there is no rainfall comes out, and the result is the same as the LRM are not using rainfall.

Significant values (Sig) if it is worth> 0.05 (not significant) and <0.05 (significant) (Widayani, 2016). Seen from a statistical standpoint, the category of classes that do not significantly removed from the logistics function to generate a better model. However, when not all the classes the Sig> 0.05 then these variables can still be included in the regression function. MLR results as shown in Table 8, showing none of the significant parameters for all the parameters have the Sig> 0.05 that is of significance for the value of 0.022 on land cover class 3 is settlement to the highest significance value of 1.0 on soil grade 3, grade lithology 1-3 and 5, the slope grade 2-6 and grade 1 and 2 textures.

Table 8. Significant Constant Value and Results Processing of Logistic Regression Using Six (6) Variables

		B	S.E.	Wald	df	Sig.	Exp(B)
				0.912	3	0.822	
	Soil (1)	-23.208	14996.195	0.000	1	0.999	0.000
	Soil (2)	**Variabel Indenpenden**	2.119	0.912	1	0.340	0.132
	Soil (3)	Soil	60753.873	0.000	1	1.000	0.000
	Soil (4)	0					
	Lithology			1.495	5	0.914	
	Lithology(1)	-20.327	58786.273	0.000	1	1.000	0.000
	Lithology(2)	-19.324	58786.273	0.000	1	1.000	0.000
	Lithology(3)	-20.764	58786.273	0.000	1	1.000	0.000
	Lithology(4)	-38.755	59626.103	0.000	1	0.999	0.000
	Lithology(5)	-22.661	58786.273	0.000	1	1.000	0.000
	Lithology (6)	0					
	Slope			0.407	5	0.995	
	Slope(2)	19.694	40192.923	0.000	1	1.000	357337838.557
Step 1[a]	Slope (3)	19.932	40192.922	0.000	1	1.000	453300979.384
	Slope (4)	20.509	40192.922	0.000	1	1.000	807352397.363
	Slope (5)	19.983	40192.922	0.000	1	1.000	476976285.725
	Slope (6)	0.939	41396.951	0.000	1	1.000	2.557
	Slope (7)	0					
	Landcover			6.155	3	0.104	
	Landcover(1)	1.025	1.689	0.368	1	0.544	2.788
	Landcover(2)	1.446	1.486	0.946	1	0.331	4.245
	Landcover(3)	3.598	1.565	5.283	1	0.022	36.520
	Landcover (4)	0					
	Texture			0.000	1	1.000	
	Texture(2)	0.440	56841.468	0.000	1	1.000	1.552
	Texture (3)	0					
	Depth			0.673	2	0.714	
	Depth (1)	-24.467	11204.883	0.000	1	0.998	0.000
	Depth (2)	-1.713	2.087	0.673	1	0.412	0.180
	Depth (3)	0					
	Constant	0.120	42899.434	0.000	1	1.000	1.127

a Variable(s) entered on step 1: Soil, Lithology, Slope, Landcover, Texture, Deepness

3.3.2. Vulnerability Analysis of Landslide Through a probability value can produce maps of landslide vulnerability depicting the extent of vulnerability at the study site and their distribution [22]. Nilia P on logistic regression of the probability calculations obtained by linear regression function of the sum of all independent variables constant longsorlahan. The result longsorlahan vulnerability can be represented in terms of probability of value 0 to 1, which is getting closer to 1, the possibility of a longsorlahan will be even greater (Figure 4).

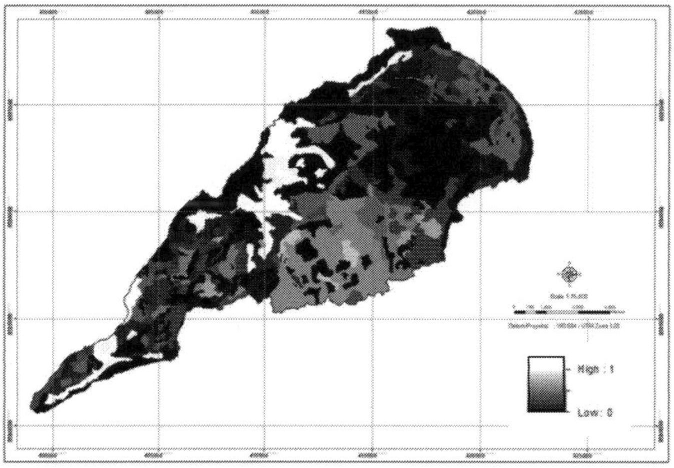

Figure 4. Probability Map of Landslide Vulnerability
Source: Logistc Regression Analyses Using ArcGIS 10.

The probability map on the next classified into three classes which are prone to low, medium and high, by using natural breaks classification method, since the distribution of the value of the probability that there is uneven so the method is suitable for reducing the frequency deviation between classes. The Landslide Vulnerability Map results at entire of the logistic regression analysis which this is presented in Figure 5.

The results show that the classification of the landslide vulnerability at the research location in Leitimur Peninsula, Ambon Island is highest prone to low level with a percentage of 65.47%, followed by prone to medium level with a percentage of 24.35% and prone to high percentage of 10.17% (Table 9). The Landslide Point scene is mostly in the high level of vulnerability.

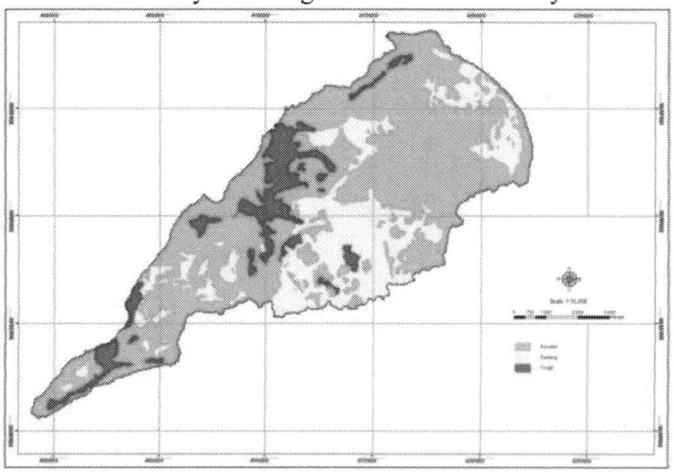

Figure 5. Logistic Regression of Landslide Vulnerability Map
Source: Logistic Regression Analyses Using ArcGIS 10.

Table 9. Relationship of Vulnerability level and its width

Vulnerability Level	Width (hectare)	Distribution Precentage (%)
Low	9.957,33	65,47
Moderate/Medium	3.703,84	24,36
High	1.546,83	10,17

Source : *Analyses Result of Logistic Regression*

3.3.3. Model Validation Testing Testing the validity of a model is important, because it concerns the appropriateness and accuracy of the model. Testing the validity of the vulnerability level of its landslide model accuracy is performed by the method success rate. This validation type will generate a comparison between the percentage probability curve of the landslide vulnerability that created from the model with the actual percentage of landlside event. The actual number of landslide events for the validation test at this time was as much as 16 area has taken randomly from the total of landlside event in the research location, is not included in the 18 area of landslide has been taken to modeling. The landslide's probability for its interval class vulnerability and probability of landslide are presented in Table 10.

Table 10. The Distribution Location of Landslide Validation at Probability of Interval Class (According to Map)

Class Interval Probability	Width Probability (ha)	Width Landslide (ha)	Cumulative Landslide (ha)	Area Precentage Landslide (%)
0-10	7.115,63	0,55	0,55	10,74
10-20	852,12	0,32	0,87	16,99
20-30	694,44	0	0,87	16,99
30-40	1.025,42	0	0,87	16,99
40-50	426,03	0,45	1,32	25,78
50-60	1.291,1	1,19	2,51	49,00
60-70	2.062,17	1,47	3,98	77,73
70-80	193,16	0	3,98	77,73
80-90	1.162,23	0,38	4,36	85,16
90-100	385,7	0,76	5,12	100,00

Form the Rate of the Success curve in Figure 6 shows that the level of accuracy of the model was calculated from the size of the area under the curve by 41%. Compared with previous studies, this value is still low, as Wahono [23] examined, which carried out a research in the watershed latitude of landslide vulnerability and model accuracy of landslide vulnerability was obtained by 81.58%.

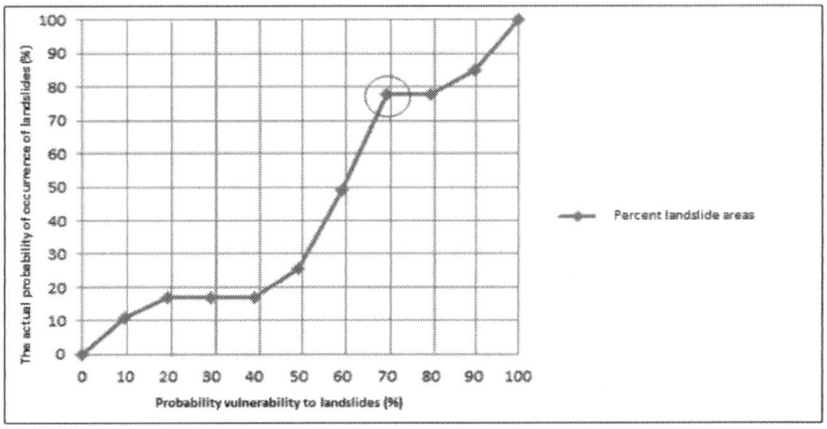

Figure 6. *Success Rate* Curve, The Level-Model of the Landslide Vulnerability

Thus, a small percentage of accuracy on a model of landslide vulnerability in Leitimur Peninsula due to extensive comparison with most locations landslide area which is very much different probability of less than 1 ha. Also, the deployment of landslide in the class probabilities is uneven and is not much proportional to the area of research location extent of 15 208 ha, a ratio of about one in three by the density of the scene and a very small area each landslide, h 3 / ha (0.0003 / km2) and 5 m / ha (0.00059 m / km2), and there are even prone to landslide probability class that was not found its landslide ie the probability classes 20-30, 30-40, and 70 - 80.

In addition to the factors above, then one factor that makes the accuracy percentage of the resulting model parameters of landlside's small value is used, or better known as the dependent variable using only 6 parameters, and in the analysis of LRM is used five parameters without precipitation because of the similarity of data has caused there is no effect on the results of the analysis. However, from the graph was obtained although most classes probability of the landslide vulnerability not indicates that the relationship of positive effect of increasing the percentage of probability of landslide occurrence in the real-time, but among them all there were one class that shows the positive effect that the probability of landslide vulnerability is 70% can affect the increasing of the probability of landslide occurrence in the real-time by 78%.

3.4. Artificial Neural Network Classification Level of Landslide Vulnerbaility

3.4.1 Artificial Neural Network Analysis The use of artificial neural network method for the classification level of landslide vulnerability is used as a method of comparison of the results obtained from the logistic regression method.

The analysis of the artificial neural networks which has been produced a classification level of landslide vulnerability, has also been done through simulation of multiple channels with network topology (hiden layer), training parameters (learning rate, momentum factor, sigmoid constanta), and stoping criteria (RMS, iterations) which varies according to design built in Multilater Perceptron network architecture. Neural network simulations performed in this study there are five, namely: (1) using a 6 channel (Landsat Band 1 - Band 6), (2) 7 channel, 6 channel spectral (Landsat Band 1 to Band 6) and 1 channel non-spectral (grade slope), (3) 8-channel, 6 channel spectral (Landsat Band 1 to Band 6), and two non-spectral channels (grade slope, land cover), (4) 8-channel, 6 channel spectral (Citra Landsat Band Band 1- 6), and two non-spectral channels (class slopes, soil), and (5) 8-channel, 6 channel spectral (Landsat Band 1 to Band 6), and two non-spectral channels (grade slope, geology).

The previous study conducted by Arif [14] and Wahyudi [24], said that the addition of the non spectral each in a simulated artificisl neural network will improve the accuracy of the results, therefore, in the simulation of artificial neural networks for the classification level of landslide vulnerability used the landslide grading-parameter are slope, land cover, soil and geology as the non spectral. The simulation results of artificial neural networks for the classification level of landslide vulnerability are explained as follows.

The simulation results ANN 6 channels shows that JST8 with hiden layer 2 and 20,000 iterations have a high overall accuracy is 54.97% with simulation time of 24 minutes. Nevertheless, eighth simulated artificial neural network has an overall accuracy increases with the magnitude of both the iteration hiden hiden layer 1 and layer 2. The amount of iterations and many hiden layer affect the duration of the simulation. Hiden layer 2 has a longer time with increasing iterations in the simulation process compared hiden layer 1. The simulation results showed that ANN 7 channels JST4 with hiden layer 1 and 20,000 iterations have a high overall accuracy is 58.83% with simulation time 17 minutes. Similarly ANN 6 eighth channel has a simulated artificial neural network that is overall accuracy increases with the magnitude of both the iteration hiden hiden layer 1 and layer 2. The amount of iterations and many hiden layer affect the duration of the simulation. Hiden layer 2 has a longer time with increasing iterations in the simulation process compared to hiden layer 1. The simulation results of ANN 8, the additional channels of data, Non spectral of slopes' class and land cover, have been shown that JST4 with hiden layer 1 and 20,000 iterations have a high overall accuracy which is not only to channel 8 with additional slopes and the classes of land cover but for all the channels simulated artificial neural network that is 64.71% with simulation time of 22 minutes.

The simulation results of ANN on various channels over the value of its accuracy, is determined by the level of landslide vulnerability traning area and ANN parameters that used. The Parameter of learning rate, momentum factor, sigmoid Constanta and RMS are the best values that used in this simulation which are based on the results of research and practice some previous investigators. It became a standard of high accuracy of the simulation results of the ANN. In order to have a better of landslide vulnerability's classification, is then used a hiden layer (hidden layer) which is different, namely 1 and 2 with nodes adapted to input data or input. However, the higher accuracy results in each channel of ANN is in hiden layer 1 with big iterations is 20,000 and the training and testing of RMS is low, and lasts a long time about duration of 14-17 minutes. Both at JST with hiden layer 1 and layer 2 has been shown that with the increasing value of the iteration, also increases the overall value of its accuracy. That means the effect is also very large of iteration for its value accuracy.

Related to the training area, its influence on the simulation results of ANN that is the sampling bit and uneven for each class of vulnerability affects the value of resulting its accuracy, following that on most maps generated residential area on the plains should enter the classes of the low vulnerability, the simulation results came into a high-grade vulnerability, it is due to sampling for high vulnerability classification according to the results of logistic regression are in a residential area in the hills or mountainous areas. These samples on Landsat 8 has a spectral reflectance / the same color as the existing settlements in highland areas with a low level of vulnerability, as well as samples taken over a lot of samples to lower the settlement's vulnerability. Caparini et al [25] mentions that the unsatisfactory results in the classification due to errors in the data rather than on its truth data, the classification of JST. Here, the matter is on the classification results of logistic regression in the making training area.

As stated in the previous section by some previous investigators for example[14] has shown that the addition of non-spectral data is going to produce a high accuracy data. Non spectral data that used are four namely landslide parameter, but not all used at once, only one additional parameter in JST 7 channels and the addition of two parameters at JST 8 channels with different combinations. The result of the addition of non-spectral data is very good because the average yield of high accuracy that is at JST4 channel 7, JST4 channel 8, as opposed to just using the spectral data on the JST 6 channels, which reached only 54.868% of its overall accuracy. The slopes used to be combined with three other parameters for consideration of the slope is a very decisive factor in landslide event, and the results of the trial were combined with a better accuracy than the slopes with lithological soil or soil with land cover.

3.4.2. Testing Results for calcification Results of Artificial Neural Network Testing the accuracy or accuracy is critical to remote sensing that is carried out by researchers or researchers who use remote sensing data. The accuracy of data interpretation results is important to know before proceeding further with the analysis based on the data. For the users of the accuracy of the data strongly influences the confidence that can be given to the data[4]. Danoedoro[6] argued that in testing the accuracy of the classification of remote sensing data is not only the final crucial classification, but starting from the specification or the type of imagery used in the modeling, the number of channels, methods of correction radiometric and geometric, base map reference, the current resolution resampling, an interpolation technique, the method of classification, the number of samples and statistical parameters for classification algorithms are applied, is also very important.

Based on the above understanding, the various components associated with the pre-processing of data to remote sensing data processing including vulnerability landslide's classification using artificial neural networks to be tested for its accuracy by the field data that has been classified by logistic regression analysis. The data that will be tested is a simulation that has the best architecture in the training data that is JST8 on 6 channels, JST4 on channel 7, JST4 on 8 channels, and JST7 on 7 channel map simulation results almost identical to map, the results of classification by logistic regression analysis. The number of 51 samples tested were also not involved in the training, which for the training sample for a class of prone to low 21 samples, 16 samples were vulnerable, and prone to high 14 samples.

Figure 7 shows that the calcification is at best accuracy accuracy value JST7_4 and JST7_7 ie 72.55. Namu because maps generated by JST7_4 severe impact, including in the areas of land use settlements on the slopes of <15% (flat-corrugated), that are used in this study as the zone map the landslide vulnerability is the result of the classification or simulated JST7_7, as presented in Figure 8.

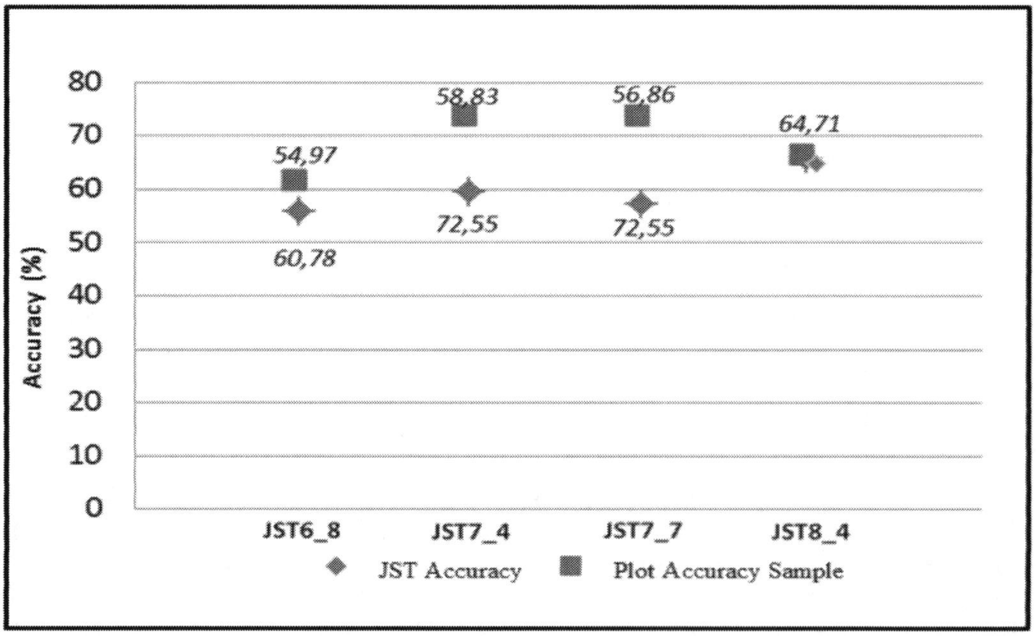

Figure 7. Comparison between the Accuracy and JST Classification and also the Plot Accuracy Sample (Classification Results of Logistic Regression)

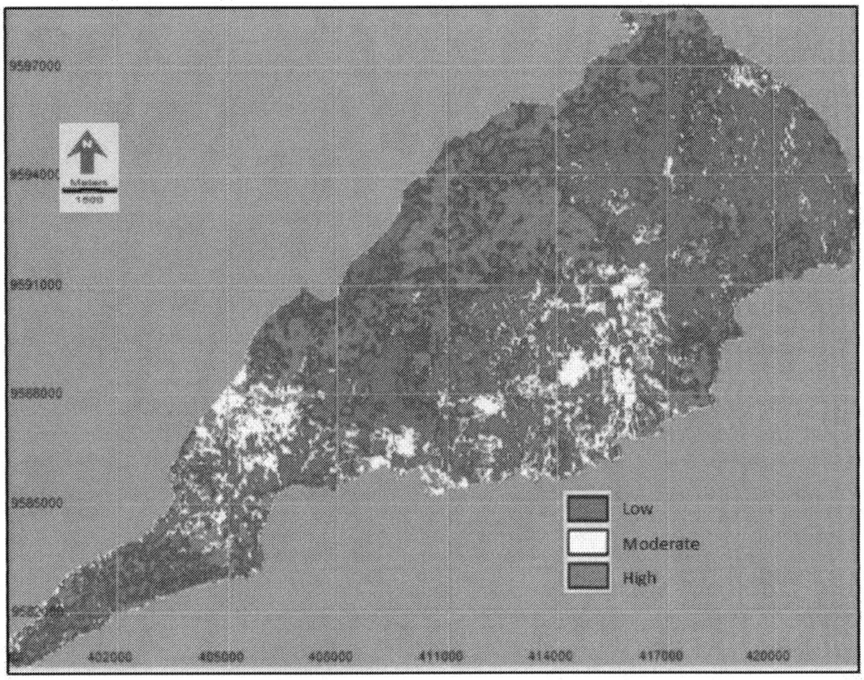

Figure 8. The Level of Vulnerability of Landslide Map, JST7_7, with Highest Value *Overal Accuracy*

This result shows that the landslide vulnerability in the research location in Leitimur Peninsula, Ambon Island is highest prone to low level with a percentage of 54.99%, followed by high-level cartilage with a percentage of 30.98% and prone to moderate/medium with a percentage of 14.33% (Table 11). The point scene of landslide is almost mostly in the high level of vulnerability. Tingginnya areal extents low level of vulnerability, is acceptable because there is still a Leitimur Peninsula forest land use of secondary and mixed farms are quite spacious compared to the use of land settlements and shrubs or reeds.

Table 11. The Relationship of the Level of Vulnerability and Its Width

Level of Vulnerability	Width (Hectare)	Precentage of Distribution (%)
Low	8364.15	54.99
Moderate/Medium	2179.61	14.33
High	4711.37	30.98

Source *: Result of Analyses, JST7_7*

4. Conclusion

Artificial neural network method which is a digital method further in remote sensing and logistic regression statistical methods used in this study, resulted in the zoning area of landslide vulnerability in Leitimur Peninsula. Based on the testing results obtained by the best accuracy is the result of the analysis of the artificial neural network with a value of 72.55% of the analytical results JST JST 7 4 7 7 channel.

Acknowledgement
This paper is part of dissertation of the author. therefore the author expressed his gratitude and appreciation to the team of promoters in this paper are included as well as an author who has guided and always provide motivation for me to be able to produce a scientific work of this valuable.

References

[1] BNPB, 2012. Data dan Informasi Bencana Indonesia. http://dibi.bnpb.go.id/DesInventar/dashboard.jsp?countrycode=id&continue=y&lang=ID. Akses 4 Mei 2012
[2] BPS Kota Ambon, 2014. Kota Ambon dalam Angka Tahun 2014.
[3] Puturuhu, F. Pengembangan Metode penginderaan Jauh dan Sistem Informasi Geografis untuk Pemetaan Kerawanan Longsorlahan di Jazirah leitimur Pulau Ambon. Universitas Gadjah Mada.
[4] Sutanto. 1994. *Penginderaan Jauh Jilid I.,* Gadjam Mada University Press.
[5] Choi J, Hyun Joo Oh, Joong Sun Won, Saro Lee., 2010. Validation of an Artificial Neural Network Model for Landslide Susceptibility Mapping. Environ Earth Sci-Springer.
[6] Danoedoro, P., 2012. *Pengantar Penginderaan Jauh Digita*l. Penerbit ANDI.
[7] Sabins, F. F. 2007. *Remote Sensing : Principles and Interpretation,* Waveland Press,
[8] Lee s, Ryu JH, Min K, Won JS. 2003. Landslide Susceptibility Analysis Using GIS and Artificial Neural Network. Earth Surf Proc Landf 28.
[9] Zhau W., 1999. Verivication of the Non Parametric Characteristics of Backpropagation Neural Networks for Image Classifications. IEEE Trans Geosci Remote Sens 37 (2).
[10] Dai F C, C F Lee, J Li, Z W Xu., 2001. Assessment of Landslide Susceptibility on The Natural Terrain of Lamtau Island, Hongkong. Environmental Geology 40 (3)-Springer-Verlag.
[11] Widayani Prima, Totok Gunawan, Projo Danoedoro, and Sugeng Juwono Mardihusodo., 2016. Application of Geographycaly Weghted Regression for Vulnerable Area Mapping of Leptospirosis in Bantul District. Indonesia Journal of Geography Nol 48, No.2.
[12] Kleinbaum. D.G., 1994. Logistic Regression a Self Learning Text. Sprenger, Berlin Heidelberg New York.
[13] Bui Dieu Tien, Tien Chung Ho, Inge Refhaung, Biswajeet Pradhan and Duy Ba Nguyen., 2014. Landslide Susceptibility Mapping Along The National Road 32 of Vietnam Using GIS-Based J48 Decision Tree Classifier and Its Ensembles. Springer-Verlag Berlin Hedelberg.
[14] Arif, N., 2011. Kajian Kemampuan Jaringan Syaraf Tiruan Berbasis Citra ALOS Dalam Identifikasi Lahan Kritis: Studi Kasus Kecamatan Dlingo Dan Sekitarnya. *Tesis (tidak dipublikasikan).* Program Pascasarjana Fakultas Geografi Universitas Gadjah Mada, Yogyakarta.
[15] Caniani D and Francesco Sdao., 2016. Neural Networks and Landslide Susceptibility: A Case Study of The Urban Area of Potenza. Natural Hazards.
[16] Pradhan Biswajeet, and Saro Lee., 2009. Landslide risk Analysis Using Artificial Neural Network Model Focussing on Different Training Sites. International Journal of Physical Sciences Vol. 4 (1)
[17] Li Y, G. Chen, C. Tang, G. Zhou, and L. Zheng., 2012 Rainfall and earthquake-induced landslide susceptibility assessment using GIS and Artificial Neural Network. Natural Hazards and Earth System Sciences.
[18] Zhang Shi Liang and Ting Cheng Chang., 2015. A Studi of Image Classification of Remote Sensing Based on Back Propagation Neural Network With Extended Delta Bar Delta. Mathematical Problems in Enginering, Article ID 178598.
[19] Dent, D., and Young, A., 1981. *Soil Survey and Land Evaluation*. George Allen & Unwin Ltd. London
[20] Saha A. K. Gupta RP, Arora MK., 2002. GIS-Based Landslide Hazard Zonation in a Part of The Himalayas. Int J. Remote Sensing, 23 (2).

[21] Clerici A, Perego S, Telini C, Vescovi P., 2002. A Prosedure For Landslide Susceptibility Zonation by the Conditional Analysis Method. Geomorphologi. 48 (4).
[22] Hadmoko .D.S., 2007. Toward GIS-Based Integrated Landslide Hazard Assessment: A Critical Overview. Indonesia Journal of Geography Volume 39. Number 1.
[23] Wahono, B.F., 2010. Application of Statistical and Heuristic Methods For Landslide Susceptibility Assessments. *Double Degree M.Sc Thesis (not published).* International Institute For Geo-Information Science and Earth Observation
[24] Wahyudi, D., 2013. Kajian Jaringan Syaraf Tiruan Berbasis Citra ASTER VNIR dan SWIR Untuk Klasifikasi Penutup Lahan dan Penggunaan :Lahan di Kecamatan Katingan Tengah, Kabupaten Katingan, Provinsi Kalimantan Tengah. *Tesis (tidak dipublikasikan).* Program Pascasarjan Fakultas Geografi Universitas Gadjah Mada, Yogyakarta.
[25] Capparini, F., Caporali E., and Casteli F., 2001. Land-cover in the Arno Basin Italy: Multispectral Classification and Neural Network, *Proceedings Of a Symposium Held at Santa Fe*. New Mexico, USA.

IOP Conf. Series: Earth and Environmental Science **989** (2022) 012014 doi:10.1088/1755-1315/989/1/012014

Assessment of the Coastal Vulnerability Index (CVI) for disaster mitigation strategies in some coastal tourism areas in Gunungkidul, Yogyakarta-Indonesia

Elsa Widura and Djati Mardiatno*

Faculty of Geography and Research Centre for Disasters, Universitas Gadjah Mada, Yogyakarta, Indonesia

*Email: djati.mardiatno@ugm.ac.id

Abstract. South coastal area of Gunungkidul, Yogyakarta-Indonesia has a high tourism development due to the unique physical conditions of coastal typology. However, during the last few years, extreme wave height associated with cyclones damaged many tourist facilities there. Coastal management for ecosystem sustainability requires disaster mitigation, which one considering the aspect of vulnerability. This study aims to assess the level of coastal vulnerability according to coastal typology of Baron-Pok Tunggal section and to determine the coastal management model based on the level of vulnerability. The Coastal Vulnerability Index (CVI) is used in vulnerability assessment. The variable formulation of CVI is divided into hydrodynamic factors (mean significant wave heights, mean tide range, and mean sea level rise) and morphodynamics factors, i.e. geomorphology, coastal slope, and average width of emerged beach. The morphodynamics data were obtained through observation, direct measurements in the field by systematic sampling, extraction from aerial photographs and digital surface model (DSM). The results of modeling and predictions by several agencies are derived to obtain the hydrodynamics data. The result of this study revealed that marine deposition coast typology have a high and moderate vulnerability (twelve and eleven coastal units, respectively). Meanwhile, a low vulnerability can be found in the cliff coast or wave erosion coast typology (24 units). The convex or concave beaches tend to be more vulnerable than beaches with the straight form. The beaches of Buluk, Drini, Kosakora, Sarangan and all beaches from Krakal to Indrayanti have a high vulnerability. Geomorphological variables, significant wave height, and sea level rise have the strongest influence on vulnerability. Coastal management in the marine deposition coast typology can be carried out either in a protective manner or by accommodative approach, such as by observing changes in natural conditions due to sea level rise.

1. Introduction

Sea level rise due to global warming can threaten the sustainability of ecosystems in coastal areas. The global average surface temperature increased by about 0.85oC from 1880 – 2012, while the global average sea level rose by about 0.11 m from 1901 – 2010 [1]. The impacts of global warming and sea level rise in coastal areas include increased risk of inundation, increased erosion and abrasion, changes in sediment deposition patterns, changes in the profile of sandy beaches, damage to coral reefs, and damage to residential infrastructure [2] [3].

Government Regulation No. 64 of 2010 concerning Disaster Mitigation in Coastal Areas and Small Islands explains that the management of coastal and small islands by the government must include disaster mitigation. Proactive mitigation strategies can be achieved through the identification of natural hazards and environmental or social characteristics as part of the vulnerability aspect [4]. Vulnerability assessment can be used as a starting point for disaster risk reduction and mitigation plans.

The coastal area of Gunungkidul Regency as a Tourism Strategic Area (KSP) has a unique natural attraction and has the potential to be better developed. The coastal area of Gunungkidul Regency has a beach that is identical to the morphology of karst rock and pocket beach. Some beaches have been designated as geo-sites by UNESCO because they have unique characteristics in terms of geodiversity to be protected. These beaches include the Baron-Kukup-Krakal Beach and Siung-Wediombo Beach clusters. Several beaches between Baron Beach and Krakal Beach have also been developed for tourism by the government and local coastal villages. These beaches include Rawe Beach, Buluk Beach, Porok Beach, Lolang Beach, Panjang Beach, Sanglen Beach, Watu Kodok Beach, Drini Beach, Kosakora Beach, and Ngrumput Beach.

Spatial information related to the level of vulnerability based on the vulnerability index in the coastal area of Gunungkidul does not yet exist. The Indonesian Coastal Vulnerability Index map made by the Marine and Fisheries Research Agency, Ministry of Maritime Affairs and Fisheries of the Republic of Indonesia in 2009 has not been able to describe the conditions in more detail in the Gunungkidul coastal area because the scale is too small. Spatial information on the level of vulnerability is very important in the management of sustainable tourism as part of disaster risk reduction efforts.

Based on the previous description, this study aims to assess the level of coastal vulnerability according to coastal typology of Baron-Pok Tunggal section. The level of coastal vulnerability can be used as a basis to propose the coastal management model.

2. Methods

The semi-quantitative vulnerability assessment method uses the Coastal Vulnerability Index (CVI) with variables grouped into two factors, i.e. morphodynamics and hydrodynamics. The morphological factors consist of geomorphological variables, beach slope, and beach width. The hydrodynamic factor consists of variables of significant wave height, tides, and sea level rise. Data from morphological factors were obtained through the interpretation of aerial photographs, images, observations, and direct measurements. Geomorphological data was obtained through the interpretation of 2017 aerial photos and images from the google earth portal, as well as field observations. The beach slope and beach width in the marine deposition coast typology were measured directly at intervals of 50 and 100 m according to the length of the beach. Meanwhile, the beach slope in the wave erosion coast typology is obtained from the Digital Surface Model data extraction.

The data from the hydrodynamic factors were obtained through the results of modeling and predictions by several agencies. Significant wave height data from Copernicus Marine, Environment Monitoring Service (CMEMS); tidal data from tides modeling by the Geospatial Information Agency; and water level rise data from AVISO+. The relative sea level rise needs to be reduced by the uplift rate in the study area. The lifting rate is 0.016 cm/year [5]. Interpolation of significant wave height data and sea level rise was carried out to obtain data close to the coastal typology unit. Interpolation was carried out using the kriging method. Vulnerability classes are made into three classes based on data percentiles as follows: <33% (low), 33 – 67% (moderate), and >67% (high). The tidal range is calculated by the following equation.

$$\mathrm{TR} = \mathrm{HHW} - \mathrm{LLW} \tag{1}$$

Where:

TR: Tidal range (m)
HHW: Highest water level (m)
LLW: Lowest water level (m)

Moreover, CVI index value is calculated by the equation as follows.

$$\text{CVI} = \frac{\sqrt{a\ x\ b\ x\ c\ x\ d\ x\ e\ x\ f}}{6} \tag{2}$$

Where:

a: Geomorphology; b: Beach slope (percent); c :Beach width (m); d: Significant wave height (m); e: Tidal range (mm/year); f: Sea level rise

3. Results and Discussion

3.1. Geomorphology and Coastal Typology

The research area is part of the Gunungsewu Karst Area (Figure 1). The development of the karst landscape is entering an advanced stage, where it is characterized by a complex series of hills [6]. The identified karst landforms include cone hills, foot slopes, dolines, uvalas, and karst valleys. Meanwhile, the identified marine landforms include cliff, headland, sea stack, gisik, shoal gisik, and coastal plains. Geomorphology is represented by landforms that determine the level of erodibility [7]. Marine landforms in the form of cliffs or rocky coasts have lower erodibility than coastal rocks. The wave erosion coast typology of coastal cliffs (cliffs) and wave erosion, and the marine deposition coast typology of coastal beaches were found in the study area. The secondary coastal typology is a structurally shaped coast primary typology development due to the uplift and fault processes in the past [8]. In addition, wave-cut platforms composed of friging reefs are part of the coastal typology built by organisms.

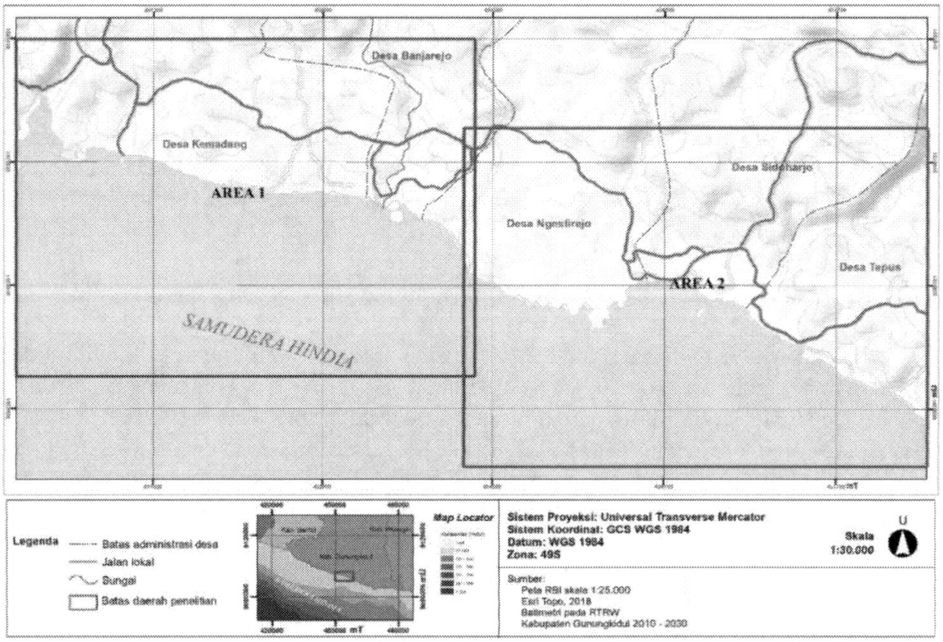

Figure 1. Research area as a part of Gunungsewu Karst Landscape

The process of destruction by waves with high intensity and tidal influences results in cliff changes. This creates erosion marks in the form of wave-cut notch, rock fall, sea stacks, and sea caves. Beaches that develop in the valleys between karst hills have material deposits in the form of boulder, gravel, and sand. Shingle beach with boulder and gravel material is found at Buluk Beach, Kosakora Beach, and Sarangan Beach (Figure 2). Meanwhile, another beach has a sandy beach with fine to coarse sand (Figure 3). The shape of the pocket beach is mostly straight, but there are some beaches that have convex and concave shapes, such as Drini Beach, Sadranan-Ngandong, and Sarangan-Krakal.

Figure 2. Shingle beach with boulder and gravel material in Kosakora Beach (*Photo by Widura*)

Figure 3. Sandy beach with fine to coarse material in Watukodok Beach (*Photo by Widura*)

3.2. Beach Slope and Width

The slope of the coast from the beach and the rocky coast tends to be contrasting. High coastal slopes are less likely to experience inundation during sea level rise [9]. The maximum slope of the coast with cliffs in each typology segment is >55°. The height of the cliffs ranges from 25 – 49 m. Most of the sandy coast has a coastal slope of 5 – 10° (undulating) and a width of 20 – 30 m. The highest beach slope is around 11.88° at Lolang Beach, while the lowest is around 1.79° at Baron Beach. The widest beach is found at Baron Beach, about 53 m and the narrowest at Buluk Beach, about 8 m. There is a tendency for wide coastal rocks to have low coastal slopes, which is related to marine depositional processes.

IOP Conf. Series: Earth and Environmental Science **989** (2022) 012014 doi:10.1088/1755-1315/989/1/012014

3.3. Significant Wave Height and Sea Level Rise

The average significant wave height in 2016 – 2019 ranged from 1.87 – 1.90 m. The significant wave heights tend to be higher than the average in June – August. This is related to the wind speed of the east monsoon (Australian monsoon). Significant wave height illustrates the potential for greater wave energy during storms so that erosion occurs [7]. Meanwhile, the relative average sea level rise from 1998 to 2018 ranged from 4.40 to 4.41 mm/year. This value is higher than the global average of sea level rise.

3.4. Tides

The average tidal range from 1998 – 2018 is 2.20 m. Meanwhile, the maximum tidal range (spring tides) is around 2,379 m and the minimum tidal range (neap tides) is around 1,719 m (Figure 4). Hurricane phenomena are more associated with low tidal ranges [10].

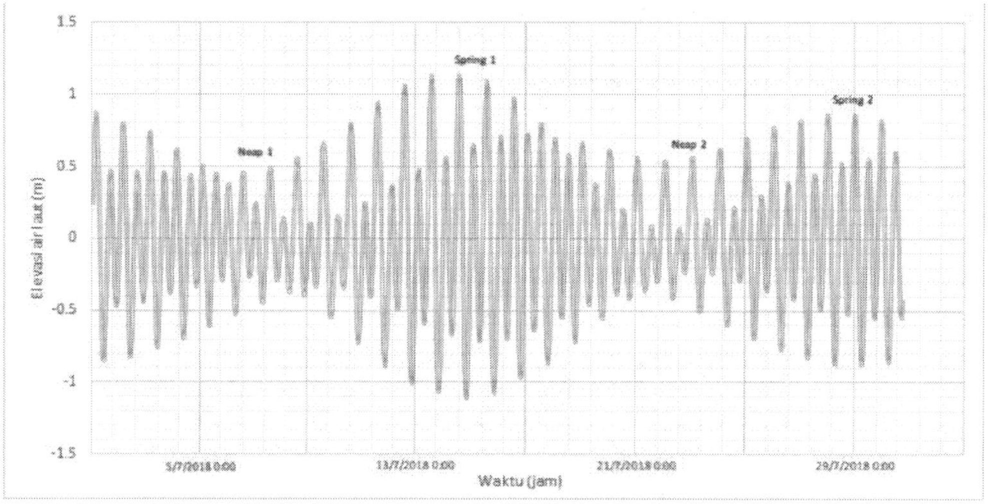

Figure 4. Tidal pattern on July, 2018 around the research area
Source: data analysis of tides model (BIG, 2020)

Based on the tidal ratio value, the research area has a mixed prevailing semidiurnal type of tidal pattern. This type of tide is characterized by high and low tides occurring twice a day, but with different amplitudes and periods. The tidal pattern in July 2018 in chart 1 shows changes in sea water elevation when the highest tide goes to the lowest tide or vice versa has a period of 6–7 hours.

3.5. Coastal Vulnerability Index

The coastal geomorphology of the research area is represented by the typology of coastal beaches and cliff coasts or eroded waves. The level of vulnerability tends to contrast with the almost balanced length of the coast, which is about ±4.6 km (Table 1). Coastal cliffs or eroded waves are included in the very low vulnerability of variable coastal slopes with a beach length of 7.67 km. Meanwhile, the shingle beach has a coastal slope with relatively high vulnerability, among others, at Rawe Beach – Kukup, Throughout, Sanglen, Watu Kodok, and Krakal Beach to Pok Tunggal Beach. The width of the beach with moderate and high levels of vulnerability is almost found on every beach, except Baron Beach and Ngrumput Beach which have a low level of vulnerability. All coasts are included in high vulnerability from significant wave height variables and sea level rise, as well as moderate vulnerability from tidal range variables. Physical process variables (significant wave height, sea level rise, and tidal range) contribute to the inundation hazard [11].

Table 1. Shoreline length distribution with vulnerability level for each variable

Variables	Shoreline Length (km)				
	Very low	Low	Medium	High	Very high
Geomorphology	4.78	2.89	0	0.42	4.63
Beach slope	7.67	0	0.1	3.95	1
Beach width	0	0.23	2.49	2.22	0.1
Sea level rise	0	0	0	0	12.72
Mean significant wave height	0	0	0	0	12.72
Mean tidal range	0	0	0	12.72	0

Source: data analysis (2020)

The value of the Coastal Vulnerability Index or CVI ranges from 9.13 to 40.84. The mean value of CVI is 22.62 and mode is 12.91. CVI value <12.91 is included in the low vulnerability level; CVI value 12.91 – 31.62 in the medium vulnerability level; CVI value >31.62 is included in the high level of vulnerability. The spatial distribution of vulnerability levels on the coast of the Baron – Pok Tunggal Beach ODTW area is shown in Figure 5 and Figure 6. There are 12 and 11 units (24.54 and 24.4%) of beaches with high and medium vulnerability. Meanwhile, a low level of vulnerability is found in 24-units (51.06%) of the cliff coast.

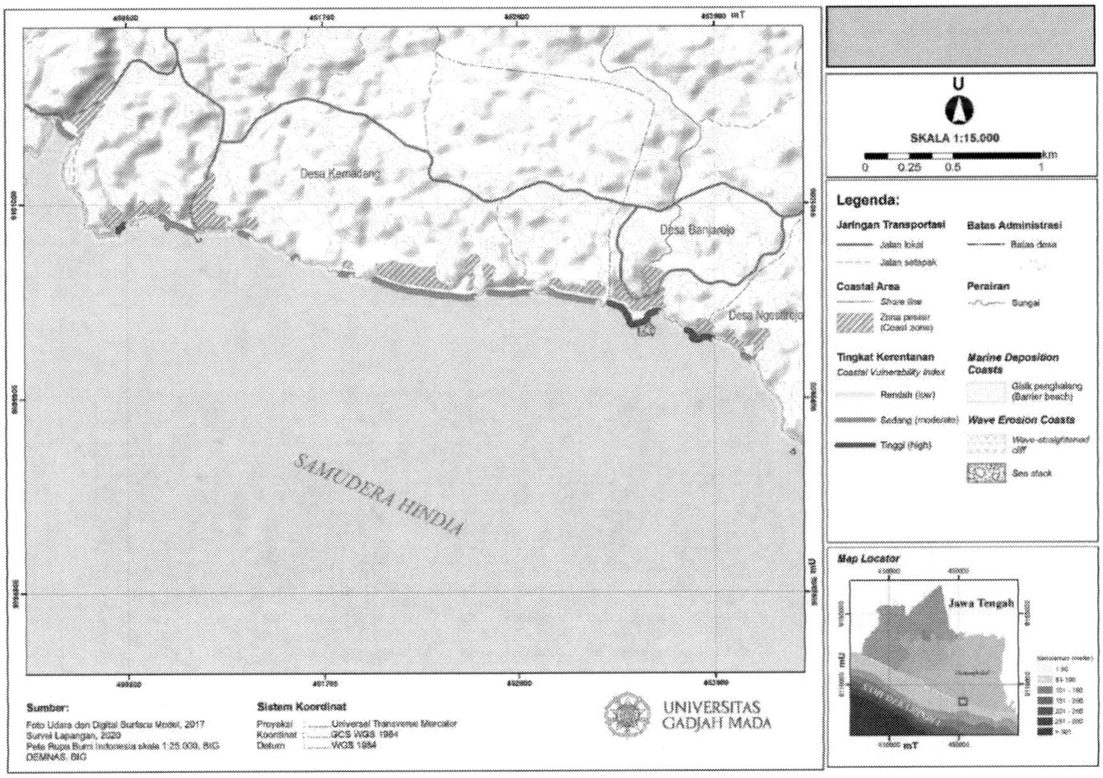

Figure 5. CVI map for the western part of the research area

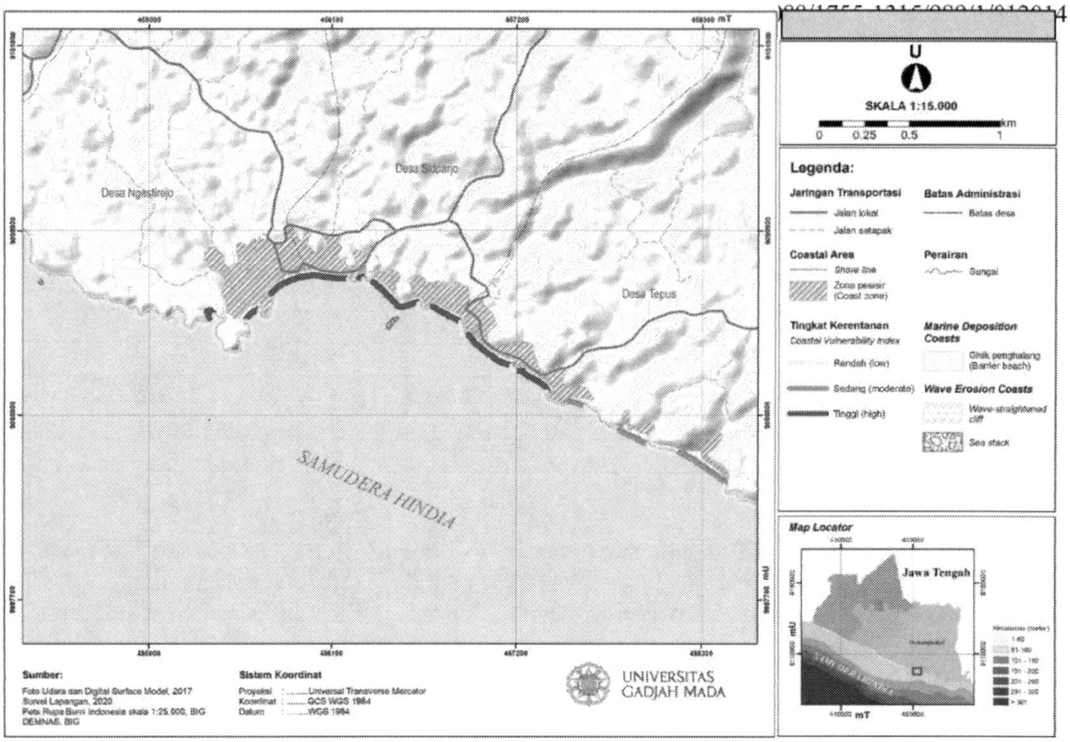

Figure 6. CVI map for the eastern part of the research area

Beaches that have a concave or convex beach tend to be more vulnerable than beaches with straight beaches. Buluk Beach, Drini, Kosakora, Sarangan, and Krakal Beach to Indrayanti have a high level of vulnerability. The level of vulnerability can be different in the same coastal typology due to differences in intensity and marine processes [12].

Most beaches with a high level of vulnerability have many tourist facilities buildings behind the beach. This condition is different from beaches with a moderate level of vulnerability, where there are relatively few buildings and quite a lot of protective vegetation such as casuarina, sea pandanus, and other shrubs. However, Sepanjang Beach has buildings along the back of the beach which is quite dense (see Figure 7). Such conditions can endanger the beach.

Figure 7. Dense building in Sepanjang Beach (*Photo by Widura*)

Some beaches that have quite dense vegetation include Porok Beach, Lolang Beach, Watu Kodok Beach and Sanglen Beach. The presence of vegetation behind the beach has a very important role in reducing wave energy and reducing the abrasion/erosion process, especially during extreme events [7]. The management of coastal areas through better spatial planning of coastal areas should be carried out. Refer to [13], coastal management in the marine deposition coast typology can be carried out protectively, i.e. making wave-retaining or accommodative buildings, namely management by looking at changes in natural conditions due to sea level rise.

4. Conclusion

High and medium vulnerability levels are in 12 and 11-unit (24.54 and 24.4%) coastal typologies with beach (marine deposition coast), while low vulnerability levels are in 24-unit (51.06%) coastal typologies cliffs and wave erosion. Coastal areas with high vulnerability include Buluk Beach, Drini, Kosakora, Sarangan, and Krakal Beach to Indrayanti Beach. Beaches with concave or convex shaped beaches tend to be more vulnerable than beaches with straight beaches. Geomorphological variables, significant wave height, and sea level rise have the strongest influence on vulnerability. This is indicated by the very high level of vulnerability that is spread more widely. Coastal management in the marine deposition coast typology can be carried out either in a protective manner or by accommodative approach, such as by observing changes in natural conditions due to sea level rise.

Acknowledgement

The authors thank to Regional Disaster Management Agency (BPBD) and Department of Land and Spatial Planning (DISPERTARU) of Gunungkidul Regency for data support.

References

[1] IPCC 2014 *IPCC Fifth Assessment Report: Climate Change 2014 (AR5)* (Intergovernmental Panel on Climate Change: Working Group II report. Impacts, Adaptation and Vulnerability Judge).

[2] Numberi, F. 2009 *Perubahan Iklim: Implikasinya terhadap Kehidupan di Laut, Pesisir dan Pulau-Pulau Kecil* (Jakarta: Citrakreasi Indonesia).

[3] UNCCS. 2019 Climate action and support trends. In *United Nations Climate Change Secretariat*. Retrieved on 8 January 2020 from https://unfccc.int/sites/default/files/resource/Climate_Action_Support_Trends_2019.pdf

[4] Ciurean, R. L. and Schröter, D. 2013 Conceptual Frameworks of Vulnerability Assessments for Natural Disasters Reduction. In J. Tiefenbacher (1st ed.), *Approach to Disaster Management Examining the Implications of Hazards, Emergencies and Disasters* (Rijeka: InTech) pp 3–32.

[5] Muchitawati, G. S., Husna, H. Z., & Marliyani, G. I. 2017 Studi Karakteristik Geologi Pantai Selatan Gunung Kidul untuk Pengembangan Pembangkit Listrik Tenaga Gelombang (*Paper in Seminar Nasional Kebumian Ke-10*) pp 1836–1848.

[6] Tjia, H. D. 2013 Morphostructural Development of Gunungsewu Karst. *Indonesian Journal of Geology*, *8*(2), 75–88.

[7] Pantusa, D., D'Alessandro, F., Riefolo, L., Principato, F., & Tomasicchio, G. R. 2018 Application of a coastal vulnerability index. A case study along the Apulian Coastline, Italy. *Water (Switzerland)*. 10, 1218.

[8] Marfai, M. A., Cahyadi, A., & Anggraini, D. F. 2013 Tipologi, Dinamika, dan Potensi Bencana di Pesisir Kawasan Karst Kabupaten Gunungkidul. *Forum Geografi*, *27*(2), 147–158.

[9] Abuodha, P. A. O., & Woodroffe, C. D. 2010 Assessing vulnerability to sea-level rise using a coastal sensitivity index: a case study from southeast Australia. *Journal of Coastal Research*, *4*, 189–205.

[10] L´opez Royo, M.; Ranasinghe, R., and Jim´enez, J.A. 2016 A rapid, low-cost approach to coastal vulnerability assessment at a national level. *Journal of Coastal Research, 32*(4), 932–945.

[11] Pendleton, E. A., Thieler, E. R., & Williams, S. J. 2010 Importance of Coastal Change Variables in Determining Vulnerability to Sea- and Lake-Level Change. *Journal of Coastal Research, 26*(1), 176–183.

[12] Mutaqin, B. W., Cahyadi, A., & Dipayana, G. A. 2012 Indeks Kerentanan Kepesisiran Terhadap Kenaikan Muka Air Laut Pada Beberapa Tipologi Kepesisiran di Propinsi Daerah Istimewa Yogyakarta (*Paper in Seminar Nasional Penginderaan Jauh Dan Sistem Informasi Geografis 2012*).

[13] Setiawan, K. P. 2016 *Kajian Pengelolaan Pesisir Berbasis Tipologi, Morfodinamika dan Kerentanan Bencana (Studi Kasus: Pesisir Kabupaten Tanah Laut, Kalimantan Selatan)*. (Yogyakarta: Universitas Gadjah Mada).

IOP Conf. Series: Earth and Environmental Science **989** (2022) 012015 doi:10.1088/1755-1315/989/1/012015

ICT application in CBDRM in Karangtengah village Bantul Regency

S T Mahendra*, A H Pratomo, and E T Paripurno

University of Pembangunan Nasional Veteran Yogyakarta, Jalan Lingkar Utara, Condongcatur, Sleman 55283

*Email: septyanteguhmahendra1990@gmail.com

Abstract. This research was conducted in Karangtengah Village, Imogiri District, Bantul Regency, Special Region of Yogyakarta. The landslide incident that occurred in the Karangtengah Village area on November 28, 2017 then prompted several groups to take the initiative to install an early warning system and several other supporting tools at the scene. This study aims to identify what Information and Communication Technology tools already exist in the Karangtengah Village area that can support landslide hazard monitoring activities, and see how residents manage these tools. This research was conducted with a qualitative method using a Participatory Rural Appraisal tool. The results of this study show that the presence of Information and Communication Technology tools requires the community to have the capacity to synergize with existing tools and manage them independently so that there is no failure of the role of Information and Communication Technology in supporting the work of Community-Based Disaster Risk Management in Karangtengah Village. Access to data generated by existing Information and Communication Technology tools will also be useful for residents to monitor phenomena or conditions around them.

1. Introduction

One of the natural disasters that often occur in the Special Region of Yogyakarta is a landslide. One of the landslide events occurred in November 28, 2017, high rainfall caused landslides in Karangtengah Village, Bantul Regency, Special Region of Yogyakarta, precisely in the the Local Transmigration area of Mojolegi (RT 06) and Karangrejek (RT 07) Hamlets. This incident caused severe and moderate damage to 2 resident's houses. There were no fatalities in this incident, but at that time 49 families had to be evacuated as a result of the incident.

In the following landslide hazard map of Bantul Regency, it can be seen that most of the Karangtengah Village area is in the medium category, but there are also a small number of areas that are in the high category.

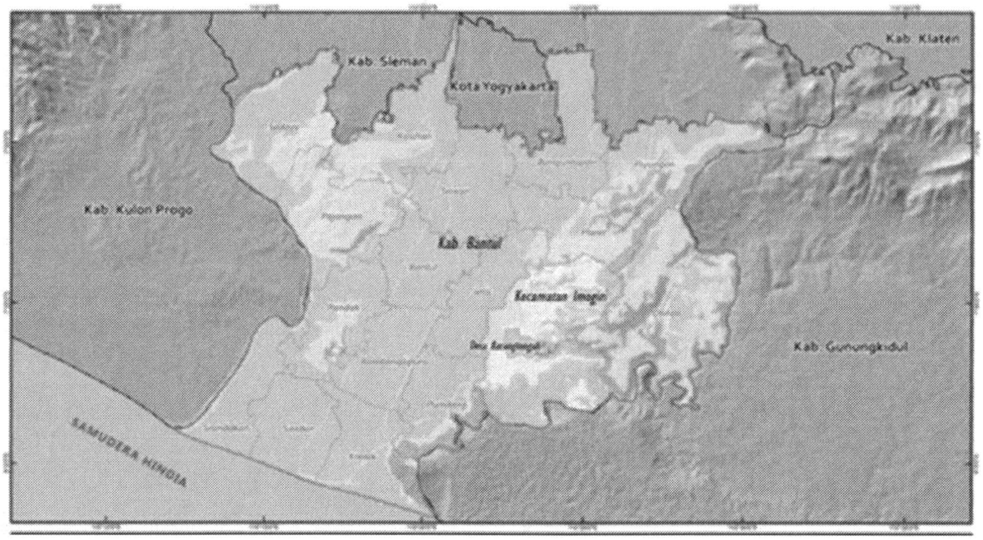

Figure 1. Bantul Regency landslide hazard map

The occurrence of natural disasters cannot be prevented, but humans can minimize losses due to disasters by managing the risks. Disaster risk management is defined as a conceptual framework which basically focuses not on managing disasters and their consequences, but on reducing threats and their potential losses.

Currently Community-Based Disaster Risk Management (CBDRM) has become an important pillar in disaster risk management affairs, using the CBDRM approach disaster risk management affairs will at least be more efficient, because by maximizing local resources and minimizing external resources, then in ideally will be able to reduce transaction costs. So far, disaster management works have shown less significant results in the field of disaster risk reduction because they still focus more on emergencies, the approach used is top-down and tends to be structural [1].

The community of Karangtengah Village itself has actually formed a Disaster Risk Reduction Forum (FPRB) as one of the efforts to implement CBDRM. In the Karangtengah Village area, several types of hazard detection tools have even been installed as a form of implementing ICT for disaster management, such as ground movement detection devices and rainfall gauges. However, it is suspected that the management of these tools is still lacking in involving the community. Therefore, a documented study is needed to identify the current application of ICT in supporting CBDRM practices in Karangtengah Village, as well as how the capacity of residents to synergize with existing ICTs.

ICT has great potential in disaster management [2]. Studies around the world show that the application of ICT by the disaster risk management agency of a region is an important aspect in relation to their disaster risk management efforts. Information is the key in disaster risk management, so ICT plays an important role in collecting and processing information dissemination on disaster risk management efforts globally, for that there must be a documented study on the application of ICT in disaster risk management efforts that have been ongoing so far [3.

In an exploratory study entitled "Redefining Community Based Disaster Risk Management (CBDRM) through Enhanced Early Warning Processes" it is stated that early warning is a trigger for actions taken by the community in responding to disasters. Therefore, it is necessary to emphasize that early warning is an important element in CBDRM. The study used a CBDRM prototype design which was systematized to be integrated with early warning tools, which was later called a community-based early warning system, the system enabled the community to take an active role in its implementation [4].

Based on the background described before, some questions can be formulated as follows:

- How is the community-based landslides risk management in Karangtengah Village?
- How is the application of ICT in community-based landslide risk management in Karangtengah Village?
- How is the community management of ICT in Karangtengah Village?

2. Methods

The approach used in this study is a qualitative method using PRA tools, the use of PRA tools is in accordance with qualitative research which basically has the aim of understanding the community with their own frame of reference [5].

PRA is a set of approaches and methods that function in learning about rural life and conditions by, together, and from rural communities, it can also be used to enable rural communities to share, improve, and analyze life knowledge and the conditions they have, to plan and act [6. This approach aims to change social reality based on deep knowledge into everyday practices obtained through participatory research by reflecting on participatory methods [7.

3. Results and Discussion

3.1. Community-based landslides risk management in Karangtengah Village

3.1.1. Landslide Disaster Risk Management in Pre-Disaster Phase. Sugito said that about a month before the landslide incident, the village held a Destana inauguration activity, in this series of activities the Destana team and residents also carried out emergency response simulations. And also coincidentally the location used for the simulation is right at the location of the landslide. However, basically the selection of the location as a simulation location was actually based on a survey conducted by the BPBD, where according to them, the location was the most prone to landslides.

Sugiman told that the Destana activities organized by BPBD Bantul Regency were overall carried out for 13 days and were attended by 30 representative participants from the Civil Service, FPRB, PKK, Karang Taruna, Linmas, and community leaders from Karangtengah Village. In this series of activities, emergency response trainings were provided to deal with earthquakes, landslides, floods and hurricanes, both in the form of training in theories and practice of evacuation simulations as well as the establishment of refugee camps and public kitchens.

3.1.2. Landslide Disaster Risk Management in the Emergency Response Phase. Sugito, who at that time served as Head of Karangtengah Village, arrived at the location when the landslide material had buried several resident's houses and roads. The situation at that time, the landslide had stopped but the rain was still falling heavily, seeing this condition Sugito immediately instructed to evacuate residents who were near the landslide point.

Evacuation was carried out by the Destana team and directly led by the Village Head, the village government provided 2 main points as evacuation sites, namely the As-Shobari Mosque in Karangrejek Hamlet which was approximately 400 meters from the location of the landslide incident, and the At-Taqorub Mosque in Mojolegi Hamlet which was located approximately 800 meters from the location of the landslide, but there were also residents who chose to flee to the homes of their relatives, who were still in the vicinity of Karangrejek and Mojolegi hamlets.

After the evacuation of the residents was under control, at around 3 pm, the Village Head and the Destana team immediately set up a public kitchen in the resident's houses which were on the south side of the As-Shobari Mosque, Karangrejek Hamlet. After all residents have been evacuated, then data collection is carried out on all displaced residents, both those who fled in evacuation sites provided by the village government or who took refuge in people's homes.

IOP Conf. Series: Earth and Environmental Science **989** (2022) 012015 doi:10.1088/1755-1315/989/1/012015

3.1.3. Landslide Disaster Risk Management in the Post-Disaster Phase. The day after the incident, a volunteer offered to conduct a study on the landslide incident which would be carried out by a team from UPN Veteran Yogyakarta. The village government accepted the offer. The next day the UPN Veterans Yogyakarta team came to the Command Post at the Village Hall and then immediately inspected the location of the landslide incident accompanied by the FPRB of Karangtengah Village. A few days later the team from UPN Veteran Yogyakarta returned to the location to conduct a further study, and produced a recommendation that approximately 6 houses were in the landslide red zone and had to be relocated because the soil conditions on the hillside on the east side of the residential area were prone to landslides, the houses belongs to (1) Hariadi, (2) Suhardopo, (3) Widodo, (4) Wahono, (5) Agus, and (6) Muji Mulyono. In addition, from the results of the study, it is also recommended for several houses in the yellow zone of landslides whose position is below the west side of the red zone to leave the location temporarily in case of heavy rains.

Figure 2. Landslide red zone map

Based on the recommendation from the UPN Veterans Yogyakarta team to relocate the residents' houses in the red zone, the village government then submitted an application to the BPBD for assistance in relocating the residents' houses. Then, in 2019 funds from the APBD were disbursed for the realization of the relocation project.

3.2. Application of ICT in community-based landslide risk management in Karangtengah Village

There are 4 ICT tools that have a function for disaster risk reduction in Karangtengah Village. These tools are (1) two landslide early warning systems, (2) soil moisture monitoring tools, and (3) automatic rainfall gauges. The flow of information processed by these tools can be seen from the working concept of these tools as follows.

3.2.1. Long Pulley Landslide Early Warning System. Mechanical linear device → analog data detector → data acquisition device → SMS gateway device → receiver & alarm.

3.2.2. Soil Moisture Monitoring Tool. Soil moisture sensor, underground temperature sensor, ground vibration sensor → Atmega 2560 16 MHz chip → Nordic chip → D-Link network module → server.

3.2.3. Automatic Rain Recorder. Tipping bucket → AVR chip → Nordic chip → D-Link network module → server.

3.2.4. Short Pulley Landslide Early Warning System. Rotary sensor → Atmega 16 MHz chip → oscillator → alarm.

IOP Conf. Series: Earth and Environmental Science **989** (2022) 012015 doi:10.1088/1755-1315/989/1/012015

3.3. Community management of ICT in Karangtengah Village

Based on field facts, it was found that the short pulley landslide early warning device was not working properly. Sugito said that the wire attached to the device is too sensitive to movement, even if it is not caused by ground movement, so that it often causes false warnings.

Widodo also said that the device often gave the wrong warning. Because incidents like this have happened several times, which caused panic among the residents, Widodo turned off the device

There were several wishes from residents regarding the management of soil moisture monitoring tools and automatic rainfall gauges that had been conveyed, Sugito wanted the village government or FPRB to have access to information obtained from these two tools so that they could also monitor soil moisture conditions and rainfall. So that when the level of landslides vulnerability increases and begins to endanger local residents, they can convey this information and prepare to take the necessary follow-up actions.

Another wish from residents for the management of ICT tools installed in the Karangtengah Village area was conveyed by Yasin, he wanted the information generated from soil moisture monitoring tools, automatic rainfall gauges, and long pulley landslide early warning tools to be compiled and presented. in a single information system to facilitate monitoring. Yasin also expressed his desire for socialization to be carried out for local transmigration residents who are around landslide-prone areas along with representatives from the village government and FPRB regarding the long pull landslide early warning tool.

Following up on the request, the researcher and the team from the Disaster Management Study Center of UPN Veteran Yogyakarta conducted a socialization activity on how the tools in Karangtengah Village work, then facilitated the community to make deals so that they were able to manage these tools independently so that they could improve its effectiveness in supporting community-based landslide risk management activities in Karangtengah Village.

The topics discussed in the community facilitation activities are inclusive early warning systems and evacuation plans. From the facilitation activities carried out, agreements were made on several things as follows.

3.3.1. Monitoring and Warning Service. Previously, residents did not have an agreement regarding monitoring and warning services, residents only relied on a few FPRB members who sometimes volunteered to monitor conditions in the field. After facilitation, residents make an agreement regarding:

- Source of danger warning
- Hazard warning form
- How to monitor danger
- How to make sure the warning is correct

3.3.2. Dissemination and Communication. In matters of dissemination and communication, residents have never made an agreement before, no one has yet been responsible for this matter. Residents only take instinctive actions spontaneously when an emergency occurs. After facilitation, residents make an agreement regarding:

- Warning sender
- Warning target
- How to send a warning
- Warning form

3.3.3. Response Capacity. Basically, in terms of the ability to respond, the village government has sufficient capacity. However, sometimes there is still misinformation so that the instructions given from the village government cannot be received correctly by the residents. From the facilitation carried out, residents made an agreement about:

- Village government actions
- Community action

3.3.4. Evacuation Plan. In the discussion of the evacuation plan, the following aspects were identified:

- Residents in vulnerable areas
- Evacuation place
- Equipment for evacuation owned
- Evacuation location capacity

3.3.5. Evacuation Map. After the evacuation plan is agreed upon, then an evacuation map is drawn. Evacuation routes are determined based on the results of the disaster risk assessment. By describing the evacuation map in a visual form, the community will be easier to understand.

4. Conclusion

4.1. ICT in Karangtengah Village:

- There are 4 units of ICT tools in Karangtengah Village, namely 2 units of tools as an early warning system for landslides, 1 unit of soil moisture monitoring tool, and 1 unit of automatic rainfall gauge.
- The four units of the tool produce data regarding natural phenomena or phenomena that occur.
- Both units of the landslide early warning system have a mechanism for disseminating information to residents around the location for the data generated. The data produced by soil moisture monitoring tools and automatic rainfall gauges have not yet been disseminated to residents.

4.2. Community management of ICT in Karangtengah Village:

- involving residents in the independent management of existing ICT tools in the form of providing access to data owned by the owner of the tool to the village party either through village officials or the village FPRB will be useful for residents to monitor natural phenomena or phenomena that occur in order to increase their awareness.
- To avoid the failure of the role of ICT that has been implemented in Karangtengah Village in realizing the success of CBDRM, the residents need to have an agreement by doing (1) Establishing hazard monitoring and warnings, (2) Determining the dissemination of hazard warnings, (3) Determining responses/actions to warnings.

References

[1] Paripurno E, Lassa J, Pujiono P, Magatani A and Parlan H 2014 *Panduan Pengelolaan Risiko Bencana Berbasis Komunitas (PRBBK)* Jakarta
[2] Kuppuswamy S and Rajarathnam S 2009 *Women, information technology and disaster management: tsunami districs of Tamil Nadu. International Journal of Innovation and Sustainable Development* **4(2/3)** 206-216
[3] Wilson J and Wilson F 2018 *The use of ICT in Disaster Risk Management : A Case Study of Nema Borno State* MAT Journals **5(1)** 44–66
[4] Pineda M V G 2015 *Redefining Community Based Disaster Risk Management (CBDRM) through Enhanced Early Warning Processes* International Journal of Information and Education Technology **5(7)** 543–548
[5] Taylor S J, Bogdan R and DeVault M L 2016 *Introduction to Qualitative Research Methods* Wiley
[6] Chambers R 1994 *The Origins and Practice of Rural Appraisal* World
[7] Bergold J and Thomas S 2012 *Participatory research methods: A methodological approach in motion* Historical Social Research **37(4)** 191–222

IOP Conf. Series: Earth and Environmental Science **989** (2022) 012016 doi:10.1088/1755-1315/989/1/012016

Flood control simulation for disaster management in Wai Batu Merah River Basin

D Fitriani[1,2], M Farid[3,4], H Mahfudz[3] and Y Suryadi[3]

[1] Water Resources and Management Study Program, Faculty of Civil and Environmental Engineering, Institut Teknologi Bandung, Jl. Let. Jend. Purn. Dr. (HC) Mashudi No.1 Jatinangor

[2] Center for Water Resources Development, Institute for Research and Community Services, Institut Teknologi Bandung, Jl. Ganesha No. 10 Bandung

[3] Water Resources Engineering Research Group, Faculty of Civil and Environmental Engineering, Institut Teknologi Bandung, Jl. Ganesha No. 10 Bandung

[4] Center for Coastal and Marine Development, Institut Teknologi Bandung, Jl. Ganesha No. 10 Bandung

*Email: djeliafitriani@gmail.com

Abstract. Ambon City, located in Maluku Island, experiences flood disaster every year. It has been considered that the flood in this city occurs more often compared to the other city in the same island. One of the river that is related to the flood occurrence in Ambon City is Wai Batu Merah River. This study aims to find appropriate structural measurement by using modelling simulation to control flood disaster in Wai Batu Merah River Basin. The modelling used HEC-RAS 5.0.7 software for coupled 1D and 2D hydraulic simulation. Flood discharge was calibrated by applying flood event in 2013. The SCS method was selected to generate flood hydrograph by using synthetic method. The records of water depth in the flood event were compared to the hydraulic model results and it showed good agreement. Based on the model, flood control scenarios i.e. normalization and dike were exercised for 25 year return period as required by government regulation. The combination of both scenarios can manage flood in Wai Batu Merah River Basin effectively. However it might need high cost to construct the structures. Therefore, it is necessary to have priority between the scenarios.

1. Introduction

Wai Batu Merah Basin is located in Ambon City, Province Maluku, Indonesia. Wai Batu Merah Basin has been flooding every single year. Extreme floods happened in 2012 and 2013 [1]. The Flood was closest happen on 13 July 2021 with losses of 324 people affected and 56 units houses and 1 unit bridge damaged [2]. The Flood has been causing the overflow discharge. Many contributing factors to floods are land-use change, soil erosion, and rainfall intensity [3].

The precipitation intensity in Ambon city had reached five years ago [4] besides that, a land-use change had increased from 966,9 ha to 1036,1 ha 10 years ago [5] and erosion from July until October 2016 specifically 2326,556 tons/ha/year [6]. The erosion made the capacity of Wai Batu Merah River decreased [7].

Planning flood control in the Wai Batu Merah basin was approximate by modeling. Several models are support modeling floods. one of the models is Hydrologic Engineering Center (HEC) has developed by the US Army Corps of Engineers (USACE) [8]. HEC-RAS can perform with coupling analysis or each other. The river was modeling by one dimension then the pond was modeling by two dimensions [9]. Flood modeling has been running with the unsteady flow. HEC-RAS can give the information from change water level and discharge by a function of time and place used Saint Venant equation [10]. The modeling result shows a water level that did the calibration dan verification with the yield data. calibration was doing to get coefficient manning's [11].

The purpose of this study is to flood control by coupling modeling in the Wai Batu Merah basin. Planning flood control Wai Batu Merah basin has two scenarios i.e., normalization and dike. Normalization river was planning by discharge 25-year return period [12]. Normalization made the river form back to nature that the water not flooded again [13] and the dikes are used to resist overflowing floods [14].

2. Methods

2.1. Study Area

Wai batu merah basin is located in ambon city, province maluku, Indonesia. Ambon city located in 3°34'4,80'' until 3°47'38,4'' Lintang Selatan (LS) and 128°1'33,6'' until 128°18'7,20'' Bujur Timur (BT) with an administrative area of 377 km2. Ambon city has an area classified as hilly and sloping. 73 % of the ambon city is hilly and sloping with a level above 20% and 20% with a sloping level under 20%. Type of soil on ambon city are regosol, aluvial, kambisol, litosol, rendzina and gleisol.

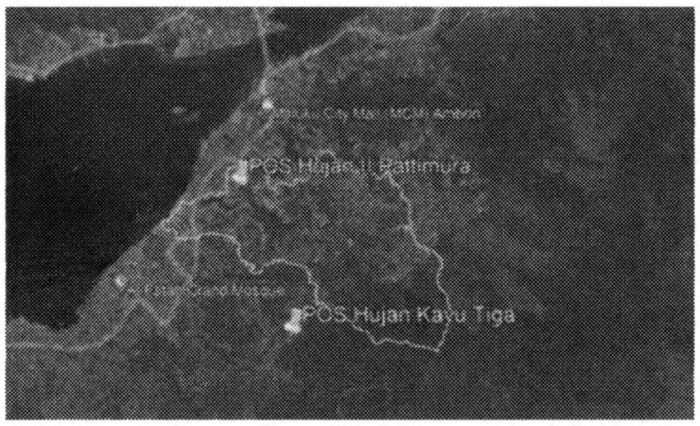

Figure 1. Location Wai Batu Merah Basin

2.2. Precipitation Data

2.2.1. The gauge precipitation stations The gauge precipitation based on stations is from 2012 to 2017. That is the gauge presipitation in 2012

Table 1. The Gauge Precipitation Data

Pos Hujan II Pattimura												
Tahun 2012												
Tgl	**Jan**	**Feb**	**Mar**	**Apr**	**May**	**Jun**	**Jul**	**Aug**	**Sep**	**Oct**	**Nov**	**Dec**
1	4	1	0	0	55	26	13	425	0	0	6	0
2	1	4	6	3	6	13	0	1	0	0	0	0
3	31	1	4	0	1	41	20	1	0	2	0	0
4	4	0	25	0	16	50	3	4	0	0	0	0
5	4	0	5	0	19	1	51	3	6	0	0	0
6	0	0	24	0	2	1	39	6	37	0	0	4
7	1	15	0	0	0	204	44	8	10	0	0	9
8	0	1	0	0	0	208	17	7	0	0	0	0
9	0	0	48	0	0	33	34	70	1	73	0	1
10	7	0	69	1	1	15	3	3	0	0	0	0
11	9	5	16	10	0	3	0	1	7	0	0	7
12	0	2	2	0	0	0	0	1	1	0	3	0
13	0	0	2	0	0	0	28	15	0	2	1	0
14	0	0	0	0	0	0	5	6	3	37	0	1
15	1	15	0	7	39	0	260	4	7	0	0	1
16	8	3	0	6	1	2	87	12	13	3	0	0
17	0	2	19	30	14	26	18	4	0	0	0	0
18	0	4	4	0	0	189	23	0	0	0	0	0
19	14	3	6	0	1	47	28	0	0	0	0	1
20	2	11	2	0	55	11	100	1	0	1	0	1
21	0	27	0	0	8	1	5	0	65	0	0	31
22	7	0	0	5	59	0	41	2	0	0	0	0
23	0	0	0	15	22	10	14	2	0	1	0	2
24	4	0	0	0	7	14	16	3	0	0	0	10
25	0	3	17	2	0	45	1	76	0	0	0	5
26	0	1	0	30	218	11	0	9	2	24	0	0
27	0	0	9	4	58	26	0	45	1	0	0	0
28	2	1	0	0	10	111	0	9	1	6	0	0
29	0	0	10	11	93	76	44	0	0	0	0	0
30	0		25	4	60	0	40	22	0	1	0	2
31	0		0		22		155	0		0		0

Pos Hujan Kayu Tiga												
Tahun 2012												
Tgl	**Jan**	**Feb**	**Mar**	**Apr**	**May**	**Jun**	**Jul**	**Aug**	**Sep**	**Oct**	**Nov**	**Dec**
1	0.0	0.5	0.0	0.4	0.0	1.0	10.1	20.4	0.0	10.3	6.1	0.0
2	0.0	0.0	0.0	0.0	20.1	0.0	9.7	0.0	0.0	20.1	0.0	0.0
3	0.0	0.0	0.0	0.0	0.4	40.0	6.5	0.0	0.0	0.0	0.0	0.0
4	0.0	0.0	0.0	0.0	20.1	0.0	10.2	0.0	0.0	0.0	0.0	0.0
5	0.0	0.0	10.1	0.0	0.0	40.1	30.3	0.4	0.0	0.0	0.0	9.1
6	0.0	0.0	0.6	0.0	0.0	20.7	10.1	0.7	10.6	0.0	0.0	0.0
7	0.0	0.0	0.0	0.0	0.0	50.1	20.6	0.0	0.0	0.0	0.0	0.0
8	0.0	0.8	0.0	0.0	0.0	60.5	0.7	10.6	0.0	0.0	0.0	0.0
9	10.1	0.0	10.1	0.0	0.0	0.9	10.7	0.0	0.0	0.0	0.0	0.0
10	0.0	0.0	10.0	0.0	0.0	10.0	0.0	0.0	8.1	0.0	0.0	10.7
11	0.0	0.0	0.5	0.0	0.0	0.0	0.0	0.0	5.0	0.0	0.0	5.1
12	0.0	0.4	0.0	0.0	0.0	0.1	10.5	0.9	0.0	0.0	0.0	0.0
13	0.0	0.5	0.0	0.0	0.0	0.0	0.0	1.0	0.0	0.0	0.0	0.0
14	0.0	0.0	0.0	0.0	0.0	0.1	10.0	0.6	0.0	0.0	0.0	0.0
15	0.0	0.0	0.0	0.8	10.6	0.0	80.1	0.0	0.0	10.1	0.0	0.0
16	0.5	0.0	0.0	0.0	0.0	10.1	0.9	10.4	5.0	0.0	0.0	0.0
17	0.0	0.7	0.9	0.9	0.0	50.0	0.0	0.0	11.0	0.0	0.0	0.0
18	0.0	0.0	0.0	0.4	0.7	40.1	0.7	0.0	0.0	0.0	0.0	0.0
19	0.0	0.9	0.0	0.0	0.0	20.1	20.0	0.0	0.0	0.0	0.0	0.0
20	0.0	0.6	0.0	0.0	20.5	0.5	0.0	0.0	0.0	6.1	0.0	0.0
21	0.0	0.0	0.0	0.0	20.7	0.0	20.5	0.0	20.7	0.0	0.0	0.0
22	0.0	0.0	0.0	0.0	10.3	0.0	9.1	0.0	0.0	0.0	0.0	0.0
23	0.0	0.0	0.5	0.0	0.8	0.0	7.1	0.8	0.0	0.0	0.0	0.0
24	0.8	0.0	0.0	0.4	0.0	0.8	9.1	0.0	0.0	0.0	0.0	0.0
25	0.0	0.0	0.0	0.9	0.0	20.4	0.0	30.6	0.0	0.0	0.0	0.0
26	0.0	0.8	0.0	8.3	50.0	0.0	0.0	0.0	0.0	0.0	0.0	0.0
27	0.0	0.0	0.7	0.5	0.0	10.1	0.0	30.0	0.0	8.1	8.1	0.0
28	0.0	0.0	0.0	0.0	41.0	20.8	0.0	0.0	0.0	0.0	0.0	0.0
29	0.0	0.0	0.0	0.0	0.0	0.5	20.0	0.0	0.0	0.0	0.0	0.0
30	0.4		107.0	0.5	40.8	0.5	20.1	0.0	0.0	0.0	0.0	0.0
31	0.0		0.0		20.5		114.0	0.0		0.0		0.0

2.2.2. TRMM precipitation TRMM precipitation based on the grid is from 1999 to 2019. That is the TRMM precipitation in 1999

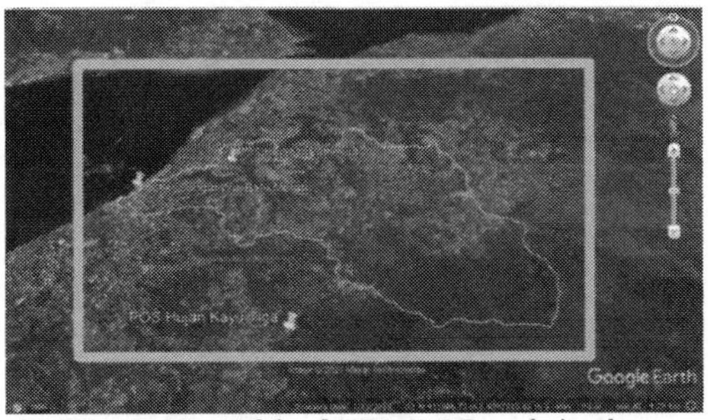

Figure 2. Grid of TRMM Precipitation

Table 2 TRMM Precipitation Data

TRMM Precipitation In 1999												
Date	Jan	Feb	Mar	Apr	May	Jun	Jul	Aug	Sep	Oct	Nov	Dec
1	1	11	0	0	0	5	1	1	0	6	1	11
2	0	0	3	0	8	13	143	107	4	11	0	3
3	5	49	23	0	3	0	78	17	8	3	0	0
4	0	5	4	17	94	1	28	0	9	11	19	5
5	40	9	6	0	0	22	70	4	14	16	0	0
6	6	3	0	1	25	0	29	41	0	0	0	0
7	24	31	1	1	3	2	42	0	2	2	0	0
8	20	14	16	0	10	0	10	46	4	35	0	11
9	9	44	2	0	5	1	22	7	5	0	34	9
10	0	9	6	0	19	0	14	0	5	24	4	65
11	1	2	1	1	0	6	28	2	16	15	20	8
12	1	0	4	2	23	7	1	41	13	0	28	13
13	4	0	26	4	17	0	1	2	14	4	4	0
14	0	3	0	1	10	33	1	0	0	0	0	0
15	0	17	2	18	1	18	0	0	0	0	0	0
16	2	0	4	3	47	12	12	2	1	0	9	6
17	2	2	16	7	85	0	17	0	37	6	0	10
18	0	0	0	2	15	0	15	1	0	0	0	70
19	0	15	0	11	34	2	4	6	0	2	0	4
20	0	0	5	61	1	0	5	0	19	2	0	0
21	6	0	2	0	47	24	2	4	0	5	0	0
22	26	0	19	0	7	3	2	0	0	2	0	3
23	2	6	30	71	0	9	0	0	0	11	0	0
24	16	0	1	1	0	86	0	0	30	9	0	16
25	0	8	9	0	19	54	12	1	10	0	5	3
26	19	5	8	0	21	72	0	1	7	16	5	0
27	0	10	14	1	1	54	0	0	4	2	1	1
28	1	0	29	51	0	30	0	0	2	0	0	17
29	1		14	21	0	31	2	0	19	5	0	1
30	2		0	0	0	2	0	0	5	0	0	0
31	2		12		0		2	0		13		0

2.3. Corrected TRMM precipitation with gauge precipitation station

The gauge precipitation based on stations is from 2012 to 2017. that has not capable of computed discharge. The precipitation distribution data be required to use TRMM precipitation. How to choose the gauge precipitation stations for corrected TRMM precipitation are:

1. The gauge precipitation stations must have daily precipitation data minimum of three years.
2. The TRMM precipitation data period has been choosing from 1998 until now.
3. choose TRMM grid was closest with the gauge precipitation station.
4. plot flow direction curve of gauge precipitation and TRMM precipitation.
5. compute the coefficient correlation of monthly gauge precipitation stations with TRMM precipitation.
6. if the coefficient correlation was under 0,6 then the gauge precipitation station was not capable compute flood discharge.

2.4. Unit hydrograph

2.4.1. Soil Conservation Service (SCS) Unit Hydrograph Soil Conservation Service unit hydrograph has developed by Victor Mockus required by a result of observation characteristic unit hydrograph on the basin in America Serikat. Soil Conservation Service unit hydrograph equation is expressed by

$$Q_p = \frac{0{,}2083\ x\ A}{T_p}$$

$$T_p = \frac{t_r}{2} + t_p$$

$$t_p = Ct(L\ x\ L_c)^n$$

Where:

Q_p : *peak discharge* (m^3/s)
A : *basin area* (km^2)
T_p : *time to peak (hour)*
t_r : *rainfall duration (hour)*
t_p : *time lag (hour)*
Ct, n : *time coefficient*
L : *length of the mainstream* (km)
L_c : *length of mainstream of the nearest watershed* (km)

2.4.2. Snyder Unit Hydrograph Snyder unit hydrograph has four parametric i.e. time lag, peak discharge, time base, and rainfall duration for unit hydrograph from the geometric basin (Gupta, 1989). Snyder unit hydrograph equation is expressed by

$$Q_p = 2{,}083 \frac{c_p\ A}{t_p}$$

$$t_p = C_t(L\ x\ L_c)^n$$

$$t_c = \frac{t_p}{5{,}5}$$

For $t_c > t_R$

$$T_p = t_p + 0{,}75(t_R - t_c)$$

For $t_c < t_R$

$$T_p = t_p + 0{,}5t_R$$

where

Q_p : peak discharge (m^3/s)
A : basin area (km^2)
c_p : peaking coefficient
t_c : time concentration (hour)
T_p : time to peak (hour)
t_r : rainfall duration (hour)
t_p : time lag (hour)
Ct, n : time coefficient
L : length of the mainstream (km)
L_c : length of mainstream of the nearest watershed (km)

2.4.3. Nakayasu Unit Hydrograph Nakayasu unit hydrograph is developed based on rivers in japan (Soemarto, 1987). Nakayasu unir hydrograph equation is expressed by

$$Q_p = \frac{1}{3{,}6}\left(\frac{AR}{0{,}3T_p + T_{0,3}}\right)$$

$$T_p = T_g + 0{,}8T_r$$

$$T_r = 0{,}75T_g$$

For $L > 15\ km$

$$T_g = 0{,}5279 + 0{,}058L$$

For $L < 15\ km$

$$T_g = 0{,}21L^{0{,}7}$$
$$T_{0,3} = \alpha T_g$$

Dimana

Q_p *peak discharge*(m^3/s)
A *basin area* (km^2)
R *effective precipitation* $(1\ mm)$
T_p *time to peak (hour)*
$T_{0,3}$ *time to peak until 0,3 peak discharge (hour)*

2.4.4. Institut Teknologi Bandung (ITB) Unit Hydrograph Institut Teknologi Bandung (ITB) hydrograph developed by D. K. Natakusumah, W. Hatmoko and Dhemi Harlan. Institut Teknologi Bandung hydrograph based on observation of work principle, structure, function, and operation compute method and calculation various unit hydrograph. Institut Teknologi Bandung-1 unit hydrograph equation is expressed by

$$Q_p = \frac{R\ x\ A_{DAS}}{3.6\ x\ T_p\ x\ A_{HSS}}$$
$$T_p = T_L + 0{,}5T_r$$
$$T_L = C_t\ x\ 0{,}81225L^{0{,}6}$$

Institut Teknologi Bandung-2 unit hydrograph equation is expressed by

$$Q_p = \frac{R\ x\ A_{DAS}}{3.6\ x\ T_p\ x\ A_{HSS}}$$
$$T_b = 10\ x\ T_p$$
$$T_p = 1{,}6\ x\ t_p$$
$$t_p = C_t\ (0{,}0394L + 0{,}201L^{0{,}5})$$

Where

Q_p : peak discharge (m^3/s)
A : basin area (km^2)
c_p : peaking coefficient
t_c : time concentration (hour)
T_p : time to peak (hour)
t_r : rainfall duration (hour)
t_p : time lag (hour)
Ct, n : time coefficient
L : length of the mainstream (km)
T_b : time base $(hour)$

2.5. HEC-RAS

The saint Venant equation consists of continuity and conservation momentum. The flowing river applied the law of physics i.e., principle/law conservation mass (continuities) and law conservation of momentum. The law is expressed mathematically in differential form by

2.5.1. Continuity equation The continuity equation of mass states that the mass flow rate into the volume must equal the mass flow rate out. the continuity equation is expressed by

$$\frac{\partial A_T}{\partial_t} + \frac{\partial Q}{\partial x} - q_l = 0$$

Where
q_l is lateral inflow (m^3/s)

2.5.2. Momentum *equation* The momentum equation is states of Newton's second law. The momentum equation states that the rate of change momentum with added external force is equal to accumulation rate momentum. the momentum equation is expressed by

$$\frac{\partial Q}{\partial t} + \frac{\partial QV}{\partial x} + gA\left(\frac{\partial z}{\partial x} + S_f\right) = 0$$

Where
t : time
x : flow path
z : water surface elevation
Q : lateral inflow
V : velocity
A : cross section area
g : acceleration of gravity
Sf : friction slope

3. Result and discussion

3.1. hydrology analysis
The precipitation distribution corrected between gauge precipitation stations with TRMM precipitation used the data from 2012 until 2014. Outlier test and Trend test do know the correction precipitation has outlier and trend or not. The correction precipitation has not been an outlier and trend.

Table 3. Outlier Test

Tahun	R24Max	ln(R24Max)
1999	364	5.898
2000	156	5.050
2001	156	5.051
2002	72	4.282
2003	261	5.565
2004	81	4.394
2005	206	5.329
2006	414	6.026
2007	301	5.709
2008	290	5.671
2009	80	4.388
2010	565	6.337
2011	442	6.092
2012	423	6.048
2013	319	5.766
2014	218	5.386
2015	156	5.051
2016	321	5.773
2017	427	6.058
2018	305	5.721
2019	188	5.235
S		0.592
x		5.468
N		21
Kn		2.407
XH		985
XL		57
Max		565
Min		72

Tahun	R24Max	ln(R24Max)
1999	153	5.032
2000	53	3.962
2001	54	3.998
2002	36	3.577
2003	107	4.672
2004	39	3.676
2005	73	4.291
2006	200	5.300
2007	112	4.716
2008	111	4.706
2009	39	3.671
2010	362	5.891
2011	229	5.432
2012	209	5.344
2013	113	4.729
2014	80	4.387
2015	57	4.049
2016	116	4.751
2017	214	5.365
2018	112	4.719
2019	64	4.159
S		0.649
x		4.592
N		21
Kn		2.407
XH		471
XL		21
Max		362
Min		36

Then the correction precipitation to analyze the Thiessen polygon. after that, the precipitation data has been using to calculate frequency analysis. There are some analyses i.e. Normal analysis, log Normal analysis, Pearson III analysis, log Pearson III analysis, and Gumbel analysis. The result of frequency analysis is the precipitation period. The frequency analysis must be Parametric test, Chi-square test, and Smirnov Kolmogorov test. the result of Parametric test, Chi-square test and Smirnov Kolmogorov test are

Table 4. Parametric test, Chi-square test and Smirnov Kolmogorov test

Uji Parameter					
	Normal	**Log Normal**	**Pearson III**	**Log Pearson III**	**Gumbel**
Rata-rata	200.87	196.85	198.03	197.58	196.47
Standar Deviasi	94.93	105.02	93.85	98.30	90.62
Cs	0.00	0.00	1.09	0.41	0.76
Ck	-0.56	0.92	-0.37	0.08	0.00
Cv	0.47	0.53	0.47	0.50	0.46
Cs/Cv	0.00	0.00	2.29	0.82	1.64
Uji Chi Square (Dkritis = 5,991)	1.970	3.840	3.640	0.350	3.310
Uji Smirnov (Dkritis = 0.286)	0.027	0.027	0.043	0.034	0.058
Uji	Cs ≈ 0	Cs/Cv ≈ 3		Any	Cs ≈ 1.1396
	Ck ≈ 3				Ck ≈ 5.4002
Keterangan	No	No	Inconclusive	Ok	Not Ok

There was a log Pearson III method as input to calculate distribution precipitation. Petunjuk Teknis Perencanaan Bendungan state that distribution precipitatin for small basin use distribution PSA-007 of six-hour. The distribution PSA-007 was added and stored in the soil that was the effective precipitation. The effective precipitation is calculated by the Soil Conservation Service method which is based on the potential for the soil to absorb a certain amount of moisture.

Table 5. The Effective Precipitation

t (jam)	P (mm)	S	Refektif (mm)
1	26.84	108	0
2	35.79	108	2
3	62.63	108	11
4	246.06	108	152
5	44.74	108	4
6	31.32	108	1

After that, effective precipitation use to calculate flood discharge. Calculate the flood discharge used the Soil Conservation Service Unit Hydrograph, Snyder Unit Hydrograph, Nakayasu Unit Hydrograph, and Institut Teknologi Bandung Unit Hydrograph. The flood discharge on 29 July 2013 has calibrated with AWLR data and obtained that Soil Conservation Service Unit Hygrograph has been a minimum error.

Table 6. Calibration Flood Discharge

Tanggal	Curah Hujan II Pattimura	Curah Hujan Kayu Tiga	Debit Sungai	Nakayasu	Snyder	SCS	ITB-1	ITB-2
29 Juli 2013	315	210	41	42	42	41	41	41
E absolute				1.427	1.009	0.280	0.317	0.540
E relative				0.034	0.024	0.007	0.008	0.013
r				1.000	1.000	1.000	1.000	1.000
RMSE				1.427	1.009	0.280	0.317	0.540

The result that Soil Conservation Service Unit Hydrograph used as a reference to calculate flood discharge. required by government regulation to control Floods use a 25-year discharge return period. Discharge 25-year return period is $138,11 m^3/s$.

3.2. hydraulic analysis

The modeling flood used the coupling method with a connection between 1d and 2d used lateral structure. The calibration was modeling HEC-RAS with discharge and tide on 29 July 2018. Then the verification with the yield data of water surface elevation on 14 until 18 Mei 2019. The results of the modeling verification are

Figure 3. Verification Modeling

Table 7. Error Verification Modeling

Titik Banjir	Tinggi	Ho	Hm	(Hm-Ho)^2	(Ho-Hobar)^2	NSE
Negeri Batu Merah	> 2 m	2	1.724	0.076	0.037	0.760
Negeri Batu Merah	30 – 100 cm	0.65	0.629	0.0004	1.339	
Waihoka	> 2 m	2	2.204	0.042	0.037	
Kel Rijali	> 2 m	2	2.494	0.244	0.037	
Kel. Amantelu	> 2 m	2	2.055	0.003	0.037	
Ahuru	> 2 m	2	2.087	0.008	0.037	
Kepala air	> 2 m	2	1.954	0.002	0.037	
Rata-rata		1.807143	Total	0.375	1.562	

The value of NSE more than 0,75 was good, that the modeling can use to planning control flood. HEC-RAS modeling has been doing by the 25-year discharge return period and HHWL tide. that is modeling HEC-RAS existing condition with 25-year discharge return period and HHWL tide

IOP Conf. Series: Earth and Environmental Science **989** (2022) 012016 doi:10.1088/1755-1315/989/1/012016

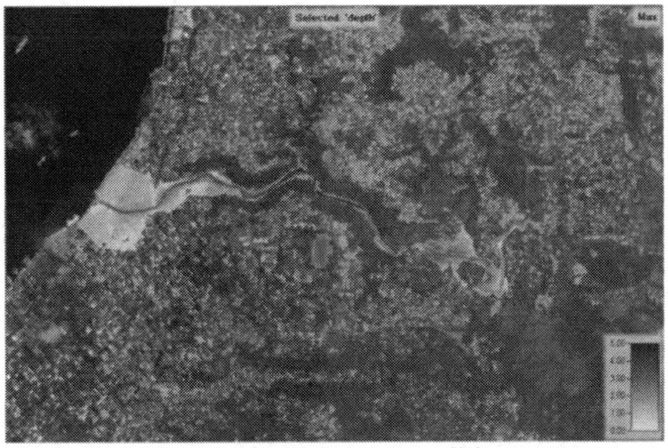

Figure 4. Existing Condition

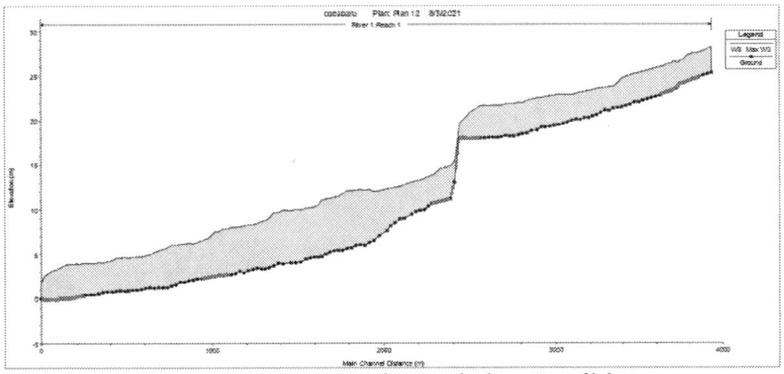

Figure 5. Long Section Existing Condition

Plan the normalization in HEC-RAS used tools channel design and modification. Normalization upstream was dredging 5 meters that the sabo dam return to capacity. that is normalization in upstream, transition, and downstream

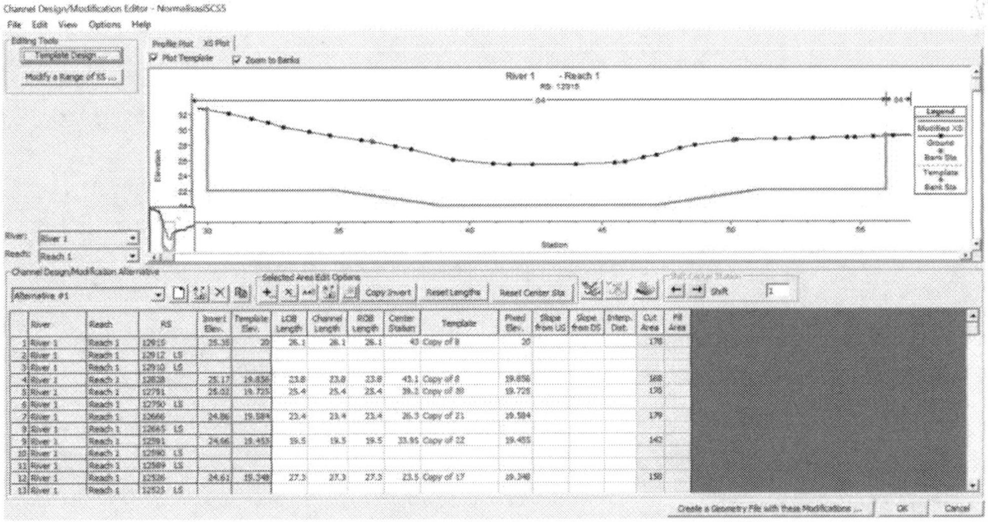

Figure 6. Normalization in Upstream

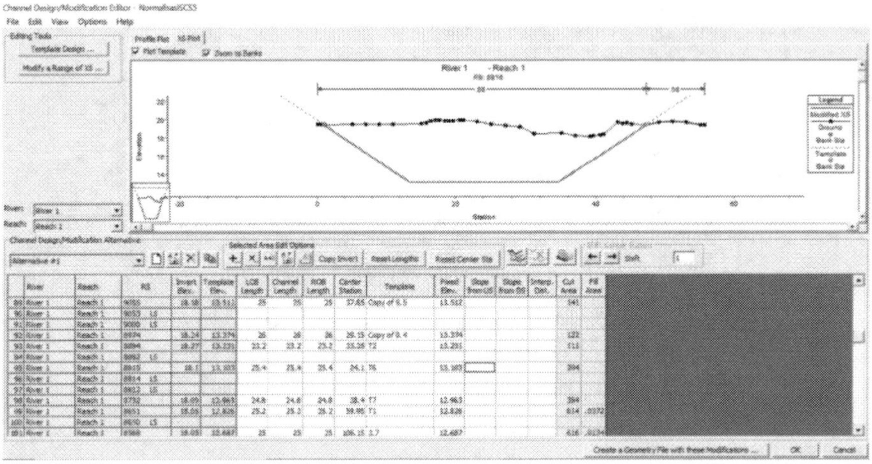

Figure 7. Normalization in Transition

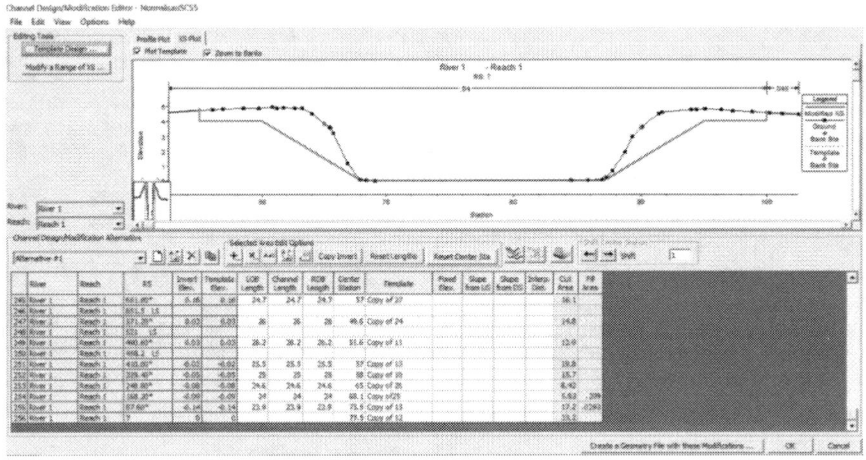

Figure 8. Normalization in Downstream

If the cross-section was still flooded then the cross-section must plan the dike. The dikes had been calculating by water surface elevation with freeboard 0,6 m. Modeling HEC-RAS after construction normalization with a dike is

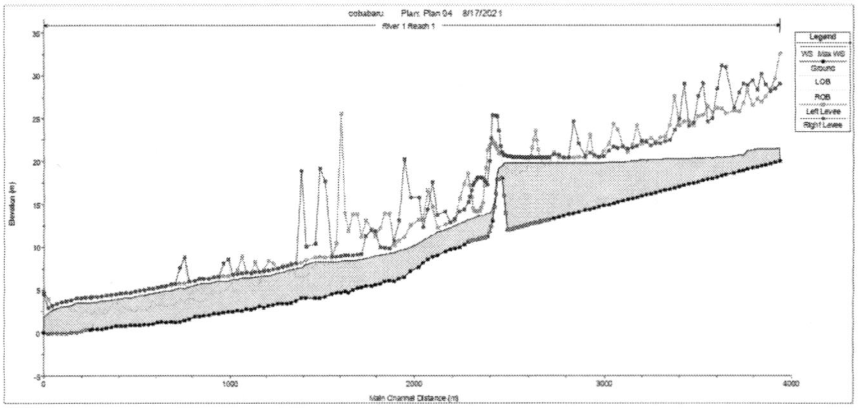

Figure 9. Modeling Normalization and Dike

The result that the normalization will be reaching 13655 m^3, besides that the dike will be assembled along 5028,76 km with land acquisition has been reaching 28779,2 m^2.

4. Conclusions

Flood in wai batu merah basin was caused by the river had not capable to accommodate water and the estuary be affected by the tide. A 25-year return period of water discharge with SCS unit hydrograph was 138,11 m^3/s. A 25-year return period is required by government regulation. A Scenario solution to control flood disaster in wai batu merah basin is normalization and dike. the result that the normalization will be reaching 13655 m^3, besides that the dike will be assembled along 5028,76 km with land acquisition has been reaching 28779,2 m^2. Cost to control flood disaster in wai batu merah for normalization and dike reach Rp.112.881.969.800,00.

References

[1] Pattipeilohy, D., Pattiselanno, A. E., & Mardiatmoko, G. 2019. Community Resilience Against Floods (Case Study in Batu Merah Village, Sirimau District, Ambon City). Agrilan: Journal of Archipelago Agribusiness, 7(1), 88-105.

[2] BNPB PUSDALOPS DAILY REPORT (Monday, 12 July 2021).

[3] Sebastian, L. 2008. Approach to flood prevention and control.

[4] Rumihin, A., Djajadi, R., & Kusumastuti, C. 2015. Flood Analysis in Wai Ruhu and Wai Batu Merah watersheds, Ambon. Journal of Civil Engineering Primary Dimensions, 4(1).

[5] Laka, B. M., Sideng, U., & Amal, A. 2017. Changes in Land Use in Sirimau District, Ambon City. Journal of Geocelebes, 1(2), 43-52.

[6] Osok, R. M., Talakua, S. M., & Gaspersz, E. J. 2018. Analysis of soil erosion factors, and the level of erosion hazard using the Rusle method in the Wai Batu Merah watershed, Ambon City, Maluku Province. Journal of Agricultural Cultivation, 14(2), 89-96.

[7] Yudono, A. R. A., Sungkowo, A., & Gomareuzzaman, M. 2020. Analysis of Erosion Impact on Dead River Capacity in Tawangsari and Sukoharjo Districts. Journal of Minerals, Energy, and the Environment, 4(1), 61-72.

[8] Muin, S. F., Boer, R., & Suharnoto, Y. 2015. Flood Modeling and Analysis of Losses Due to Flood Disasters in the Upper Citarum Watershed. Journal of Soil and Climate, 39(2), 75-84.

[9] Siregar, R. I., & Indrawan, I. 2017. Comparative Study of 1-D (One Dimension) and 2-D (Two-Dimensional) Modeling in Modeling Upper Citarum Watershed Floods. Educational Building Journal of Civil and Building Engineering Education, 3(2 DECEMBER), 31-37.

[10] Syahroni, M., Amri, K., & Afrizal, Y. ANALYSIS OF PEAK DEBT USING SYNTHETIC UNIT HYDROGRAPH METHODS ITB 1 AND HEC-RAS VERSION 5.0. 7.

[11] Zakaria, A. 2013. Manning Coefficient Sensitivity Analysis for 1-D Steady Flow Using HEC-RAS Program. In Proceedings of the National Seminar on Mathematics and Natural Sciences and Applications (ISBN: 978-602-98559-1-3) (Vol. 3, No. 3).

[12] Regulation of the Minister of Public Works and Public Housing Number: 28/PRT/M/2015

[13] Hernanda, R. Y., Nainggolan, R. M., Darsono, S., & Kurniani, D. 2016. Serayu River Embankment Downstream From the Estuary to the Serayu Motion Weir. Journal of Civil Engineering Works, 5(1), 135-145.

[14] Ginting, S. H. 2010. Analysis of River Water Level Profile Using HEC-RAS.A reference

This reference has two entries but the second one is not numbered (it uses the 'Reference (no number)' style.

IOP Conf. Series: Earth and Environmental Science **989** (2022) 012017 doi:10.1088/1755-1315/989/1/012017

School resilience policy development and strategic steps to disaster-resilient communities

S I A Dwiningrum [1,*], N Sitompul [2], A A Nisa [3], and D R S Sumunar [4]

[1]Faculty of Educational Sciences, Yogyakarta State University Indonesia
[2]Student Magister of Educational Research and Evaluation, Yogyakarta State University Indonesia
[3]Student Undergraduad of Education, Yogyakarta State University Indonesia
[4]Faculty of Social Sciences, Yogyakarta State University Indonesia

[*]Email: siti_ireneastuti@uny.ac.id

Abstract. Mitigation education policies have not been implemented optimally in schools because they have not received synergistic social support. This study will articulate four main points, which are: 1) social support for disaster mitigation education policy; 2) school resilience profile; 3) development of school's resilience policy, and 4) factors influencing the success of policy implementation. The research method is descriptive analysis approach to describe the respondents' opinions as they are regarding the draft policy brief and the indicators of school resilience policies for disaster mitigation. The study sample consisted of the heads of the education office and the principals of upper secondary schools in Indonesia. Sample selection was done by purposive sampling, from the regions in 18 disaster-prone provinces, 24 heads of department and 24 principals of each disaster-prone area at the high school level in Indonesia are taken as respondents. The data collection technique used in the study was the use of online media through Google Forms and focus group discussion through Zoom meetings. The results of the study concluded that: 1) disaster mitigation education policies require social support from the central government, regional governments, higher educational establishments, and students who work synergistically; 2) the school's resilience profile is still relatively low and requires school policies strengthening; 3) the development of school resilience policies requires strategic and proactive steps from the school; 4) the success of the policy implementation is largely determined by the school's strategy in dealing with the supporting and hindering factors. By successfully implementing school resilience policies in schools in Indonesia, the role of schools will be more optimal in disaster mitigation.

1. Introduction

Disaster is an event or series of events that threaten and disrupt people's lives and livelihoods caused by both natural and/or non-natural factors as well as human factors, resulting in human casualties, environmental damage, property losses, and psychological impacts. This is in line with the results of the research conducted by Zaim, et al. [1] which states that natural disasters have an impact of damage and destruction as well as loss of life. Natural disasters, according to Law Number 24 of 2007, are disasters caused by events or a series of events caused by nature, including earthquakes, tsunamis, volcanic eruptions, floods, droughts, hurricanes, and landslides. Natural disaster is a process that damages aspects of human life which is promoted by nature and causes harm to humans both physically and psychologically.

Indonesia has the potential for disaster that should not be overlooked. Indonesia is the largest disaster-prone area in Southeast Asia. The National Disaster Management Agency (BNPB) noted that 148.4 million people live in earthquake-prone areas; 3.8 million in tsunami-prone areas; 1.2 million people in areas prone to volcanic eruptions; 63.7 million people in flood-prone areas; and 40.9 million people live in landslide-prone areas; 11.1 million are vulnerable to high waves and abrasion. In addition, there are 386 regencies/cities in the medium to high earthquake hazard zone; there are 233 districts/cities located in tsunami-prone areas; 75 districts or cities vulnerable to volcanic eruptions; and 315 regencies/cities are in moderate to high flood hazard zone [2].

As handling of disaster risk is yet optimized, Indonesia is still faced with two main problems, namely: 1) the performance of disaster management is still low; and 2) the attention to disaster risk reduction is lacking. These two problems are a challenge for Indonesian people to be able to earnestly design disaster risk management creatively and proactively. Disaster management is a science related to efforts to reduce risk of disaster, which includes actions to prepare, support, and rebuild communities when disasters occur. In general, disaster management is a continuous process carried out by individuals, groups, and communities in managing hazards in an effort to avoid or reduce the impact of disasters. In line with Mabon, Leslie [3] mentions implementing disaster management policies by empowering the role of communities and the environment helps in the adaptation to the changes and will reduce losses. The actions taken depend on the perception of the risks faced. Mitigation is an action taken to reduce the impact caused by a disaster. The mitigation phase is focused on long-term actions to reduce disaster risk. According to Paunga & Lassa [4], Disaster Risk Reduction (DRR) has a significant impact and thus, every country in the world and the international community is encouraged to develop DRR into a policy framework as an investment in development goals and community resilience. Mitigation strategies implementation can be seen as part of the recovery process if mitigation actions are taken after a disaster occurs. However, even though the implementation action is a recovery effort, the actions taken to eliminate or reduce future risks are categorized as mitigation actions [5].

Schools have a strategic role in disaster mitigation, this is in accordance with the policy of the Sendai Framework for Disaster Risk Reduction 2015-2030 [6]. Therefore, the development of school resilience policies is needed for disaster mitigation and is in line with the results of Dwiningrum's research [7] which states that building school resilience for disaster mitigation cannot be separated from the involvement of the school community. The development of school resilience policies will be effective if it is built based on the values of local wisdom. With local wisdom as the base, which aimed at developing the personality and cultural identity of the community in the form of values, norms, ethics, beliefs, and customs that are taught and practiced in schools for the purpose of disaster mitigation, namely providing protection and safety to students, educators, and staffs from disaster risk and to ensure the continuity of education services in disaster-affected education units, a regulation from the Minister of Education and Culture regarding the Implementation of the Disaster Safe Education Unit Program is needed.

Disaster mitigation efforts are dynamic, so disaster mitigation and prevention is the best strategy in respect to people's livelihoods to survive, adapt and recover from disasters [4]. Interpretation and adaptation to events is determined by the knowledge possessed by the community. However, the current knowledge and community responses still require help from comprehensive education and socialization. Related to the issue of disaster mitigation education, schools have a strategic role in disaster mitigation. This is because the locations of the schools are mostly located in disaster-prone areas. The development of school resilience will not be the same in each region because each region has a different context of disaster events. Therefore, education policy needs to be studied based on its level [8] namely, Macro Policy (National Level), Meso Policy (State Level), and Micro Policy (Local Level). The development of resilience policies is a micro policy (Local Level) that should be implemented by all schools in Indonesia. This fact is supported by the opinion of the heads of offices and school principals in Indonesia as depicted as follows:

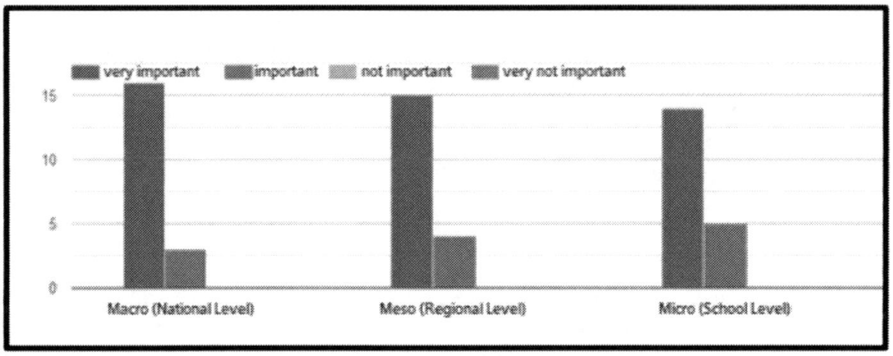

Figure 1. Opinions on the importance of mitigation education policies in schools

Schools have a strategic role in disaster mitigation, this is in accordance with the policy of Sendai Framework for Disaster Risk Reduction 2015-2030 [6]. Therefore, the development of school resilience policies is needed for disaster mitigation at schools. The reason schools are selected to build resilience policies for disaster mitigation is based on the fact that school community's role is not just as subjects but as agents who translate the policies into practices [9] and also that the policies can be incorporated in teaching and leadership. This paper will discuss the development of school policies and strategic steps to build a disaster-resilient society, specifically related to: 1) why social support for disaster mitigation education policies is needed; 2) how is the school's resilience profile; 3) how is the development of school resilience policies, and 4) factors influencing the success of policy implementation.

2. Theoretical review

2.1. Disaster Mitigation

Mitigation is an action taken to reduce the impact caused by a disaster. The mitigation phase focuses on long-term actions to reduce disaster risk. Implementation of mitigation strategies can be seen as part of the recovery process if mitigation actions are taken after a disaster occurs. However, even though the implementation is a recovery effort, the action taken to eliminate or reduce the risk in the future is categorized as a mitigation action. Mitigation actions include structural mitigation and non-structural mitigation. Structural mitigation is an action to reduce or avoid the possible physical impact of a disaster. Examples of structural mitigation actions are the construction of earthquake-resistant houses, infrastructure development, construction of embankments on the banks of rivers, and so on. Non-structural mitigation is action related to policies, awareness building, knowledge development, public commitment, and implementation of methods and operations, including participatory mechanisms and information dissemination, which are carried out to reduce risks related to the impact of disasters. Mitigation is the most efficient action to reduce the impact caused by a disaster [5].

Disaster management is a science related to efforts to reduce risk, which includes actions to prepare, support, and rebuild communities when disasters occur. In general, disaster management is a continuous process carried out by individuals, groups, and communities in managing hazards in an effort to avoid or reduce the impact of disasters. The actions taken depend on the perception of the risks faced. The effectiveness of disaster management depends on the integration of all elements, both government and non-government. Activities in each hierarchy (individual, group, and community) exert influence at different levels.

Disaster mitigation education through schools is quite effective, because schools as formal institutions have an important role in building a disaster culture effectively. With the knowledge and awareness built by schools, the role of schools is greater in disaster mitigation. On the other hand, disaster response culture of the school will become more responsive and proactive concerning disaster events.

2.2. Disaster Mitigation Education Policy

Policies are needed to solve problems. The school's resilience policy is aimed at disaster mitigation. This is because schools have not been taking their role optimally for disaster mitigation and thus, social impacts continue occurring when disaster events happen in Indonesia. This is because school resilience policies have not been implementable in most schools in Indonesia. The process of policy-making is not instantaneous, for it is built through political processes that involve many social roles in the community, until the policy can be implemented. The formulation of a policy starts from accumulation, then articulation, and leads to accommodation. In general, a policy is carried out through at least two stages, namely, the articulation stage and ends with accommodation, and the stabilization stage [10]. Each stage has a different process, sometimes policy formulation is discussed through a complicated and complex political and analytical process which transpire dialectically so that it often does not have a clear beginning and end. There are experts who say that the policy formulation stage is actually a transformation from input into political output [11] quoted by Rohman [12].

The formulation of disaster mitigation education policy requires a social process to articulate and define the problem, the formulation of possible answers to all political demands, the submission of all these demands into the system, the imposition of sanctions or legitimacy on the selected actions, as well as the ratification of implementation, monitoring, and feedback. While in the second stage, namely the implementation stage, a policy that is implemented is basically not only related to the mechanism of elaborating political decisions into routine procedures through several existing bureaucratic channels, but further, it also involves issues of conflict, decisions of choice, and who gains from a policy, including in this case is the school's resilience policy for disaster mitigation [12].

2.3. School resilience for disaster mitigation

School resilience policy as a process of transformation from input to political output, is not only to realize the interests of the mechanism for elaborating political decisions into routine procedures through several existing bureaucratic channels, and decision of choices [12]. School resilience policies for disaster mitigation require an effective approach model that can be applied in schools in Indonesia.

Resilience is quite important to build in school life, especially in Indonesia. School resilience has an important role and is an important aspect in disaster mitigation. School resilience will play a role in dealing with critical conditions, especially during disaster events. With resilience, schools will be able to help the community in facing critical conditions and be able to provide a protective factor for all internal and external assets in the school. In addition, resilient schools will have the adaptive capabilities needed to deal with disasters. Resilience as a social process will be able to improve the social capacity needed to face challenges and problems. Schools will be able to provide sufficient provisions for students to become resilient individuals. As explained by the American Psychological Association, the broad definition of resilience is a process of well adaptation in facing hardship, trauma, threat, or a significant stressor such as health, family and relationship problems.

School resilience is a social process built by two principals, namely: mitigating risk factors in the environment, and building resilience in the environment [13]. Henderson & Milstein [13] also explains that the stages in building school resilience are determined by six variables, namely: “increase bonding”; “set clear and consistent boundaries”; “teach life skills”; and “provide caring and support”; “set and communicate high expectations”; and "provide opportunities for meaningful participation". In building school resilience, teachers have an important role in mobilizing all aspects to make students become more resilient [13], [14].

IOP Conf. Series: Earth and Environmental Science **989** (2022) 012017 doi:10.1088/1755-1315/989/1/012017

3. Research methodology

The research uses a descriptive approach using qualitative methods to describe what respondents think about the policy brief and indicators of school resilience policies for disaster mitigation. The qualitative method aims to describe the perceptions and inputs of the stakeholder, in this case the education office and high school principals in Indonesia, on policy briefs and school resilience policy indicators for disaster mitigation. The research was carried out through Zoom meetings involving a number of samples, namely the heads of the education offices and the principals of upper secondary schools in Indonesia. The sample selection was carried out by purposive sampling, wherein 18 provinces considered as disaster-prone areas were selected along with 24 heads of department and 24 principals of each disaster-prone area at the high school level. The data collection technique used in the study is the use of online media through Google Forms, which can be accessed via (https://forms.gle/7YhASYLfGEQeCBaA), and focus group discussions (FGD) through Zoom meetings. As for what is highlighted is the perception of the heads of education office and senior high school principals who are in disaster-prone areas of the policy brief and indicators of school resilience policies for disaster mitigation.

Table 1. Techniques, Respondents and Target Information

No	Technique	Respondent	Information Target
1.	Questionnaire (through online media)	Heads of education offices and heads of senior high schools.	▪ Perceptions and inputs on policy briefs and indicators on school resilience policies for disaster mitigation
2.	FGD	Heads of education offices and heads of senior high schools.	▪ Feedback and suggestions for policy brief and school resilience policy indicators for disaster mitigation

The process of data analysis was carried out at the time of data collection and after data collection. The process of analyzing the qualitative data is done by using qualitative analysis according to Milles and Huberman [15] with the help of the application of the Latent Dirichlet Allocation (LDA) method which looks for the semantic structure in a collection of corpus (collection of documents) based on hierarchical bayesian analysis. Data analysis in qualitative research includes the following stages: First, data reduction, which is the process of selecting, separating, paying attention to simplification, abstracting and transforming rough data that emerges from written notes in the field and then transferred into a complete and detailed report. Second, presentation of data, is an attempt to compile a collection of information that has been obtained in the field to be presented clearly and systematically so that it will be easier to draw conclusions from. Lastly, drawing conclusions and verification. At this stage, the researcher tries to understand, analyze, and find meaning from the data collected, and finally after the data is collected a conclusion will be obtained. The education profile of the respondents is as follows:

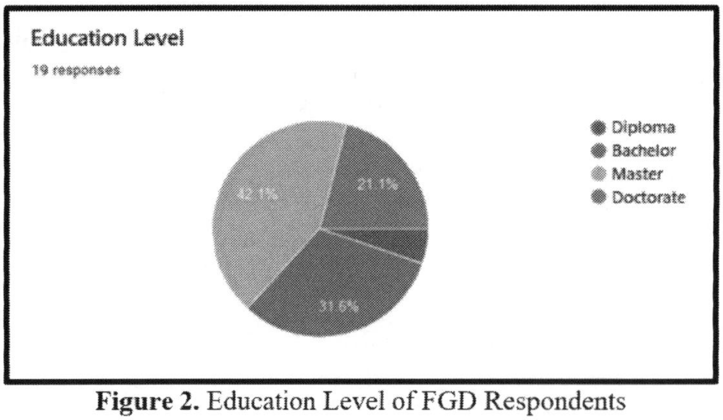

Figure 2. Education Level of FGD Respondents

IOP Conf. Series: Earth and Environmental Science **989** (2022) 012017 doi:10.1088/1755-1315/989/1/012017

4. Results and Discussion

The development of school policies and strategic steps to build a disaster-resilient society is not an easy job, because it requires a strategic and synergistic design. To provide a comprehensive picture, the main findings are presented as follows:

4.1. Social Support for Disaster Mitigation Education Policy

Mitigation education policies are not easy to implement in schools. Policy as a social process requires stages in its implementation as a public policy. As explained by Van Meter and Van Horn [16], the policy implementation process model has several factors that form the link between policies, namely (1) standards and policy objectives; (2) policy resources; (3) interorganizational communication and implementation activities; (4) characteristics of the implementing agency; (5) economic, social and political conditions; (6) the attitude of the implementers. According to Edward III, there are four factors that are important requirements in reviewing and researching the success or failure of policy implementation, which are "communication, resources, disposition, and bureautic structure" [17]. A policy in its implementation process is very dependent on the ability of policy implementers to interpret or translate the policy in accordance with the procedures, systems, and mechanisms, so that the policy objectives can be realized effectively. In other words, the success of the implementation process of a policy is largely determined by the honesty, competence, and high commitment of the implementers to achieve its goals, and high attentiveness to government regulations [18].

Likewise, the process of forming school resilience policies requires a social process carried out in various stages of research, resulting in a policy brief that is expected to be able to be used by the schools to decide on school programs that are expected to improve school resilience. Based on the literature review and primary data as well as qualitative data that is systematically processed, a conceptual picture of the importance of implementing policies for disaster mitigation in schools is pictured as follows:

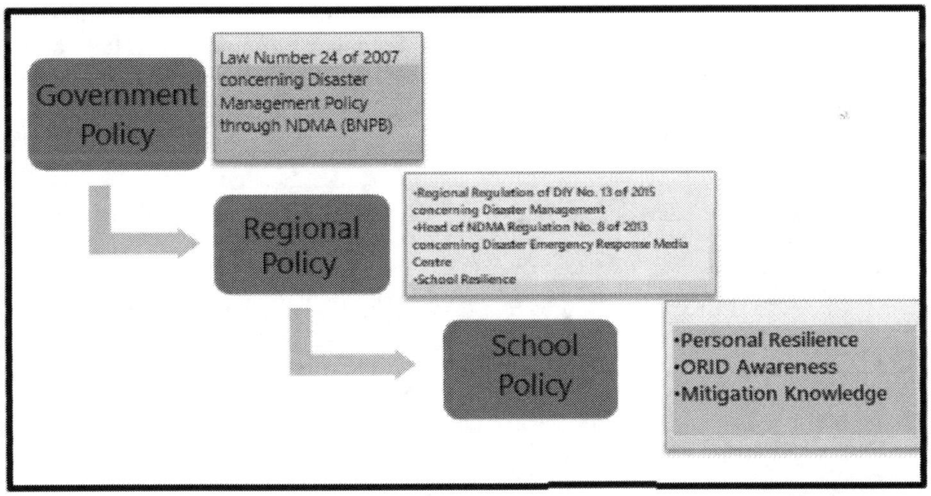

Figure 3. Hierarchy of School Resilience Policy for Disaster Mitigation

The policy hierarchy for disaster mitigation becomes the basis for schools to design strategies to achieve disaster mitigation goals. Systemic work and synergy are the basic principles for schools to make a truly collaborative work program between various stakeholders to support a more optimal disaster mitigation. The dynamics in social relations between classes, schools, communities, and governments must be built with the strength of high social capital. This kind of awareness becomes very important, because a policy is not understood partially, but must be understood holistically and comprehensively, so that the social dynamics that continue to occur can be better understood. Based on the results of the FGD, the following results were obtained:

IOP Conf. Series: Earth and Environmental Science **989** (2022) 012017 doi:10.1088/1755-1315/989/1/012017

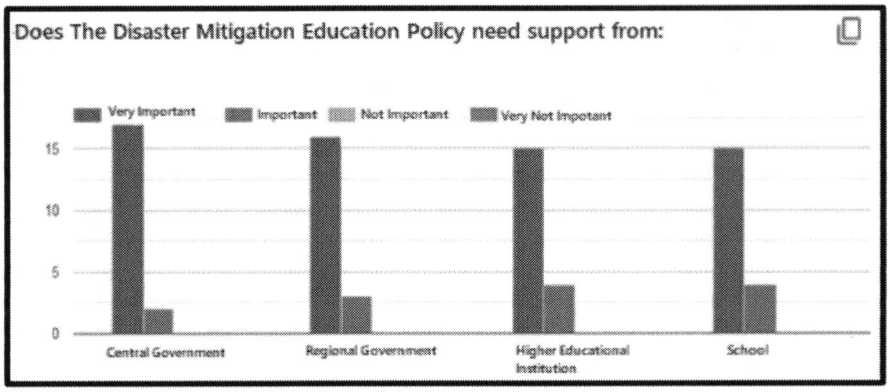

Figure 4. Social support for disaster mitigation education policy

The figure above proves that the implementation of disaster mitigation education policies requires synergistic support from the central government, regional government, higher educational institution, and school. The data above can be interpreted that the success in disaster mitigation is not a partial work, but one that requires systematic and synergistic work for disaster mitigation purposes. Schools as formal institutions are considered to have a strategic role in socializing and educating about disaster mitigation.

4.2. School Resilience Profile

The school resilience policy for disaster mitigation is one of the decisions that is expected to support the realization of Law No. 24 of 2007 concerning Disaster Management Policy through the National Disaster Management Authority (BNPB) in which each region's application is adjusted to their needs and conditions. For example, D. I. Yogyakarta, which is vulnerable to natural disasters such as tsunami, earthquake, and volcano eruption, issued the Regional Regulation of DIY No. 13 of 2015 concerning Disaster Management and DIY Regulation No. 8 of 2013 concerning the Media Center for Disaster Emergency Response. The results of the research related to school resilience policies are based on the data described in table 3.

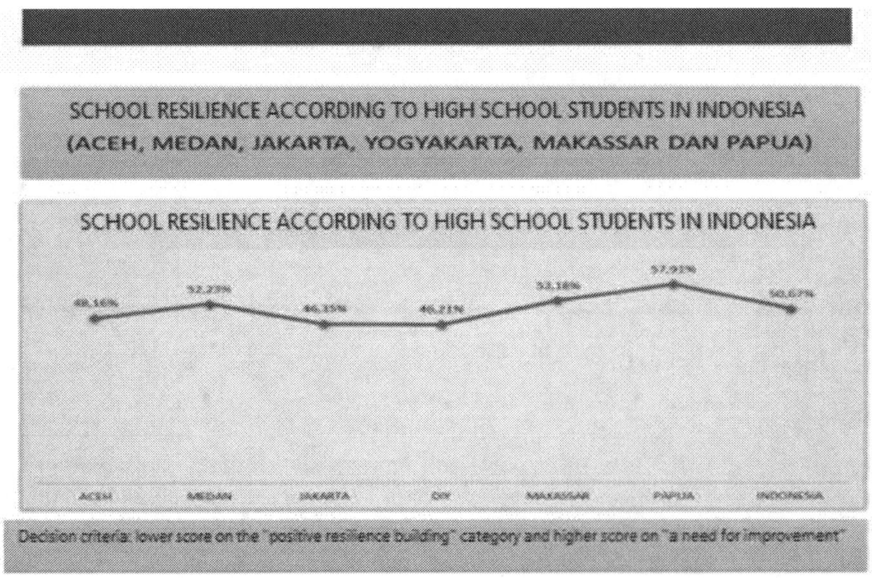

Figure 5. Profile of school resilience in Indonesia

Based on Figure 5, it can be concluded that the level of school resilience in several schools in Indonesia is relatively yet to be high. Therefore, the implementation of school resilience policies must be enacted by the government as a form of the school's concern in realizing macro policies in which are the government provisions.

4.3. School Resilience Policy Development

The school's resilience policy is not an easy work, because it requires very strong support from various components that must be optimally prepared by the school. As explained by Bahtiar [19] that the analytical framework needed in the policy-making process is related to four elements that must be considered, namely as follows: (1) Elements of Problem. The element of problem is related to the fields of government work, such as education, public welfare, public health, regional development, foreign relations, defense and security, taxation, population, and others. These elements are better known by the fields of ideology, politics, socio-culture, defense and security. (2) Elements of Purpose. The element of purpose relates to the targets that were to be achieved through programs that have been determined by the State. (3) Elements of Working Method. The element of working methods is related to the procedures and systems based on applicable laws and regulations. (4) The Elements of Authority. This element is related to the authorities that were given the trust to carry out government activities.

Based on several main principles in the process of forming school resilience policies, the concept being developed requires dialogue with the Head of Offices, School Principals, Teachers, and Community Leaders. Based on the results of the FGD related to the mitigation education policy, the following conclusions were obtained:

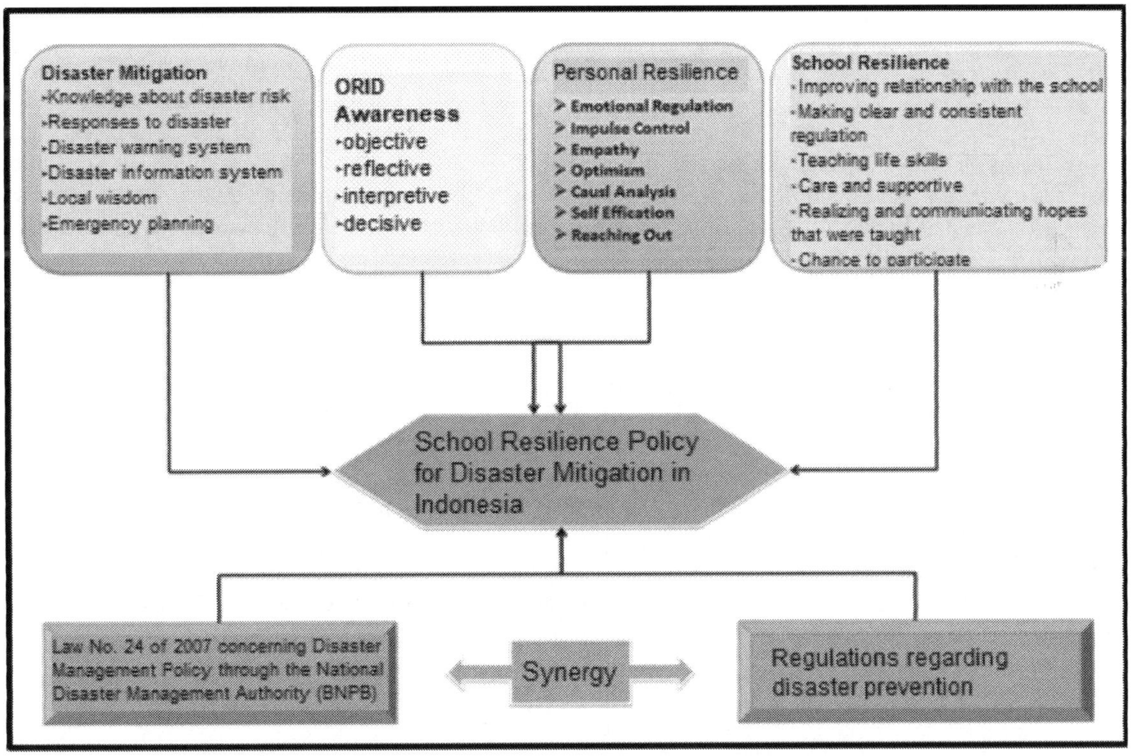

Figure 6. School Resilience Policy System Development

Figure 6. Illustrates that the policies that have been formulated and determined to be implemented have achieved goals and targets, and to find out whether these targets have been realized, the policies must be implemented in schools. Policy implementation is an important aspect of the policy process.

Thus, implementation implies that policy implementation can be seen as a process of carrying out policy decisions [12]. Based on the development of a school resilience policy system, the policy decisions usually take form of laws, presidential instructions, government regulations, court decisions, ministerial regulations and so on. As for the benchmark for the success of the policy is largely determined by its implementation. Because whatever the policy formulates, if it is not implemented, it will not be useful. On the other hand, any simple policy that has been implemented will be more useful, whatever and however useful it is [20]. To understand respondents' opinions about the importance of implementing mitigation policies in schools, the following responses were obtained:

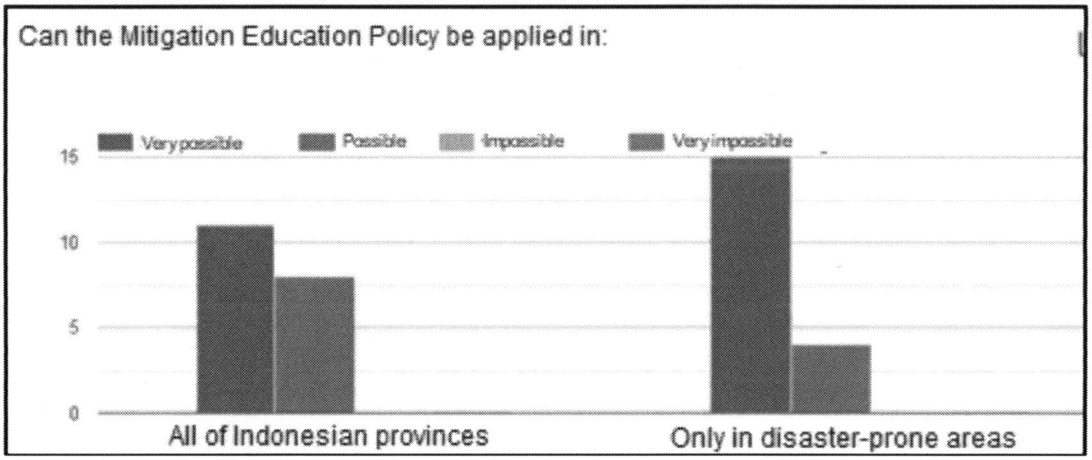

Figure 7. Implementation of school resilience policies

Based on the opinion of the respondents regarding the implementation of school resilience policies in general, the policy is very likely to be applied in all provinces in Indonesia and also in disaster-prone areas. The social facts reinforce the initial assumption that the implementation of school resilience policies can be applied in schools, and the benefits can be assessed if policy decisions have been implemented in schools. Therefore, based on the data above, a concept recommendation by the government is needed to be socialized to schools in Indonesia.

4.4. Factors Affecting Policy Implementation

The level of success or failure is in accordance with the policy objectives that are wanted to be achieved and is very dependent on the use of resources both from policy makers and policy implementers, namely school organizations. According to Imron [20], the success or failure of policy implementation is largely determined by the following factors: 1) The complexity of the policies that have been made. The more complex a policy is, the more complex and difficult it is to implement; 2) If the policy formulation and alternative policy problem solving proposed in the formulation are not clear; 3) Factors of potential source that can support the policy implementation; 4) The expertise of policy implementers; 5) Support from the target audience of the implemented policies; and 6) the factors of effectiveness and efficiency of the bureaucracy. As a determinant of the success of policy implementation, a mechanism with strategic steps is needed to operationalize it, so that policy objectives can be successful. Based on the results of the FGD, the following data were obtained:

Table 2. Factors supporting and inhibiting policy implementation

Supporting factors	Inhibiting factor
1. The existence of regulations made by the central and regional governments regarding disaster management. 2. A law that integrates knowledge, attitudes and values in disaster education. 3. The existence of a disaster mitigation education policy leaflet/guide to education units, school community and the education office are pro-active in fundraising and orderly reporting the liability, high cooperation spirit, competence and understanding of disaster mitigation is quite good. 4. Facilities, infrastructure that integrate knowledge, attitudes and values in disaster education. 5. Government's political will, NGO support, public awareness and concern, financial support, facilities and infrastructure for the implementation of Disaster Mitigation Education.	1. Heterogeneous school community, in which not all of the residents made efforts of disaster risk reduction. 2. Lack of awareness and consistency of rule enforcement as well as unsynchronized handling policies between regions and the central government. 3. The community's ability to predict is still low and having a lack of support, stakeholders and various parties are not in sync. 4. Regional policies and the lack of participation of relevant agencies, funding, as well as the area width. 5. Limited human resources, infrastructure that has not been improved, coordination with stakeholders in disaster management has not been optimized. 6. Limited human resources, infrastructure that has not been maximized, coordination with stakeholders in disaster management has not been maximized. 7. Lack of participation of relevant agencies. 8. Political will of regional government, synchronization and synergy between related institutions, awareness and concern of community members that still need to be improved, financial support and infrastructure. 9. Lack of coordination between disaster mitigation training participants and disaster mitigation training providers. 10. There is no clear regulation yet. 11. Socialization is still lacking, few socialization targets, the awareness of school community is not high and financing factors. 12. Selective non-compliance with the law, where there are several laws and regulations or public policies that are less binding on individuals.

Based on table 2, it can be concluded that the success in implementing mitigation education policies has not been optimal, because there are still inhibiting factors in the process. Therefore, it is still necessary to study effective strategies to overcome problems in the implementation of school resilience policies. This is in line with Siagian's [21] opinion which explains that policy implementation is a strategy and design for managing various systems that apply in schools. Meanwhile, to achieve a high level of integration of all elements, the synergy is needed to be considered, such as the human aspects, structure, administrative and management processes, funds and resources to achieve disaster mitigation goals. In this context, managerial activities related to policies that should be interpreted as a management process that involves all elements in the school. Therefore, the use of resource strategies owned by the school can be utilized to achieve disaster mitigation goals. In field application, a policy that were to be implemented requires cooperation or synergy between the elements determining and implementing the policy [22]. In other words, policy implementation is not a stand-alone entity, but is actually a part of a whole system of interdependent problems and require a holistic approach, an approach that looks at the inseparable parts of the whole system that binds them together. Because of the solution chosen in overcoming the school's resilience policy, several varied strategies must be given so that schools can choose the most appropriate method, in which is in accordance with the situation and conditions of the school.

4.5. Supporting Factors and Inhibiting Factors for Disaster Mitigation Education Policies at the Meso Level

Table 3. Suggestions and Feedback

Suggestions and Feedback
1. Emphasis on sanctions against violators of regulations and massive socialization.
2. There is very little awareness and concern regarding disasters, there is aneed for socialization.
3. A Governor's Regulation concerning disaster mitigation education is needed.
4. Training of resource persons at the regional level.
5. A book concerning disaster and disaster mitigation to further this program.
6. The issuing of leaflets, guidelines and standardization program, the forming of task force and the existence of supervision.
7. Increasing socialization in all lines of society.
8. Awareness and consistency of rule enforcement.
9. A well-planned structured socialization in each school as the smallest unit in handling disaster mitigation.
10. A plan of action and policies based on local elements is prepared.
11. Appropriate implementation of stakeholder support policies in the environment.
12. Good preparation starting from the budget to human resources.
13. There is socialization to all levels of society about the need for disaster mitigation and management. Regional leadership policies, especially those that are prone to disasters, for anticipation are included in the curriculum or lesson in all levels of education units.
14. Implementing policies and regulations made at the macro level in a harmonious and synergistic manner, developing a more technical and operational implementation system by taking into account the carrying capacity and potential that exists at the micro level.
15. There is education to the community.
16. Synergy and collaboration of relevant agencies in dealing with disasters by the regional regulations. From BPBD, Youth and Sports Affairs Offices, Health Service, Social Service, Child Protection Service and other related agencies.
17. Policies applied at the public policy level must really be carried out in a balanced manner, meaning that the implementation process must be fair, on target, objective and accountable.

The implementation of a policy is sometimes a complicated and complex phenomenon, which can be understood as a process, and as a result, the level of success depends on and requires cooperation and mutual coordination between implementers in order to be carried out. The elements of cooperation and coordination are very important in the process of implementing a school resilience policy for disaster mitigation, which must appear as concrete actions of the policy implementers. In terms of involvement (cooperation and coordination), there are many components that can be used in the process of implementing a policy according to Edwards III [17] which states that policy implementation can include a broad set of actions that place and implement instructions, determine funds, lease, giving help, agreement approval, collect data, disseminate information, analyze problems, raise and review, form organizational units, determine alternatives, plan the future of and consult with the community, with interest groups, with business activities, with legislative commissions, with bureaucratic units, and even with other countries. To achieve the successful implementation of a policy, it involves many components including operational policies or technical guidelines, budget funds, resources, work units, activity planning, benefits obtained, desired changes and community participation, as well as interest actors (rulers). In general, a policy that originates from the top management level and then passed down to be implemented at the lower level always runs smoothly and effectively, in which the impact must be real for society. The success or failure of policy implementation has many components or elements that are correlated with each other. This condition indicates that policy implementation is made as a frame of reference or strategic step because a policy needs to be implemented. Therefore, in order for implementation to be carried out in accordance with its objectives, it is necessary to have policies that have been operationalized in the form of activity programs, have target groups as objectives, and have actors who are responsible for the implementation process [23]. The process of implementing education policies requires an analysis of factors that can determine and ensure the implementation process can run effectively and efficiently as an indication of success that must be considered by policy makers and implementers in the field. There are several theories that become models in policy implementation that pay close attention to the factors that will determine the success or failure of a policy, including school resilience policies for disaster mitigation.

5. Conclusion

The results of the discussion explained that not all provincial offices were aware of the regulation of Law Number 24 of 2007 concerning disaster mitigation education. The results of the answers from the education office participants stated that there were still several agencies that did not know the regulations of the law on mitigation education.

Out of the 19 provincial-level education offices involved, only about 12 provincial-level education offices in Indonesia mentioned the governor's regulation and local regulations referring to Law No. 24 of 2007. Meanwhile, the other provincial-level education offices gave various answers, some said they did not know so they can not mention and explain. There are those who say there are but have never received socialization from the authorities.

The inhibiting factors are emphasizzed more on culture, geographical location, lack of synchronization of regulations made, lack of awareness and consistency of regulatory actors, and the heterogeneous communities.

The supporting factors are marked by the existence of legal regulations that support disaster mitigation education programs, the availability of facilities and infrastructure to support the program, the availability of human resources, the commitment to run the program and the existence of a solid team.

Suggestions/inputs include the preparation of clear and firm rules, the need for socialization of mitigation education programs, collaboration between mitigation education and resilience in accordance with the conditions and culture of each region, mitigation and resilience education materials are integrated into relevant subjects, and the existence of awareness and consistency in implementation.

School's excellence program plan. From the results of the FGD, it was stated that some schools located in disaster-prone areas already had excellent programs, but they were only limited to discourse, and some said that schools did not have any excellent program plans related to disaster mitigation and resilience. Essentially, schools located in disaster-prone areas do not yet have a disaster program that is in accordance with disaster mitigation, even when it is known that Indonesia is a disaster-prone area. Thus, from the results of the FGD, it was agreed that collaboration in disaster mitigation and resilience education is needed as initial capital for the society, and encompasses both knowledge and experience related to disaster mitigation and resilience in vulnerable situations. This combination is expected to make the society more resilient.

Acknowledgment
This research was supported by Ministry of Research, Technology and Higher Education of the Republic of Indonesia who have provided research development funds 2021 in accordance with the contract number: T/13.4/UN34.21/PT.01.03/2021 and Lembaga *Penelitian dan Pengabdian kepada Masyarakat* (LPPM) with Prof. Dr. Samsul Hadi, M.Pd., M.T as chairman.

References

[1] Zaim M Putra R R and Rosita N 2021 Language as a Tool for Disaster Mitigation Management: Analysis of Warning System Text in Language and Institutional Framework *J. Disaster Res* **vol 16**, no 5, pp 817–826
[2] bbc Indonesia 2011 Negara Rawan Bencana https://www.bbc.com/indonesia/berita_indonesia/2011/08/110810_indonesia_tsunami.
[3] Mabon Leslie 2020 A historical approach to understanding governance of extreme urban heat in Fukuoka, Japan Disaster Prevention and Management: An International Journal, ahead-of-print(ahead-of-print)," *J. Disaster Res* **vol 16** no 5 pp 817–826 doi: 10.1108/dpm-01-2020-0010.
[4] Paunga F and Lassa J A 2020 Setting an Baseline Agenda for Enterpreneurial Governments: A Global Baseline Assessment of Disaster Risk Reduction Investment *Int. J. Disaster Manag* **vol 3** no 1
[5] Pribadi Krisna S 2008 Penyusunan Masterplan (Rencana Induk) Penanggulangan Bencana Di Propinsi Jawa Barat Bandung
[6] A D and H R Jayakody R R J C 2016 The 6th International Conference on Building Resilience in *The 6th International Conference on Building Resilience* pp 1022–1030
[7] Dwiningrum S I A 2017 Developing school resilience for disaster mitigation: a confirmatory factor analysis *Disaster Prev. Manag. An Int. J* **vol 26** no 4 pp 437–451
[8] Rohman A 2009 *Memahami pendidikan dan Ilmu Kependidikan* (Yogyakarta: Laksbang Mediatama)
[9] Bergmark Ulrika and Hansson Kristina 2021 How Teachers and Principals Enact the Policy of Building Education in Sweden on a Scientific Foundation and Proven Experience: Challenges and Opportunities *Scand. J. Educ. Res* **vol 65** no 3 pp 448–467 doi: 10.1080/00313831.2020.1713883.
[10] Silalahi O 1989 *Beberapa Aspek Kebijaksanaan Negara* (Yogyakarta: Liberty)
[11] Wahab Solichin Abdul 1997 *Analisa Kebijakan* (Jakarta: Bumi Aksara)
[12] Rohman A 2012 *Kebijakan Pendidikan: Analisis Dinamika Formulasi dan Implementasi* (Yogyakarta: Aswaja Pressindo)
[13] Henderson N and Milstein M M 2003 *Resiliency in schools: making it happen for students and educators* (California: Corwin Press)
[14] Poliner R A and Benson J 2013 *Teaching the Whole Teen: Everyday Practices that Promote Success and Resilience in School and Life* (New Delhi: SAGE Publication)
[15] Miles M, Huberman M, and J Saldaña 2013 Qualitative Data Analysis: A Methods Sourcebook: An Expanded Sourcebook p 408 [Online]
[16] Van Meter Donald S and Van Horn Carl E 1975 The Policy Implementation Process: A Concentual Framework in *Adm. Soc* **vol 6** no 4 pp 445–485
[17] Edward III George C 1980 *Implementing Public Policy* (Washington DC: Congressional Quarterly Press)
[18] Keban Yeremias T 2008 *Enam Dimensi Strategis Administrsai Publik Konsep, Teori dan Isu* (Yogyakarta: Gava Media)
[19] Bahtiar 2009 Hubungan antara Kecerdasan Emosional dengan Prestasi Belajar Siswa Kelas II SMA Negeri 2 Mataram: Mataram *J. Pemikir. Altern. kependidikan Insa* **vol 14** no 2 pp 254–268
[20] I Ali 1996 *Kebijakan pendidikan di Indonesia: Proses produk dan masa depannya* (Jakarta:

IOP Conf. Series: Earth and Environmental Science **989** (2022) 012017 doi:10.1088/1755-1315/989/1/012017

Bumi Aksara)
[21] Siagian S P 1995 *Teori Pengembangan Organisasi* (Jakarta: Sinar Grafika Offset)
[22] Dunn William N 2017 *Analisis Kebijakan Publik* (Yogyakarta: Gadjah Mada University Press)
[23] Tjilen Alexander Phuk 2019 *Konsep, Teori, dan Teknik Analisis Implementasi Kebijakan Publik* (Yogyakarta: Penerbit Hikam Media Utama)

IOP Conf. Series: Earth and Environmental Science **989** (2022) 012018 doi:10.1088/1755-1315/989/1/012018

Assasment of flood hazard reduction in DKI Jakarta: Bendungan Hilir Village

S R Taufik[1] , M Yatrib[1], A N Harman[1], T N A Kesuma[2], D Saputra[2], and M S B Kusuma[2]

[1] Water Resource Mangement Master Program, Faculty of Civil and Environmental Engineering, Institut Teknologi Bandung, Jawa Barat, Indonesia
[2] Water Resources Engineering Research Group, Institut Teknologi Bandung, Jawa Barat, Indonesia

*Email: taufikrfdh@gmail.com

Abstract. Jakarta has an essential function for Indonesia, apart from being the capital city of the country, Jakarta is also the center of the nation's economy. However, the problem of flooding that occurs every year causes enormous material and immaterial losses. One of the metropolitan areas that is often hit by floods is Bendungan Hilir Village. This village is the meeting point between the Ciliwung River and the Krukut River. One of the causes of flooding is when the water level in both rivers is high, the water from the Krukut River cannot flow, so it overflows into the surrounding area. For this reason, this study will analyze the level of flood hazard and the effect of normalization of the Krukut River. The analysis was performed using the HEC-RAS, combining 1D and 2D modeling. In the upstream part, the Krukut watershed flood hydrograph is applied for a return period of 25 years with a peak discharge of 169 m3/s, and the normal depth will be used as the downstream boundary. The modeling is carried out in two conditions, they are the existing channel and the channel after normalization. Based on the model results it is found that normalization could reduce the flood hazard of Krukut river around 75%.

1. Introduction

The Jakarta Flood Canal is a collector water channel which is an integral part of the Ciliwung River. The collector channel will accommodate overflowing water from the Krukut River and Cideng River which will then flow into the sea. This water system is the lifeblood of Jakarta's flood control. Kusuma et al [1], Formanek et al [2], Farid et al [3], conclude that as most of the flood management in the urban area of Indonesia, the flood management in Ciliwung River facing erosion, sedimentation, and land-use change problems. The land-use change in the upper part of the Ciliwung River tends to increase the rate of not only its extreme discharge but also its erosion and sedimentation. Several previous studies demonstrate very well the correlation of land-use change to the increasing extreme discharge as is shown by Papagiannaki [4] and Kuntoro et al [5]. Febi et al [6] discussed how to enhance the resiliency of the community of Jakarta to reduce the risk generated by the influence of climate change on increasing the extreme discharge of the Ciliwung River with flash floods occurring more frequently than in the past [18-21]. Meanwhile, lack of field data, of rainfall and discharge distribution, remains the main concern to get eligible results of climate change influence to the extreme discharge of Ciliwung as is indicated by Farahnaz et al [7]. Based on neural network methods Gunawan et al. [8] demonstrate the application mathematical model for river sediment load estimation, however, this model requires a good record data measurement that is not applicable for Ciliwung River data.

Suprayogi et al [9] conclude that land subsidence is one of the most important parameters that should be addressed in decreasing flood risk in Jakarta where freshwater availability for the local people is the key solution. Furthermore, several reservoirs are developed in the upper part of the Ciliwung river and one estuary dam is proposed in Jakarta Bay. The upstream reservoir is developed not only to increase the freshwater availability during the dry season but also to reduce the extreme discharge during the wet/rainy season as indicated by Kusuma et al [10]. However, Sengara et al [11] concluded that as an archipelago country located in a very active tectonic plate and volcano, Indonesia has a very high probability of earthquake hazard that will generate a high risk of dam failure of any reservoir in its surrounding area. This dam break will generate collateral hazards to its downstream area that usually become a favor urban area development. The sudden release of a large water mass of the reservoir might cause massive destructions either in life or property particularly in that urban area as indicated by Kusuma et al [10], Haltas et al [12], Kusuma et al [13], and Kusuma et al [14]. Sarminingsih et al discussed the influence of the increasing flood depth on the increasing flood disaster risk [15].

Based on the above discussion and regarding the current condition of the river as water system (physical characteristic, data availability, potential previous research results, and hydraulic capacity), this paper discuss the contribution of river normalization as one of the most probable short-term effort to decrease the flood risk in the study area. The analysis is supported by the mathematical model using HEC-RAS 5.0.7.

2. Study Area

The Krukut River with a length of 40 km and an average river slope of 30% is one of the rivers that empties into the Ciliwung River. The Krukut watershed has an area of 89 km2 covering Situ Citayam, Bogor, Depok, Jagakarsa, Cilandak, Pasar Minggu, Kemang, Mampang Prapatan, Gatot Subroto, Setiabudi, Tanah Abang, Glodok Chinatown, branching under the Pancoran Bridge, until the end in the West Flood Canal (integrated with the Ciliwung River). The scope of this research is limited to the location on the Krukut river section from the Bendungan Hilir Village intersection to the downstream of the river which empties into the West Flood Canal. The location can be seen in Figure 1.

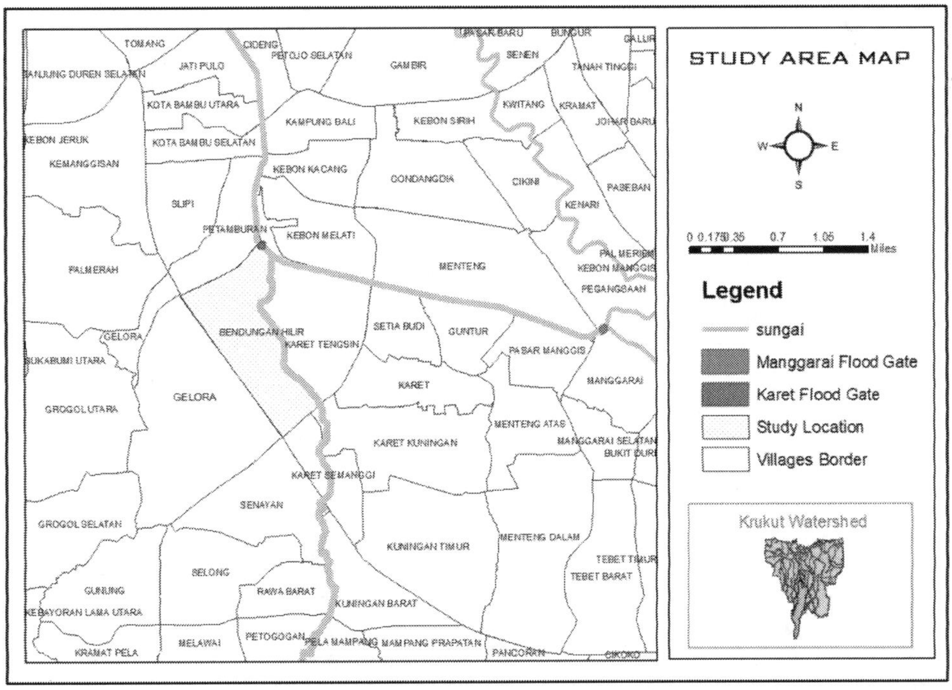

Figure 1. Study Area in Krukut River

3. Methods

3.1. Hydrology Analysis
Hydrological analysis is carried out to determine the planned flood hydrograph using a statistical approach according to the applicable reference standards. The results of the flood hydrograph analysis will be used in flood modeling as boundary conditions in the upstream. The requires data such as maximum daily rainfall data for a period of 35 years from 1985 - 2019, DEM data, and soil data. The first step is to test the consistency of the data and data outliers, then determine an average rainfall. This analysis is performed to a watershed affected by more than 1 rain station using the Thiessen polygon method that is relatively simple yet accurate.

Table 1. Thiessen Area for each Raingage.

Raingage	Thiessen Area (km2)	Percentage (%)
Kemayoran	2.69	30
Tanggerang Selatan	14.74	17
Bogor	0.00	00
Kampus UI	53.52	60
Cawang	18.13	20
Total	89.08	100

Furthermore, rainfall frequency analysis was carried out using Normal, Log-Normal, Pearson III, Pearson III, and Gumbel methods. After that, the distribution was tested by Chi-Square and Kolmogorov Smirnov method, the result was rain with Gumbel distribution was the best. According to the Technical Guidelines from the Ministry of Public Works and Public Housing of Indonesia, the return period rain should be divided into 6 hours using the PSA 007 method, and the amount of effective rain is determined which is equal to the difference between the falling rain minus the infiltration that occurs. the amount of infiltration was calculated using the SCS-CN method. In this study, the synthetic unit hydrograph calculated using Nakayasu unit hydrograph technique that can be seen in Figure 2.

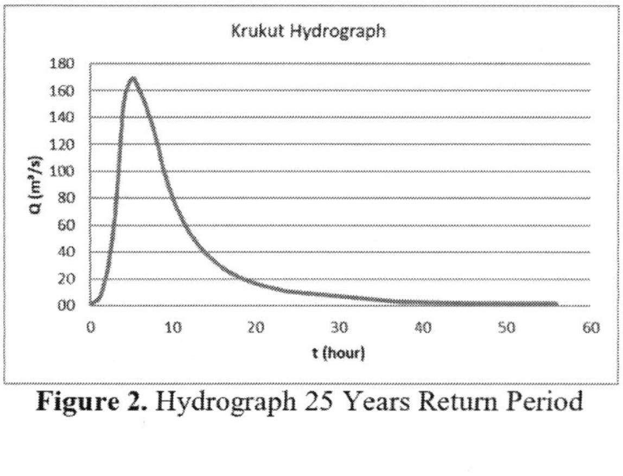

Figure 2. Hydrograph 25 Years Return Period

3.2. Existing Hydraulic Model
Hydraulic analysis with a combination of 1D and 2D techniques is modeled to determine the area and height of flood inundation at the study site. The modeling was done using Hec-RAS 5.0.7 software. The boundary conditions are set in the Unsteady Flow Data menu. For upstream boundary conditions, a Flow

Hydrograph is applied by entering hydrograph data. Meanwhile, in the lower reaches of the Krukut River, the normal depth is applied. In this model, calibration is carried out first to produce a more accurate analysis.

Calibration is done by comparing the location, total area and the height of the inundation in the flood events that occur according to the disaster map with the results from the modeling. Comparisons were made using the flood incident on February 25, 2020, the results showed that the floods that occurred had similar inundation areas and heights. So the modeling parameters that can be used to model floods with a return period of 25 years are the modeling time is 1 second, the normal depth is 0.01 and the Manning value is 0.03. The model calibration can be seen in Figure 3.

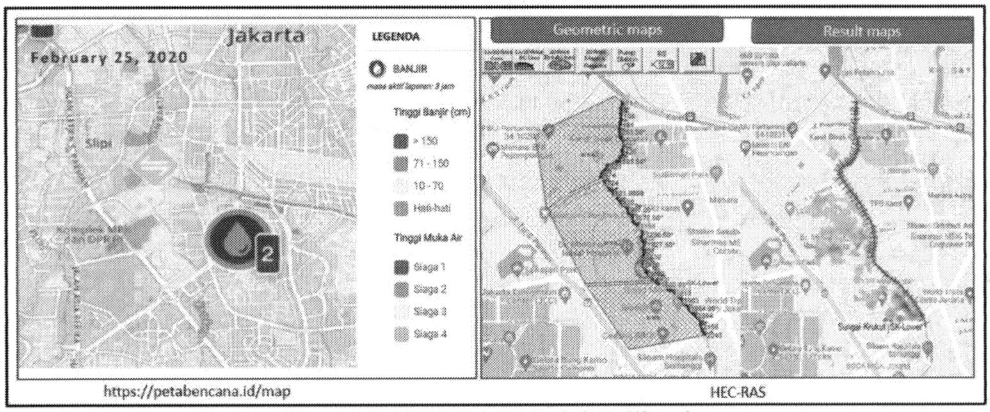

Figure 3. HEC-RAS Model Calibration

3.3. Normalized Hydraulic Model

Normalization in the river increases the capacity of the river by dredging the riverbed with a typical width of 15 to 20 meters. Normalization activities usually include river reinforcement structures, such as piles and boxes, to protect against scouring. Normalization implemented along 1.5 kilometer. The following is an example of normalization in the river section (RS) 2109 (Figure 4).

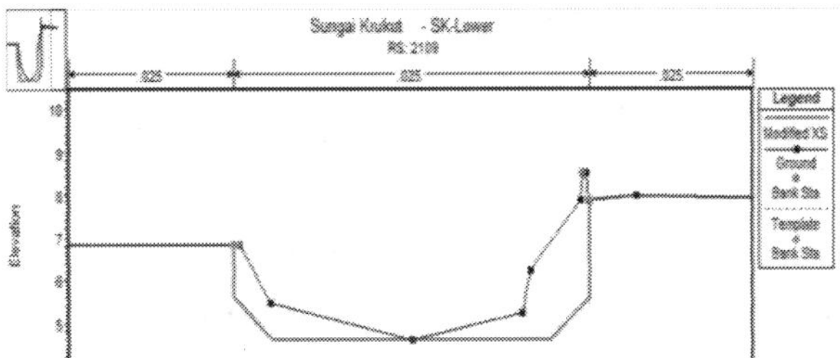

Figure 4. HEC-RAS Model Calibration

3.4. Hazard Map

Hazard level is the frequency (probability) of certain disasters that tend to occur with a certain intensity at a certain location. Hazard maps define areas where certain natural events occur with a certain frequency and intensity, depending on the vulnerability and capacity of the area, which can cause disasters. In the analysis of flood hazard level, the parameter used according to the regulations of the National Disaster Management Agency is the inundation height. Based on the provisions, the inundation height is classified into low, medium, high (Figure 5). Furthermore, the level of danger is generalized to the administrative boundaries of the Neighbourhood.

IOP Conf. Series: Earth and Environmental Science **989** (2022) 012018 doi:10.1088/1755-1315/989/1/012018

Depth (m)	Class	Score
< 0.76	Low	1
0.76 - 1.5	Medium	2
> 1.5	High	3

Figure 5. Hazard Level Index

4. Result and Discussion

The results of 1D and 2D modeling of flood events with a return period of 25 years on river condition before normalization shows that the flood inundation is quite extensive and the flood depth is high. Meanwhile, the area of the inundation was significantly reduced after the normalization was implemented (Figure 6), although the flood did not completely disappear. this happens because dredging is able to add the cross-sectional area of the river and effectively increase the river's storage capacity.

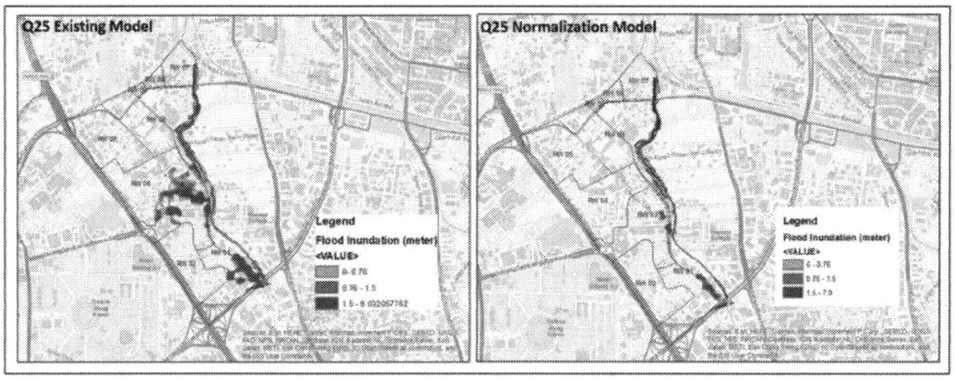

Figure 6. Modeling Result for Existing and Normalization Scenario

The inundation map is then converted to a hazard map using the administrative boundaries and the provisions of the National Disaster Management Agency. The results of the existing condition shows that RW 01, 03, and 06 have a high level of hazard. This can be interpreted that the level of disaster risk and losses that will be experienced is also high in the area, this can be also an indicator of which areas need to be treated first related to flooding. Meanwhile, RW 02, 04, and 07 are medium and RW 05, 08, and 09 are low. After the river was normalized, RW 02, 04, 07 decreased to a low level because there was no flood in the area. The hazard map can be seen in Figure 7.

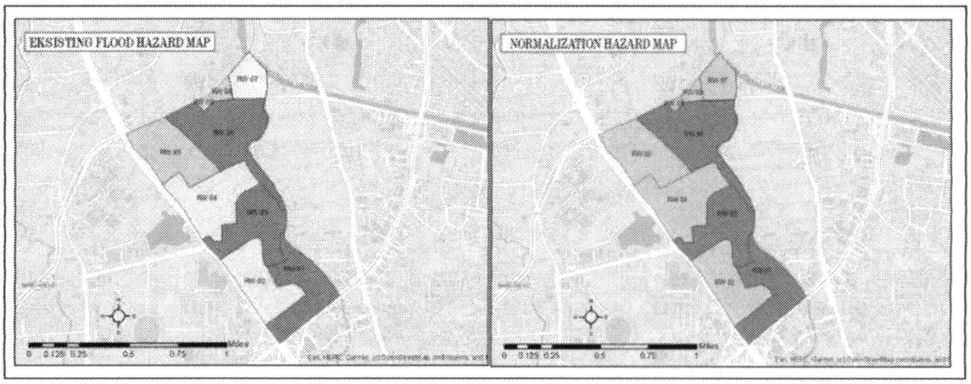

Figure 7. Hazard Map for Existing and Normalization Scenario

5. Conclusion

Based on the simulation results, normalization of river channels can increase channel capacity and could reduce the flood hazard of Krukut river around 75% but cannot overcome backwater when the water level in the West Flood Canal (Ciliwung) is high. Therefore, it is necessary to reduce the load on the channel by pumping water from the Krukut River using a polder system as an alternative. This polder functions to pump excess water when the flood from the Krukut river cannot flow by gravity, while when the flood can flow naturally, the floodgate will regulate the water to be released and the water level upstream of the flood gate.

Acknowledgement

The authors acknowledge the support given by the Republic of Indonesian Ministry of Research and Technology through 2020 Applied Research and the support given by cooperation between Institut Teknologi Bandung and Huddersfield University through the United Kingdom's Natural Environment Research Council and Economic and Social Research Council. The author's thanks also to the Indonesian Agency for Meteorological, Climatological and Geophysics (BMKG), Authority for Ciliwung and Cisadane River Watersheds (BBWS CiliCis), and National Planning and Development Agency-Directorate (BAPPENAS) for providing datasets used in this study.

References

[1] Kusuma M S B, Rahayu H P, Farid M, Adityawan M B, Setiawati T and Silasari R 2010 Studi Pengembangan peta indeks resiko banjir pada kelurahan Bukit Duri Jakarta/ *Jurnal Teknik Sipil* **17** 2 123-134

[2] Formánek A, Silasari R, Kusuma M S B, Kardhana H 2013 Two-dimensional Model of Ciliwung River Flood in DKI Jakarta for Development of the Regional Flood Index Map *Journal of Engineering and Technological Sciences*, Vol. 45, Issue **3** 307-325

[3] Farid M, Pusparani H H, Kusuma M S B and Natasaputra S 2017 Study on effectiveness of flood control based on risk level: case study of Kampung Melayu Village and Bukit Duri village *MATEC Web Conf.* **101** 05003

[4] Papagiannaki K, Lagouvardos K, Kotroni V and Bezes A 2015 Flash flood occurrence and relation to the rainfall hazard in a highly urbanized area *Nat. Hazards Earth Syst. Sci.* **15** 1859–1871, www.nat-hazards-earth-syst-sci.net/15/1859/2015/, DOI: 10.5194/nhess-15- 1859-2015

[5] Kuntoro A A, Putro A W, Kusuma M S B, Natasaputra S 2017 The effect of land use change to maximum and minimum discharge in Cikapundung river basin *the 3rd International Conference on Construction and Building Engineering (ICONBUILD)* **1903** p 100011-1,100011-7, DOI: https://doi.org/10.1063/1.5011621

[6] Dwirahmadi F, Rutherford S, Phung D and Chu C 2018 Understanding the Operational Concept of a Flood-Resilient Urban Community in Jakarta, Indonesia, from the Perspectives of Disaster Risk Reduction, Climate Change Adaptation, and Development Agencies *International Journal of Environmental Research and Public Health*.

[7] Farahnaz N, Kuntoro A A, Kusuma M S B 2020 Statistical Downscaling for the Projection of the Keetch Byram Drought Index in the Barito Basin *Int J. on Advanced Science Engineering Information Technology* pp 873-879, Vol 10 (2010) 2. http://dx.doi.org/10.18517/ijaseit.10.2.10102

[8] Gunawan T A, Kusuma M S B, Cahyono M and Nugroho J 2017 The Application of Backpropagation Neural Network Method to Estimate the Sediment Loads *MATEC Web of Conferences* **101** 05016.

[9] Suprayogi H, Rudyanto A, Bachtiar A and Limantara L M 2020 Critical-phase Sea dike construction of NCICD program in Jakarta as national capital city *IOP Conf. Series: Earth and Environmental Science.* **162** 2018 012020. doi :10.1088/1755-1315/162/1/012020

[10] Kusuma M S B, Rahayu R A, and Cahyono Development Study of Turbulent κ-ε Model for Recirculation Flow III: Two Dimension Recirculation Flow in a Reservoir *ITB Journal of*

Engineering Science, Vol. 41, Issue B (1), 2009, pp. 1-16

[11] Sengara I W, Latief H, Kusuma M S B 2008 Probabilistic seismic and tsunami hazard analysis for design criteria and disaster mitigation in rehabilitation and reconstruction of a coastal area in city of Banda Aceh *Geotechnical Engineering for Disaster Mitigation and Rehabilitation - Proceedings of the 2nd International Conference GEDMAR08* pp. 224–230

[12] Haltas I, Tayfur R and Elci S 2016 Two-dimensional Numerical Modeling of Flood Wave Propagation in an Urban Area due to Urkmez Dam-break, Izmir, Turkey *Natural Hazards.* **81** 2103-2119.

[13] Kusuma M S B, Setiawati T and Farid M 2019 Experimental Model of Dam Break Flow around several Blockages Configurations *International Journal of GEOMATE.* **16** Issue 58 26-32

[14] Kusuma M S B, Rahayu R A and Cahyono 2009 Development Study of Turbulent κ-ε Model for Recirculation Flow III: Two Dimension Recirculation Flow in a Reservoir *ITB Journal of Engineering Science.* **41**, Issue B (1) 1-16

[15] Sarminingsih A, Soekarno I, Hadihardaja I K, Kusuma M S B 2014 Flood Vulnerability Assessment of Upper Citarum River Basin, West Java, Indonesia *International Journal of Applied Engineering Research* **9** Issue 23 pp 22921-22940

[16] Kandilioti G and Makropoulos C 2012 Preliminary flood risk assessment: the case of Athens *Nat. Hazards.* **61** 441–468, doi:10.1007/s11069-011-9930-5

[17] Tsakiris G 2014 Flood risk assessment: concepts, modelling, applications *Nat. Hazards Earth Syst. Sci.* **14** 1361–1369, doi:10.5194/nhess-14-1361-2014

[18] Barrera-Escoda A and Llasat M C 2015 Evolving flood patterns in a Mediterranean region (1301–2012) and climatic factors – the case of Catalonia *Hydrol. Earth Syst. Sci.* **19** 465–483, doi:10.5194/hess-19-465-2015

[19] Ruin I, Lutoff C, Boudevillain B, Creutin J-D, Anquetin S, Rojo M B, Boissier L, Bonnifait L, Borga M, Colbeau-Justin L, Creton-Cazanave L, Delrieu G, Douvinet J, Gaume E, Gruntfest E, Naulin J-P, Payrastre O and Vannier O 2014 Social and Hydrological Responses to Extreme Precipitations: An Interdisciplinary Strategy for Postflood Investigation *Weather Clim. Soc.* **6** 135-153, DOI: https://doi.org/10.1175/WCASD-13-00009.1, 2014

[20] Papagiannaki K, Kotroni V, Lagouvardos K, Ruin I and Bezes A 2017 Urban area response to flash flood–triggering rainfall, featuring human behavioral factors: the case of 22 October 2015 in Attica *Greece Weather Clim. Soc.* **9** 621-638, DOI: https://doi.org/10.1175/wcas-d-16-0068.1

[21] Llasat M C, López L, Barnolas M, and Llasat-Botija M 2008 Flash floods in Catalonia: the social perception in a context of changing vulnerability *Adv. Geosci.* **17** 63–70, http://www.adv-geosci.net/17/63/2008/.

[22] Natakusumah D K, Hatmoko W, and Harlan D 2011 Prosedure Umum Perhitungan Hidrograf Satuan Sintetis (HSS) dan Contoh Penerapannya Dalam Pengembangan HSS ITB-1 dan HSS ITB-2, *Journal Teknik Sipil ITB, Vol. 18 No. 3*

IOP Conf. Series: Earth and Environmental Science **989** (2022) 012019 doi:10.1088/1755-1315/989/1/012019

The analysis of coastline changes of maruni beach in Manokwari during 1995 – 2021 period

Maryrose E. Tapilatu[1,2,*], Rima H. S. Siburian[1,3] and Ricardo F. Tapilatu[1,4]

[1]OurConservaSea Community Group, Manokwari 98314 – Papua Barat Indonesia
[2]Marine Science Graduate Program, University of Diponegoro, Semarang 50275 – Indonesia
[3]Faculty of Forestry, University of Papua, Manokwari 98314 – Papua Barat Indonesia
[4]Faculty of Fisheries and Marine Science, University of Papua (UNIPA), Manokwari 98314 – Papua Barat Indonesia

*Email: marytapilatu@gmail.com

Abstract. The coastal area of Maruni at Manokwari Regency is known for its potential as a fish-producing area, industry, settlements, ports, and others. However, some parts of Maruni coastal areas have been damaged due to the development. The coastal area experienced changes caused by both natural processes and human activities. The presence of natural and artificial processes in coastal areas results in accretion and abrasion. The dynamics of oceanographic pressures such as sea tidal, sea waves, ocean currents, and tides occurring in the Maruni Beach area cause changes in the movement of sediments affecting the coastal area. Continuous changes can be a threat to Maruni coastal line as the Trans Papua Barat Road passed this route. This research was conducted with the aim to quantify the rate of coastline changes and oceanographic dynamics of Maruni coastal line at the south part of Manokwari Regency, Papua Barat Province from 1995 to 2021. The utilization of remote sensing interpretation method from Landsat 5 TM, Landsat 8 OLI, and oceanographic data providers (BIG, ECMWF) accommodated data source. The analysis result concluded that changes in the coastline at Maruni Coastal line during the 1995-2021 period predominantly encountered abrasion. The magnitude rate of change experienced a very small accretion, but was replaced by a very high abrasion event that occurred during two observation periods between 1995-2021.

1. Introduction

The coastal area has been a pivotal area supporting multiple economic activities in the Manokwari Regency area. The coastal area of Manokwari Regency is known for its significant role as a fish-producing area, industry, settlements, and ports. However, some coastal areas have been damaged due to the lack and even absence of appropriate development plans considering the nature of protection and preservation in coastal areas. Coastal area experienced significant changes caused by both natural processes and human activities [7]. The presence of natural and artificial processes in coastal areas results in accretion and abrasion. The dynamics of oceanographic pressures such as ebb and flow, ocean currents, and tides occurring in the Maruni Beach area cause dynamic changes in the movement of sediments that support this area. It is known that a large sediment transfer rate is able to carry large amounts of material every time.

Human activities overlapped in order to fulfill daily needs on Maruni Beach, especially construction, sand mining, and settlements that have been lasting while ago would also change the balance of the Maruni Beach area and may cause damage. This phenomenon causes a decrease in the condition of the coastal area leading to a decrease in the function as protection for another ecosystem nearby such as lagoon located next to it. Unprevented changes would immediately have an impact on the lagoon ecosystem located in the Maruni area and the ecosystems contained therein.

The dynamic process occurring at Maruni Beach continuously needs to be observed as a provision of data for the government in carrying out regulations related to the development taking place on Maruni Beach. Coastline mapping would be conducted by direct field measurements, analysis of aerial images and analysis of remote sensing and satellite imagery [3,4]. The use of satellite image data would be a way of observing and analyzing shoreline changes. Baskoro observed the coastline changes of Paciran District using the Landsat image overlap technique which was then analyzed using the Digital Shoreline Analysis System (DSAS) method [1]. Marques and Khakim used Landsat TM in 1994, 2004 and 2014 to identify the dynamics of the coastal coastline of Semarang City [6].

Based on the situation taking place in the coastal area of Maruni District, it is necessary to conduct research to quantify changes in the coastline. This study utilizes Landsat satellite imagery data from 1995 – 2021 to identify the changes. This research is expected to provide information in the consideration of the formulation of regulations for the development and maintenance of coastal areas along the coastline of Maruni District, Manokwari Regency.

2. Methods

2.1 Time and Location

The research was conducted by collecting field data along the coast of Maruni Village in Maruni coastal line in January 2021. Data processing and the process of working on the coastline change model was carried out in May 2021. The research location is on the coast of Maruni, South Manokwari District, Manokwari Regency, West Papua Province. This village is located at coordinates 0°58'38"S and 134°01'09"E (Figure 1).

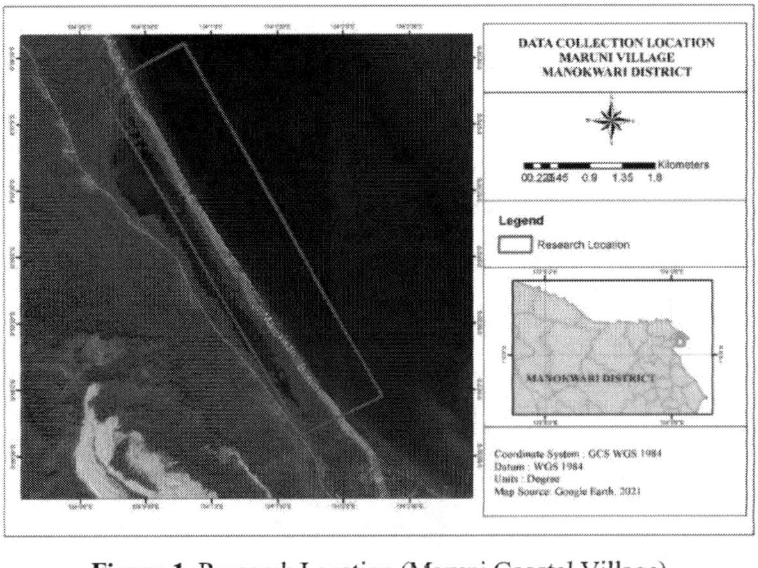

Figure 1. Research Location (Maruni Coastal Village)

2.2 Data Collection
Field data collection included coastal slope. wind direction and speed, wave height, and tide data. Oceanographic data taken to support this research are in the form of current velocity, wave height, and tidal data formats. Current speed and wave height data were obtained through the ECMWF website (https://www.ecmwf.int/). While tidal data were obtained from the provider of tidal data in the form of the Indonesia Geospatial Information Agency (BIG). The observation data were used as a validation of tidal data on the MIKE 21 Model.

2.3 Data Analysis
Coastline changes were analyzed by using ENVI 5.2 and ArcGIS 10.4 software to digitize Landsat 5 TM and Landsat 8 OLI image data in order to generate the required data and information. The calculation of shoreline changes in Maruni Village uses the DSAS application with the Net Shoreline Movement (NSM) method which measures the distance of shoreline changes based on a transect that cuts 2 coastlines and the End Point Rate (EPR) method which calculates the rate of shoreline change by dividing the distance between the coastlines. beach with time [8].

3. Results and Discussion

3.1 Shoreline Changes of Maruni Beach
Observing the situation, the Maruni Village area is a lowland with an altitude between 0-100 meters above sea level. The village is located on the crossroads between Manokwari Regency and South Manokwari Regency with such conditions supporting a variety of livelihoods for local residents. Soil types in Manokwari Regency consist of alluvial soil (18.70%), Mediterranean soil (2.44%), red-yellow podzolic soil (10.41%), gray-brown podzolic (7.57%), ultisol soil (49 .21%), latosol soil (4.49%), and organosol soil (7.17) [6].

The coast of Maruni Village consists of two beach characteristics: sandy and rocky beaches, and overall is dominated by sandy beaches. The basic geomorphology of the coastal area is influenced by wave energy originating from the Pacific Ocean considering the Manokwari Regency area gets the effect due to its open position facing the Pacific Ocean .

The results of the DSAS process to detect the average distance of shoreline change, the average rate of shoreline change, and determine the area with the most significant changes at Maruni beach (Figure 2).

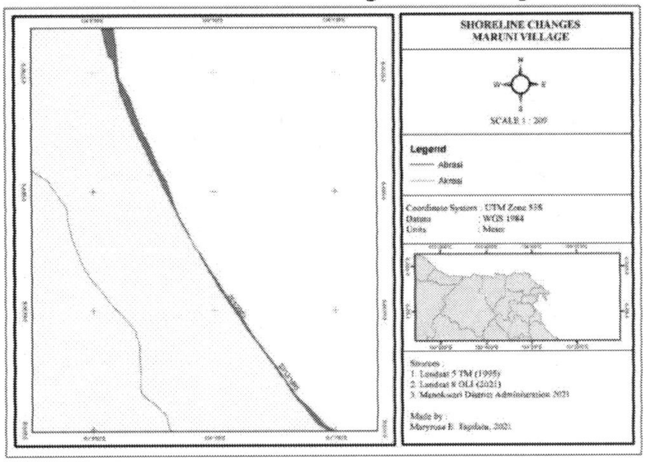

Figure 2. Shoreline Changes of Maruni Village

IOP Conf. Series: Earth and Environmental Science **989** (2022) 012019 doi:10.1088/1755-1315/989/1/012019

The results of the interpretation of Landsat images in 1995 and 2021 showed significant changes in the coastline area in the Coastal Area of Maruni Village. There has been a significant change in terms of land reduction (abrasion) which has a visible change from this research (Figure 4). The deforestation resulted in a significant reduction. Without the mangrove forest to withstand the incoming waves, it will cause damage to the barrier wall (Figure 3) which had been previously built and the possibility of a cause of road damage between these districts.

Figure 3. Abrasion Cause Damage to Barrier

The calculation of shoreline changes in Maruni Village uses the DSAS application with the Net Shoreline Movement (NSM) method which measures the distance of shoreline changes based on a transect that intersects 2 coastlines. Fauzie observed changes in two coastlines in 1998 and 2015 and found the abrasion rate in the Karawang Regency area of 15ha/year [2]. Maruni Beach since 1995 - 2021 has experienced an abrasion distance of -49.80 m and an accretion of 7.11 m. End Point Rate (EPR) method which calculates the rate of shoreline change by dividing the distance between the shoreline by the time. In 1995 - 2021 the abrasion rate that occurred was -1.18 m/year while the accretion rate that occurred was 0.28 m/year. The results of the interpretation of Landsat images in 1995 and 2021 which have been digitized, show that changes in the coastline area in the Coastal Area of Maruni Village there are significant changes in terms of land reduction (abrasion) that is visible from the research area as shown in Figure 2.

Changes shown in Figure 4 can occur both by anthropogenic factors and oceanographic dynamics. Human activities such as logging of mangrove forests for road construction, building houses for residents, and so on. Logging results in significant reductions. Without the mangrove forest to withstand the incoming waves, it will cause damage to the barrier wall that has been built and the possibility of a causeway between these districts.

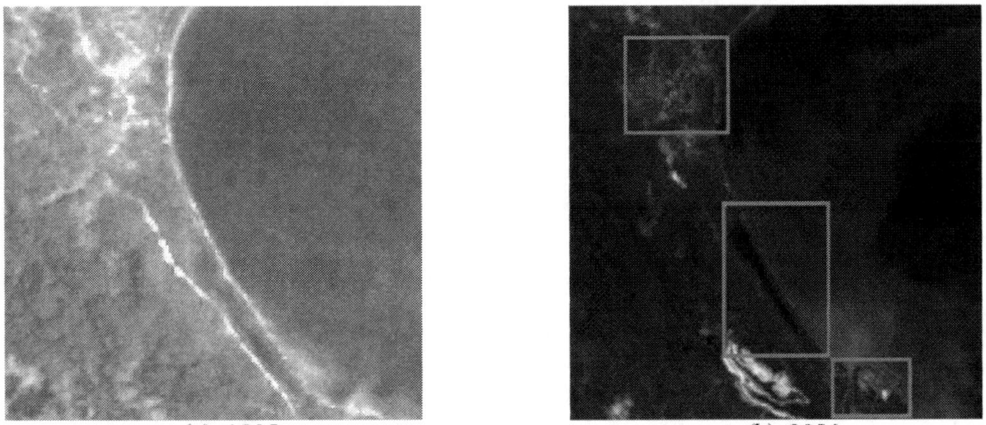

(a) 1995 (b) 2021

Figure 4. Difference Condition Between 1995 - 2021

3.2 Oceanographic Dynamic of Maruni Beach

Based on oceanographic changes in wind direction and speed that occurred in the period of December and January at Maruni Beach, namely the West monsoon (Asian Monsoon) where the wind blows from the west to the east. The duration of the gust of wind also affects the height of the resulting wave, the longer the duration of the gust of wind, the higher the wave formed. The wave height in the Maruni Beach area is quite high due to the large wave formation area. Based on Table 1, it can be seen that the significant wave value (Hs) was found in 1995, which was as high as 0.89 m. If viewed based on its class on the Beaufort Scale, the significant wave height at Maruni Beach belongs to class 4 where the waves are generated by weak winds which are indicated by the appearance of waves starting to break in this area. But in 2021 the value reaches 1.48 m, which means it is raised by moderate wind gusts. The results of the calculation of the significant wave period also experienced a slight change from 1995 to 2021.

Table 1. Characteristics of Maruni Beach Sea Waves

No	Year	Hmax (m)	Hmin (m)	Hs (m)	Tmax (s)	Tmin (s)	Ts (s)
1	1995	0.89	0.41	0.62	10.61	8.15	8.88
2	2021	1.48	0.71	1.03	10.21	7.55	8.80

Source : Data Analysis, 2021

The waters at Maruni Beach have a double daily type with M2 (the main component of lunar semi-diurnal) as the main component in the distribution of tides due to its largest value compared to the values of other harmonic components during the observation period. This tide type as same as in the Jeen Womom Turtle Nesting Coastal Park area with type of tide (semi-diurnal) as indicated by the type of tide in the Pacific Ocean [9]. This tidal cycle consists of two times the tide conditions and two times the tide conditions showed a semidiurnal tidal pattern in the meridional (north-south) direction. Based on the value of the harmonic components listed in Table 2, it was found that the MSL value in 1995 was 2.16 m while in 2021 it was greater with a value of 2.35 m. Due to sea level rise causing difference each year, people need to minimize the impact same as people in Lemon Island [10].

IOP Conf. Series: Earth and Environmental Science **989** (2022) 012019 doi:10.1088/1755-1315/989/1/012019

Table 2. Maruni Beach Harmonic Components

No.	Year		Harmonic Components										MSL
			S_0	M_2	S_2	N_2	K_1	O_1	M_4	MS_4	K_2	P_1	
1.	1995	A	2.16	0.50	0.19	0.1	0.21	0.15	0	0	0.05	0.07	2,16
		g°	0	230	320	177	72	357	91	183	320	72	
2.	2021	A	1.95	0.49	0.19	0.09	0.21	0.14	0.00	0.00	0.05	0.07	2,35
		g°	0.00	243	321	192	74	12	181	130	321	74	

Source : Data Analysis, 2021

According to the analysis of the direction and speed of the current, it shows that the current is moving from the West to the East. The dominant movement of water masses towards the East and West-Southeast is also indicated by the blue clockwise-shaped line in Figure 3. This line is a symbol of the distribution of the velocity and direction of the currents that dominate the Maruni Coastal area.

The results of data analysis of the speed of tidal currents moving in 1995 were dominated by the speed class 0.5 - 0.9 m/s at 4.03% towards the southeast, while in 2021 the dominant current speed was in the class >2.4 m/s. by 3.49% moved to the East.

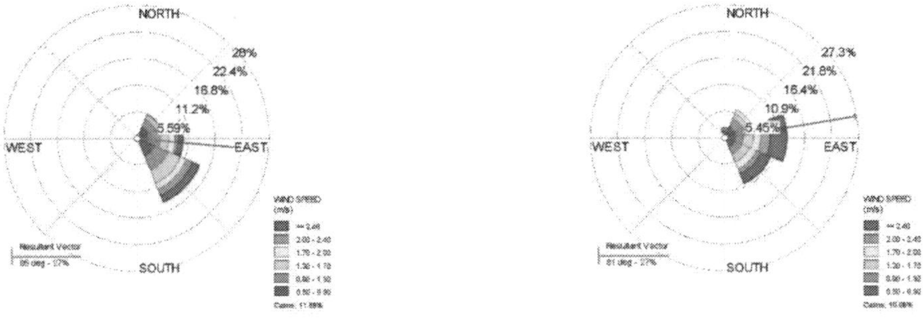

(a) 1995 (b) 2021

Figure 5. Sea Currents's Velocity and Direction

4. Conclusion

The results of the analysis carried out in this study concluded that changes in the coastline at Maruni Beach during the period 1995 - 2021 were dominantly experiencing abrasion events.

Significant wave heights and sea level values have increased with a significant period that is getting faster giving different impacts in 1995 and 2021. The values of the large group of current velocity in 1995 and 2021 also have an impact on the Maruni Beach area. In addition to natural factors, the increasing human activity in the Maruni Beach area has also brought major changes.

References

[1] Baskoro N C Joesidawati M I dan Sukma R N 2018 Perubahan Garis Pantai Kecamatan Paciran Kabupaten Lamongan, Menggunakan Citra Landsat Dengan Metode Digital Shoreline Analysis System (DSAS) Prosiding SNasPPM 3(1) 200-208

IOP Conf. Series: Earth and Environmental Science **989** (2022) 012019 doi:10.1088/1755-1315/989/1/012019

[2] Fauzie A K 2017 Analysis of Short and Medium Term Coastal Abrasion and Accretion Rates Using GIS in Karawang-West Java CR Journal Vol 03 No 02 Desember 2017 91 - 104.
[3] Guariglia A Buonamassa A Losurdo A Saladino R Trivigno M L Zaccagnino A and Colangelo A 2006. A Multisource Approach for Coastline Mapping and Identification of Shoreline Changes Annuals of Geophysics 49(1) 295-304
[4] Kasim F 2012 Pendekatan Beberapa Metode dalam Monitoring Perubahan Garis Pantai Menggunakan Dataset Penginderaan Jauh Landsat dan SIG Jurnal Ilmiah Agropolitan 5(1) 620-623
[5] KLHK 2014 Surat Keputusan Menteri Kehutanan dan Lingkungan Hidup Republik Indonesia. Nomor 783/II/2014 tentang Penetapan Kawasan Hutan dan Perairan di Papua Barat Jakarta Indonesia.
[6] Marques J N and Khakhim N 2016 Kajian Perubahan Garis Pantai Menggunakan Citra Landsat Multitemporal di Kota Semarang Jurnal Bumi Indonesia 5(2)
[7] Niya A K and Alesheikh A A 2013 Shoreline Change Mapping Using Remote Sensing and GIS International Journal of Remote Sensing Applications 102-107
[8] Suhana M P 2016 Analisis perubahan garis pantai di Pantai Timur Pulau Bintan Provinsi Kepulauan Riau (Doctoral dissertation, Bogor Agricultral University (IPB)
[9] Tapilatu R F Bonka A N Wona H Tapilatu M E Rumbiak R Ampnir T Siburian R H Wibbels T and Grady K C 2021 The Dynamic of Shoreline Change Affect Sea Turtle Nesting Sites at North Coast of Bird's Head – Papua Indonesian Journal of Science & Technology 6 (1) (In Press)
[10] Siburian R S Faidiban and Tapilatu R F 2021 The Climate Change Resilience of Lemon Island Community West Papua Indonesia Eco Env & Cons 27 2021 pp (S297-S302) ISSN 0971–765X

Protection of the elderly in disaster situations in Indonesia

Tukino*
Bandung Polytechnic of Social Welfare`
*Email: tukinoktn@gmail.com

Abstract. The territory of Indonesia is an area that is prone to disasters, both natural, non-natural and social disasters. Thus, the elderly in Indonesia live in a situation of threat of disaster. This study uses a secondary data research method with locus is focused on areas of Indonesia where disasters have occurred in the last five years from 2016 to 2020, namely; earthquake in Pidie Jaya, earthquake and tsunami in Lombok, earthquake and tsunami in Palu, social disaster in Mempawah-West Kalimantan, flash floods in Garut, and the Covid-19 disaster. The data collected was then analyzed using document analysis techniques.
Our analysis result concluded that the absence of caregiver who specifically provide programmed assistance, makes the elderly feel lonely, stressed and traumatized. Social workers contribute to the handling of the elderly in disaster locations focused on psychosocial support, and advocacy of legislative policies.

1. Introduction

Population aging is one of the issues faced by many countries in the world today. In 2015, the number of elderly people in Indonesia reached 21.6 million. This number will increase to 28.7 million, 49.6 million, and 61.4 million in 2020, 2035, and 2045 respectively (BPS: 2015-2025). The geographic location of the Indonesian territory which is between the Eurasia-Australia Plate, with the plate rate is 52 mm /year indicating the area with the plate; fractures, folds, and ridges. As a result, Indonesia is prone to earthquakes, tsunami, volcanic eruptions and landslides.

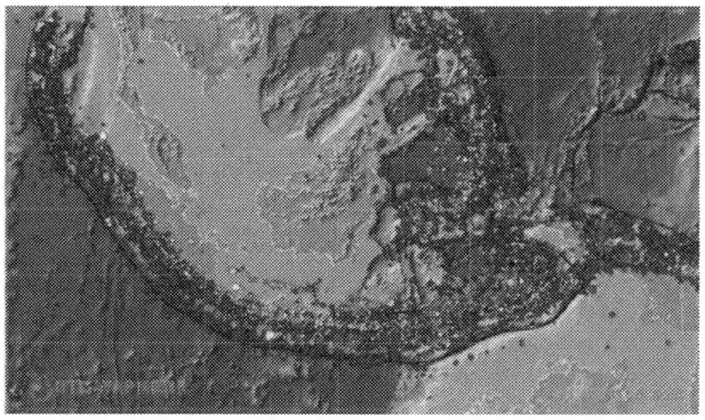

Figure 1. map: Indonesia is prone to disaster

The earthquake and tsunami that occurred at the end of 2004 in Nanggroe Aceh Darussalam, the eruption of Mount Merapi in Central Java (2010), the earthquake in Pidie Jaya Aceh (2016), the earthquake which was accompanied by the tsunami in Lombok (2017), The earthquake and tsunami disaster in Palu (2018) was a national scale natural disaster that occurred in the territory of Indonesia.

The occurrence of various disasters in Indonesia resulted in casualties and property, including the elderly. The disaster victims who survived were forced to flee to a safe place. They occupy Internally displaced persons (IDPs) camps with limited clothing, assets and food. The opportunity to fulfill their basic needs is very limited. In addition, some elderly refugees also experienced traumatic conditions as a result of the disaster which was seen as a devastating event, namely with life and death, which may still leave a profound mark on the psyche. Table 1 below is an overview disasters in Indonesia over the last 5 years

Table 1. Number of disasters victims in Indonesia last 5 years

Year	Location and type of disaster	Number of Deaths	Number of survivors / IDPs
2016	earthquake in Pidie Jaya – Aceh	104	43.529
2016	flash floods in Garut district	41	6.361
2016	social conflict of Gafatar in West Kalimantan	-	1,124
2017	earthquake and tsunami in Lombok	436	352.793
2018	earthquake, tsunami and liquefaction in Palu –Central Sulawesi	2.073	82.775
2020	non-natural disasters Covid-19	2.138 (16,5%) october, 2020	

Source: results of study, 2020

In social work research, disasters are seen as a type of collective stress situation, in which many individuals fail to have their needs met through societal processes [15]. Disasters are distinguished from other types of collective stress because, first of all, disasters are crisis situations [15]. In the social work perspective, the environment is included among the physical, biological and social factors influencing the welfare of individuals, groups, and populations. Since the late nineteenth century social workers have intervened in the microenvironments of people to improve their health status, residential living environment, workplace conditions, and social and psychological functioning [15].

There is a lack of studies explaining assistance for elderly people in disaster situations in Indonesia. Therefore, this study will contribute to the development of literature so that the literature gap on disasters and elderly people will be small. In the topic of research on the protection of the elderly in disaster situations, social workers can have a significant role both in policy advocacy and in developing methods and techniques for psychosocial support so that the elderly in disaster situations can still carry out their social functions. The purpose of this research is to broaden knowledge about the protection of the elderly in disaster situations, and the role of social workers in providing assistance to the elderly, especially during their time of displacement. This study explores the contribution of social workers to assisting the elderly in obtaining fulfillment of basic needs which include food, clothing and shelter needs, health services, and psychosocial support. The main research question is how the role of social workers in protecting the elderly in disaster situations.

2. Literature Review

2.1. Elderly in a biopsychosocial-spiritual perspective

Old age is one of the main events in a person's life span. Age 65 years is generally the middle age between middle age and old age [10]. Gerontologists who specialize in medical care for the elderly, divide old age into 2 groups, namely middle old age (65-74 years) and old age (75 years and over). The biopsychosocialspiritual is the perspective most used and accepted in various professional humanitarian service activities, especially social work (Hooyman and Kiyak (1999) cited by Adi Fahrudin (2000) [1]. To understand humans as a whole; thoughts, behavior, feelings, hopes, aspirations, desires, needs, illnesses, disturbances, it is necessary to understand the human being in the context of the situation / environment in which he is in the past and present Hooyman and Kiyak (1999) cited by Adi Fahrudin (2000) [1].

The biological aging process is a physical change that reduces the efficiency of the human organ systems, such as the heart and the circulatory system. Some physical signs of old age are changes in physical form such as; slow response, loss of motor and sensory functioning, tendency to more rapid fatigue, decreased energy and some or all. This is sometimes combined with chronic or progressive diseases due to a trait of disability. The psychological aging process refers to changes in sensory processes, perceptual processes and mental functioning (such as memory, learning and intelligence), adaptive capacity, and personality. The behavior of older people is often compared to younger individuals. The difference that appears between the two groups is seen in the psychological aspects of the elderly in such aspects as their intellectual ability to learn, solve problems and be creative. Presumably, each age group has different events and different social situations. This suggests that intellectual acceptance is more related than the illness and disease of the elderly.

Like the biological and psychological aging processes, social aging process their social role and their relationships also diminish. Most people who lived in the last century through the fourth decade of life, and because their lives are rooted in an economic system that depends on the family as the unit of life, there is always a place of old age in family and community life, including as parents and grandparents, grand mother. However, with the changes taking place through organization and industrialization, there have been family breakdowns. As a production unit. With the increasing number of people living into the next decades of life, there has been a change in the age structure of a society.

Physiological, psychological and social changes contribute to changes in the religious dimension. Older people who can accept the essence of their aging consider old age an opportunity to fill with religious life. But not a few of them, especially the drastic physiological, psychological and social changes caused them to lose faith in their God. This also has implications for changes in aspects of their life expectancy. Their life motivation also changes. In the religious dimension, important factors that need to be understood by social workers are how the philosophy of life, peace in life, meaning of life, life goals, the spirit of life in the elderly and how strong the faith they show when facing trials in their lives [1].

2.2. Protection of the elderly in disaster situations

International conventions in the Vienna International Plan of Action on Aging discuss health and nutrition, protection of the elderly, housing and environment, family, social welfare, income and employment security, education, as well as research data collection and analysis. In the United Nations Principles for Older Persons (United Nations, 1998), there are important things for the elderly, which are related to independence, participation, treatment, self-fulfillment and dignity. Likewise with the Madrid International Plan of Action on Aging (United Nations, 2008) which emphasizes on improving the health and well-being of the elderly, and increasing the capacity and support of the environment.

Zakour argues that social work disaster research has contributed new findings on the uses and effectiveness of debriefings, disaster volunteers and service delivery, vulnerable populations in disasters, organizational and interorganizational behavior in disasters, environmental disasters, cross-cultural and international aspects of disaster response, and improved measurement and theory. The scope of these contributions reveals a generalist approach to disaster response, such that social work research and intervention focuses on systems of various sizes and levels of abstraction. Improved measurement and theory has resulted from research on volunteerism, organizations, interorganizational networks, communities, and vulnerable populations [15].

In an emergency situation due to a disaster, Law no. 24 of 2007 concerning disaster management, in particular Articles 48 and 55 have emphasized that in carrying out disaster management it is necessary to protect vulnerable groups such as children, persons with disabilities, and the elderly. Protection for the elderly in a disaster emergency situation includes:

a) Rescue and evacuation of elderly disaster victims to temporary shelters by emphasizing efforts to save and protect

b) Recovery of physical conditions in the form of providing food and side dishes, and clothing. Food assistance and physical care are urgently needed by elderly people who experience physical, mental, physical and / or mental wounds for ease of activity and efforts to improve their physical and psychological abilities,
c) Recovery of elderly psychological conditions, through counseling, psychosocial and trauma healing.

2.3. Social work with the elderly in disaster situations
The International Federation of Social Work (IFSW) defines social work as follows: "The social work profession promotes social change, problem solving in human relationships and the empowerment and liberation of people to enhance well-being. Utilising theories of human behaviour and social systems, social work intervenes at points where people interact with their environments. Principles of human rights and social justice are fundamental to social work."(IFSW, 2000). Therefore, Social work is a profession that provides help to people who have difficulty carrying out their life tasks. As the incidence of natural disasters particularly earthquakes, has increased in the last two decades, studies on disaster have began to emerge and expand (Seroka, C.M., Knapp, C., Knight, S., Siemon, C.R., and Starbuck, S., 1986; Dufka C.L., 1988; Banerjee & Gillespie, 1994; Webster, 1995; Dodds and Nuehring, 1996). These early studies mainly focused on the roles and functions of social workers in the rescue stage. After the earthquake in Iran, Mohammad Reza Iravani listed techniques social workers can apply in the disaster relief including situational support, creating hope, consoling, assuring, concentrating and solution developing.

Meanwhile, Barush, Amanda (2011:348) say that physical aging and social indifference increase the physical vulnerability of elders, they may be more resilient than young adults to the psychological effects of disaster. Following his extensive review of literature in the area, Ehren Ngo described the "protective effect of older age on psychological health in the event of disaster exposure ..." (p. 83), citing studies that found elderly disaster victims report less anxiety, show less stress, are less anxious, and are less likely to suffer from long-term mental health problems than younger victims (Ngo, 2001). Of course, it is also possible that older adults manifest their emotional reactions in ways that are less easy to detect. Fuerthermore, Amanda Barush (2011:349) say that social work practitioners also contribute to disaster response efforts to provide housing assistance or emergency food, as well as counseling and mental health care [3]. The latter may be less relevant to the needs of older adults, who may be inclined to cope independently with disasters' psychological sequelae. There is some evidence to suggest that in the aftermath of disaster, older adults are as willing as younger groups to receive tangible assistance, but less willing to access mental health services. This preference has been attributed to stigma associated with counseling, as well as concerns about loss of autonomy, both of which should be taken into account in disaster response (Huerta & Horton, 1978)

Wong HC Johnston (2014:9) say that in Asia and the Pacific Region, The Kobe earthquake on Jan 17, 1995 and the 921 earthquake in Taiwan that occurred in 1999 sparked a number of studies on the roles and functions social workers play in restoring survivors mental health [14]. The need to develop work manuals was recognized and the danger of inappropriate assessments of the mental states of the survivors by untrained volunteers was raised. Social work practitioners and educators have also tried to develop post disaster actions guidelines from rescue stages to reconstruction stage. The need of an action model to guide disaster social work, which was fully recognized by the social work profession after the 911 incident, was addressed by Albert R Roberts who outlined social work tasks and methods in great details in his article "The Assessment, Crisis intervention and Trauma Treatment: the Integrative ACT Intervention Model" (2002).

Table 2. The Albert Robert's ACT Model

A	Assessment
	- Assessment/Appraisal of Immediate Medical Needs, Threats to public safety and property damage - Triage Assessment, Crisis Assessment, Trauma Assessment and Biopsychosocial and Cultural Assessment
C	Crisis Intervention
	- Connecting to support groups, the Delivery of Disaster Relief and Social Services, and Critical Incident Stress Debriefing (Mitchell & Everly's CISD Model) Implemented - Crisis Intervention (Roberts' Seven-Stage Model) Implemented, Through Strengths Perspective and Coping Attempts Bolstered
T	- Traumatic Stress Reaction, Sequelue, Posttraumatic Stress Disorders (PTSD) - Ten Step Acute Trauma and Stress Management Protocol (Lerner & Shelton) - Trauma Treatment Plan and Recovery Strategies Implemented (Brief Treatment and Crisis Intervention/ 6:1, Spring 2002)

His ACT Model is consistent with the Zunin and Myers' Phases of Disaster which is widely adopted in the mental health profession. Unfortunately Robert has left two key questions unanswered. Both Roberts and Myers are concerned about how to help the victims and survivors at the Impact stage and the Disillusionment stage. However, what differences are there between the roles of social workers and mental health professionals? Another question concerns the reconstruction phase. Roberts has only focused on treatment. What roles are there for social workers to play then in the redevelopment of the disaster sites and the prevention of future disasters?

3. Methods

The research was conducted using secondary data research methods. Heaton which is followed by Andrews, et.al., (2012:12) states that: "secondary data research method is a research strategy that utilizes existing quantitative or qualitative data to find new problems or test the results of previous research". In this study, the researcher aims to collect a number of data from various parties regarding disaster events in Indonesia in the last 5 years from 2016 to 2020 and victims affected by disasters, including the elderly, for analysis so that conclusions. As for the secondary data research steps are as follows; determine the data source, collect data that is already available, data normalization, and data analysis [13].

The data source of this research is based on social facts that in the last five years, natural disasters have occurred in Indonesia, namely; earthquake disaster in Pidie Jaya-Aceh (2016), earthquake and tsunami disaster in Lombok-West Nusa Tenggara (2017), earthquake and tsunami disaster in Palu-Central Sulawesi (2018). Furthermore, a social disaster was in the form of displacement of ex Gafatar families in Pontianak-West Kalimantan (2016). Local scale natural disasters such as those that occurred in flash floods in Garut Regency (2016), and non-natural disasters, namely Covid-19 (2020). Secondary data from the areas where the disaster occurred can illustrate how the elderly live in disaster situations.

The data collected was then analyzed using document analysis techniques, namely analyzing data from various documents to answer questions about the protection of the elderly in disaster situations from the aspect; fulfillment of basic needs, health, and psychosocial support services.

IOP Conf. Series: Earth and Environmental Science **989** (2022) 012020 doi:10.1088/1755-1315/989/1/012020

4. Results and Discussion

4.1. The problems of the Elderly in disaster

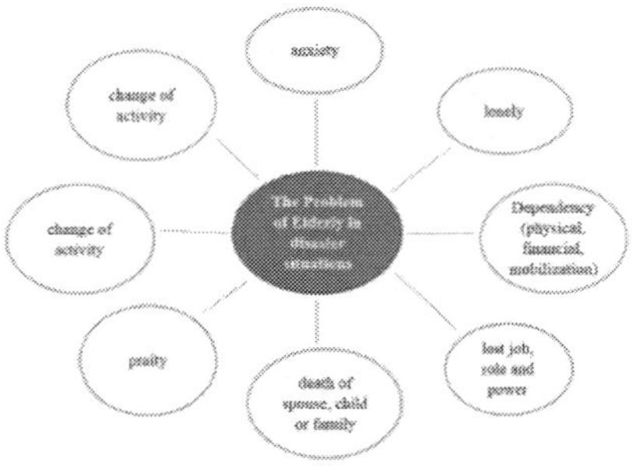

Figure 2. The problems of elderly in disaster situation

The problems of elderly in the six disaster locations, namely in Pidie Jaya, Garut, West Kalimantan, Lombok, and Palu-Central Sulawesi include; anxiety, loneliness, physical dependence, financial and mobilization, loss of job, role and power, death of spouse, children or family, praity, changes activity, and absence of caregiver.

4.2. The fulfillment of basic needs of the disaster survivor

Table 3. Fulfillment of basic needs for disaster survivors

Year	Location and type of disasters	fulfillment of basic needs
2016	earthquake in Pidie Jaya – Aceh	• Food, tent post 10 pieces, genset / 2,800 watts 10 units, • family kit, community donators in the form of: food aid, snacks, drinking water, clothing. • Assistance comes from the Ministry of Social Affairs, BNPB, BPBD, NGOs and donations from the community. Data source: • Tribune News.com: 08/12/ 2016 • Muksalmina Fadri and M. Zuhri: Syiah Kuala Banda Aceh University
2016	flash floods in Garut	Ready meals, communal kitchen, Data source: https://www.liputan6.com VIVA.co.id. 09/22/2016
2016	social conflict of Gafatar in West Kalimantan	Food, tent, bedding, communal kitchen Data source: Sindonews, 01/20/2016
2018	earthquake and tsunami in Lombok	Ready meals, communal kitchens, foodstuffs in the form of rice and side dishes, tents. Data source:*Antara news* 12/10/ 2018 Ragam Institute
2018	earthquake, tsunami and liquefaction in Palu – Central Sulawesi	Food, communal kitchen, foodstuffs in the form of rice and side dishes, tents. Data source: Kompas.com, 09/10/2018 Ragam Institute, 30/10/2019
2020	non-natural disasters Covid-19	Bantuan sosial berupa bahan makanan Source: Ministry of Social Affairs, December 2020

IOP Conf. Series: Earth and Environmental Science **989** (2022) 012020 doi:10.1088/1755-1315/989/1/012020

Table 4. Health services for disaster survivors

Year	Location and type of disasters	Forms of healthcare
2016	earthquake in Pidie Jaya – Aceh	• psychiatric specialists, medical personnel, • assistance with medicines, • medical examination, treatment of injured Data source: BPBD Pidie Jaya, Liputan6.com, 12/13/2016
2016	flash floods in Garut district	• Handling of injured • health services through mobile health centers in the camp
2016	social conflict of Gafatar in West Kalimantan	• medicalexamination of survivors • prepare a number of medicines • mosquito repellent lotion
2017	earthquake and tsunami in Lombok	• Health services for severely injured, postoperative, outpatient patients • health checks and data collection of disease history of disaster victims during the evacuation • provision of clean water and sanitation
2018	earthquake, tsunami and liquefaction in Palu – Central Sulawesi	•health services at evacuation posts for survivors including 18,634 elderly, • physiotherapy of 551 people • supplementary feeding for pregnant women and toddlers • tracking cases of malnutrition at evacuation posts
2020	non-natural disaster Covid-19	• prevention of transmission for the elderly at the community level • health services for the elderly in health facilities • blood pressure test and blood sugar test

Table 5. Psychosocial support for survivors

Year	Location and type of disasters	Forms of psychosocial support
2016	earthquake in Pidie Jaya – Aceh	• religious lectures, targhib (encouraging) for disaster survivors • psychosocial support services by the community service center team of the Bandung College of Social Welfare (STKS) in the form of; rapid assessment in biopsychosocial perspective, psychosocial interventions with various techniques including catharsis, initial psychological assistance, support, life review therapy
2016	flash floods in Garut district	STKS team and other volunteers conduct of psychosocial support
2016	social conflict of Gafatar in West Kalimantan	• The Ministry of Social Affairs Psychosocial Support Team conducts psychosocial support especially for children to restore psychological and social conditions • The psychosocial support team of The Bandung College of Social Welfare (STKS) in collaboration with HIMPSI West Kalimantan conducts psychosocial support, starting with rapid assessment and psychosocial inrvence with various techniques including social conversation, counseling, emotion freedom technique
2017	earthquake and tsunami in Lombok	• Psychosocial support in the form of medikasi and psychotherapy include cognitive and behavioral therapy, body-based interventions, eye movement desensitization and reprocessing (EDMR) Source: Zurriyatun Thoyibah, Dewi Nur Sukma Purqoti, Elisa Oktaviana (2019) • The psychosocial support team of The Bandung College of Social Welfare (STKS) in collaboration with HIMPSI West Kalimantan conducts psychosocial support, starting with rapid assessment and psychosocial inrvence with various techniques including social conversation, counseling, emotion freedom technique
2018	earthquake, tsunami and liquefaction in Palu – Central Sulawesi	• The Indonesian Red Cross (PMI) has conducted trauma recovery for vulnerable groups in accordance with the results of a quick review. Then do psychological first aid, psychoeducation on things that are considered to have a negative impact on the affected communities • STKS Bandung team conduct psychosocial supprot, starting with rapid assessment and psychosocial inrvence with various techniques such as social conversation, counseling, emotion freedom technique
2020	non-natural disasters Covid-19	• The Ministry of Health conducts mental health guidance for the elderly • The Ministry of Social Affairs through the center of social rehabilitation of the elderly conducts psychosocial guidance for the elderly both in the elderly and those in the community.

Researchers when they came to disaster sites in Pidie Jaya (2016), Garut (2016), West Kalimantan (2016), and Palu-Central Sulawesi (2018), provided psychosocial support to the elderly through several techniques including; ventilation, therapeutic communication, social conversation, emotional freedom techniques (EFT), and other techniques. However, the researchers saw firsthand that the elderly in the disaster area did not receive intensive assistance from volunteers, they saw that the elderly had received assistance of basic needs and health services. Volunteers are more focused on providing psychosocial support for children.

Regarding the 2020 Covid-19 pandemic specifically for the elderly as the group most at risk, researchers did not have the opportunity to do further research considering the limited access due to the social distancing and physical distancing policy. What researchers can do is study the number of elderly people who died until October 2020 as many as 2.138 people (16.5%) of the total deaths, and government policies in handling covid-19 for the elderly. The Ministry of Health has issued Guidelines for Elderly Health Services in the 2020 Covid-19 Pandemic Era. This guide to be a reference for health workers in health facilities in carrying out health services for the elderly. The Ministry of Social Affairs has implemented programs and activities to deal with the impact of Covid-19 for the Elderly in the form of; food assistance, temporary shelter equipment such as blankets, cooking utensils, kits for the elderly, and psychosocial support services.

4.3. Discussion

Disasters, both natural disaster, non-natural disasters and social disasters have prompted various parties to be involved in paying attention, mobilizing aid, handling emergency and post-emergency operations. In the last five years, earthquakes in Pidie Jaya (2016), earthquakes and tsunamis in Lombok (2016), earthquakes, tsunamis and liquefaction in Palu, Central Sulawesi, local scale disasters such as flash floods in Garut (2017), and social disasters in the form of social conflicts between community members and the Gafatar community in West Kalimantan (2016).

Various central government agencies and local governments, the world of education, business institutions to various community organizations are moving to help disaster victims, either directly or indirectly. Handling and assistance for disaster victims can be in the form of; fulfillment of basic needs which include food, clothing, and shelter, then health services, and psychosocial support services or trauma healing.

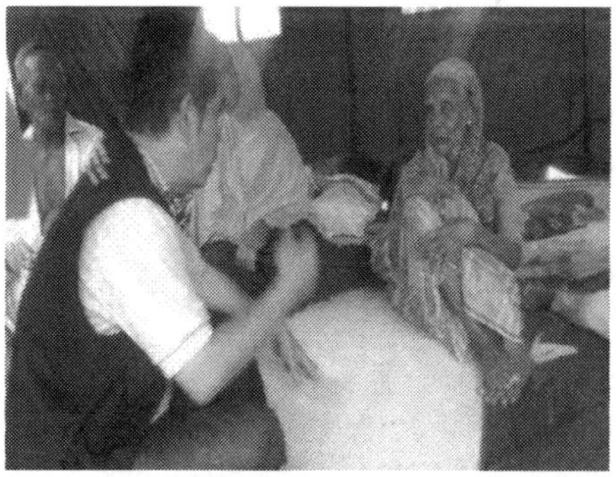

Figure 3. Reseacrcher: Tukino at Pidie Jaya

Based on the results of the secondary data study as described above, the handling and assistance of disaster victims for the elderly is still very limited, especially services to deal with social and psychological impacts.

The results of observations by researchers in the field, volunteers or caregivers who came to disaster locations were more focused on children, and did not pay special attention to the elderly. Thus, it can be said that the elderly in the disaster area did not receive assistance from the volunteers. Though, like children, the elderly also need a touch of the soul as long as they are in a disaster situation, so that the elderly can feel safe, comfortable, and still have fun.

5. Conclusion

Efforts to protect the elderly as a vulnerable group in disaster situations are still not optimal. The implementation of Article 55 paragraph 2 of Law Number 24 of 2007 concerning Disaster Management which provides protection for vulnerable groups, does not seem to have been done by the government, both central and regional, even volunteers. The elderly as a vulnerable group have not become a priority in disaster management.

This paper recommends incorporating specific materials about the elderly in disaster situation into the draft amendments to the law on the elderly that is being drafted. Disaster material in the draft law or in the future the new law on the elderly, will strengthen the mainstreaming of the elderly in disaster management. Mainstreaming the elderly in disasters should also target sustainable social protection programs for disaster victims so that they do not fall into poverty due to disasters. Other things that must also be included in the bill are gender issues, attention to elderly women as well as issues of empowerment and protection of the elderly which were previously not explored in the current law, namely Law Number 13 of 1998 concerning Elderly Welfare.

Finally, the authors hope that the new law on the elderly adopts a human rights approach (right-based approach) that encourages the obligation holders, namely the state, to carry out their obligations to fulfill and protect vulnerable groups. In the context of disaster management, these obligations include providing special treatment to the elderly during disaster situations, both pre-disaster, during disaster, and post-disaster.

Acknowledgements

I would like to thank the Director and Head of the Research Center of Bandung Polytechnic of Social Welfare, who have allowed and provided financial assistance for this research.

References

[1] Adi Fahrudin. 2000. *Gerontologi dan perkhidmatan sosial kepada warga tua. Laporan Penyelidikan. Kota Kinabalu*: Universiti Malaysiaaysia Sabah.
[2] Ahmed, Mohiuddin 2019. Brief communication: Use of Mind Stimulation Therapy Techniques as Adjunct to Clinical Practice with Elderly Psychiatric Patients (Aging Medicine and Healthcare 2019;10(2):88-93). https://www.researchgate.net/publication/334088738
[3] Barusch, Amanda 2011. Disaster, Vulnerability, and Older Adults: Toward a Social Work Response. Journal of Gerontological Social Work. https://www.researchgate.net/journal/Journal-of-Gerontological-Social-Work-1540-4048
[4] Benjamins, Maureen Reindl. 2004. Religion and Functional Health Among the Elderly Is There a Relationship and Is It Constant? Journal of ageing and Health vol. 16 no. 3 355-37
[5] Beck, Judith S. 2011. Cognitive-Behavior Therapy: Basic and Beyond (second edition). New York: The Guilford Press.
[6] Cooper, M.G. & Lesser, J.G. 2005. Clinical social work practice: An integrated approach *(2nd edition)*. Boston: Pearson Education, Inc.
[7] Edwards, P. 2002. Active Ageing: a Policy Framework. Geneva: WHO
[8] Sanders, et.al 2004. Disaster Preparedness for Older Adults and the Elderly. Journal of Gerontological Social Work. https://dmsw540.wordpress.com/jennifer/
[9] Johnston, Melisa P. 2014. Secondary Data Analysis: A Method that which a Time Has Come. Quantitative and Qualitative Methods in Library (QQML) 3.
[10] Santrock, J. W. 1997. Life-Span Development, 6th ed. Chicago: Brown& Benchmark Publisher
[11] Sheafor, B.W. & Horejsi, C.R. 2003. Techniques and Guidelinea for Social Work Practice. Boston: Allyn and Bacon.

IOP Conf. Series: Earth and Environmental Science **989** (2022) 012020 doi:10.1088/1755-1315/989/1/012020

[12] Yung-Chin Tsao, Chun-Chieh Shu, and Tian-Syung Lan 2019. Development of a Reminiscence Therapy System for the Elderly Using the Integration of Virtual Reality and Augmented Reality. Taiwan: Tatung University, Taipei City 10491. https://econpapers.repec.org/article/gamjsusta/v.3a11.3ay_3a209_3ai-3a17_3ap_3a4792-_3ad_3a263403.htm
[13] Wallace Foundation 2014. Workbook B: Conductiong Secondary Research. (Other Information restricted). Retrieved June, 2014 online from http://wallacefoundation.org/
[14] Wong HC Johnston. 2014. Social Work Interventions at Different Stages of Disaster Illustrated by Asian Experiences. Beijing Normal University
[15] Zakour, J. Michael. 2010. Hong Kong Baptist University United International College, China. https://www.researchgate.net/publication/281366547_SocialWork_Intervention_at_Different_Stages_of Disaster_Illustrated_by_ Asian_ Experience
[16] Social Work and Disaster. Tulane University
[17] Zastrow, H. Charles & Kirst-Ashman, K. Karen. 2004. Understanding Human Behavior and the Social Environment. University of Wisconsin-Whitewater: Thomson Publishing Company

IOP Conf. Series: Earth and Environmental Science **989** (2022) 012021 doi:10.1088/1755-1315/989/1/012021

Factors Determining the Welfare of the Sunda Strait Tsunami Victim's Family. A Mechanism Disaster Impact on SDGs in Family Level

Euis Sunarti[1*], Irwan Meliano[2], Cecep Pratama[3], Setya Haksana[4], Aliya Faizah Fithriyah[1], Nikmatul Khoiriyah[1], and Winny Novyanti[1]

[1]Department of Family and Consumer Sciences, Faculty of Human Ecology. Bogor Agricultural University, Indonesia
[2]Geodesy Research Group Bandung Institute of Technology Bandung – Indonesia.
[3]Department of Geodetic Engineering Gadjah Mada University Yogyakarta – Indonesia.
[4]Integrated Disaster Risk Reduction Research Group, Universitas Airlangga. Surabaya – Indonesia

[*]Email: euissunarti@apps.ipb.ac.id

Abstract. The Sunda Strait Tsunami in December 2018 has prolonged impact on the welfare of victim's families. This research aims to analyze factors determining family welfare after 27 month of Sunda Strait tsunami. Research involved 123 families consisted of two groups: refugees who living in shelter F(LS) and non-refugees who living in non shelter (FLNS). Almost two-thirds of the families stated that the impact of the tsunami is still disrupting their work, livelihood, and daily lives. Acceptance and gratitude lead to higher subjective well-being, although objectively, family welfare has not returned to how it was before the tsunami. The results of different tests showed that FLS families (compared to FLNS) had better conditions and the health of their homes as well as their livelihood strategies. The objective welfare of the family is negatively affected by economic pressures, while the subjective welfare of the family is negatively affected by both subjective and objective economic pressures. This research shows the importance of external parties helping the victim's family to reduce their economic pressure, by providing access and opportunities for family income patterns. Subjective economic pressure is a direct mechanism of disaster disruption to welfare as one of the goals of sustainable development.

1. Introduction

The 2018 Sunda Strait Tsunami caused damage to facilities and infrastructure, also in the loss of life. Disasters cause many families to lose their homes and property. Housing reconstruction is one of the most difficult challenges in post-disaster recovery [1][2], especially in developing countries [3]. Disasters disrupt various aspects of family life such as reducing welfare and disrupting the lives of individuals, families, and communities [3]. Disasters are always related to humans, disruption of life, damage, and loss. The tsunami caused families to become poor because family incomes declined sharply [5]. The destruction of productive assets (such as boats, ponds, marketing facilities, and docks) due to the disaster has eliminated people's livelihoods and incomes so that economic rehabilitation for coastal communities is needed [5][6][7].

The strategy for the rehabilitation of livelihoods and assets affected by disasters should reduce vulnerability and other socio-economic risks [8]. Several studies confirmed that disasters increased poverty [5][9][10]. Based on this background, it is important to examine the effect of economic pressures (subjectively and objectively), livelihood strategies, work skills, housing conditions and health, and family characteristics, on the welfare of families (subjectively and objectively) of victims of the Sunda Strait tsunami, as well as to analyze differences in the variables studied between displaced families (FLS=families living in shelters) and those who were not displaced (FLNS= families living in their original homes).

2. Methods

2.1. Research design, location, and Sample

This research design is a cross-sectional study. Primary data were collected through a survey in March 2021 (27 months after the tsunami). The research location was chosen in Pandeglang Regency, as one of the affected areas by the Sunda Strait Earthquake and Tsunami at the end of December 2018. Research samples were 123 families of victims of the 2018 Sunda Strait Tsunamy, consisted of 60 refugee families (FLS) and 63 non-refugee families (FLNS) who were selected by random sampling method.

2.2. Research variables and Data collection

The variables of this study were family economic conditions (income and poverty), livelihood strategy, house health and condition, economic pressure (subjectively and objectively), and family welfare (objectively and subjectively). Objective family welfare was measured by income, poverty, and food coping strategies. Subjective family welfare consisted of economic, social, and psychological aspects. All variables were measured using the instrument developed by the researcher and had met the eligibility requirements of the research instrument. Cronbach's Alpha of family livelihood strategy, house health and condition, subjective economic pressure, and subjective family well being (economic, social, and psychological) are 0.784, 0.797, 0.795, and 0,784, 0,659, and 0,691 consecutively. The types of data used in this study were primary and secondary data. Primary data were collected through direct interviews to measure family conditions at a year post-tsunami on March 2021.

2.3. Data Analysis

The data that had been collected through questionnaires were input and processed through the editing, coding, scoring, and cleaning processes using Microsoft Excel. The ordinal data were converted into percentages and indexes with the lowest value of 0 and the highest value of 100. The index was obtained by calculating the score obtained from the measured variable, divided by the result value of the subtraction of the maximum score by the minimum score, multiplied by 100. Descriptive analysis and different test T-Test using the Statistical Package for Social Science (SPSS) 2.5, while the effect test was using SEM-PLS. Descriptive analysis and t-test were used to make the distribution of data, averages, standard deviations, and to identify differences in the variable index between FLS and FLNS.

3. Results and Discussion

3.1. Family Characteristic

Family characteristics measured in this study include age (wife and husband), length of education (wife and husband), length of marriage, age of the youngest child, stage of family development, number of family dependents, and income per capita. The average value per indicator as well as the difference test between the two family groups are presented in Table 1.

Table 1. Different test and distribution of average of family characteristics

Family Characteristic	Mean ± Sd.Deviasi		p-*value*
	FLS (n=60)	FLNS (n=63)	
Wife's age (yr)	39.60 ± 9.87	42.19 ± 11.69	0.188
Husband's age (yr)	45.73 ± 11.86	47.54 ± 12.75	0.418
Wife's length of education (yr)	5.53 ± 2.84	5.40 ± 3.38	0.809
Husband's length of education (yr)	5.78 ± 3.23	5.59 ± 3.41	0.745
Long marriage (yr)	20.47 ± 10.23	23.35 ± 11.85	0.152
Youngest child's age (yr)	9.03 ± 7.73	10.57 ± 7.27	0.258
Family development stage	5.27 ± 1.35	5.44 ± 1.08	0.422
Dependents family members (person)	1.40 ± 0.616	4.94 ± 1.62	**0.000**[2]
Income/capita (IDR thousand)	900796.37 ± 986851.93	840100.86 ± 1027493.34	0.739

The results of the different test showed that there was only one indicator that showed a significant difference between the two groups, namely family dependents. FLNS has a larger number of family dependents than FLS. There were no significant differences in other family characteristics. Family dependents include the number of family members who live together in one house and whose lives are borne by the head of the family. Based on the results of the different test, the dependents of FLS were smaller than FLNS. All families in permanent shelter (FLS) were categorized as small families where the number of families is less than 4, while FLNS had more varied number of dependents and family size, namely p 47.6% are categorized as small families (less than 4 people), 50.8% are categorized as medium families (6-8 people), and 1.6% are categorized as large families (more than 8 people).

3.2. *Difference test of variables*

The average research variable index between FLS) and FLNS is presented in Table 2. Of the 8 research variables, the lowest index was work skills. This shows that the family's skills are not diverse to find alternative work. With economic conditions and an index of meeting basic needs around 66-67, families showed relatively high satisfaction with basic needs fulfillment, with an index ranging from 78.5-79.7. The results of the different tests showed that FLS has a higher livelihood strategy and better health and housing conditions than FLNS.

Table 2. Distribution of average index and result of research variables different test

Variabel	Mean		p-*value*
	FLS (n=60)	FLNS (n=63)	
Objective economic pressure	48.33	51,41	0.213
Subjective economic pressure	43.23	45.05	0.669
Livelihood strategy	52.33	41.11	**0.013**[1]
Earning skills	23.07	19.41	0.097
Objective welfare	66.33	67.46	0.721
Subjective welfare	79.70	78.57	0,527
House health and condition	58.67	45.31	**0.001**[2]
Impact tsunami per march	46.47	49.38	0,544

3.2.1. *Family livelihood strategy*

There were two types of livelihood strategies that are widely used by families of victims of the Sunda Strait tsunami in order to continue to work, which are (1) using all their abilities to earn a living and (2) being ready to do whatever work is available or not being picky about work. The results of the different test showed that the livelihood strategies carried out by FLS are more diverse than FLNS (Figure 1).

The loss or damage of boats due to the tsunami, as well as the distance from the location of permanent housing assistance from the government (formerly temporary shelters), made it almost impossible for FLS to return to being fishermen, so that their livelihood strategies were forced to change and be more diverse.

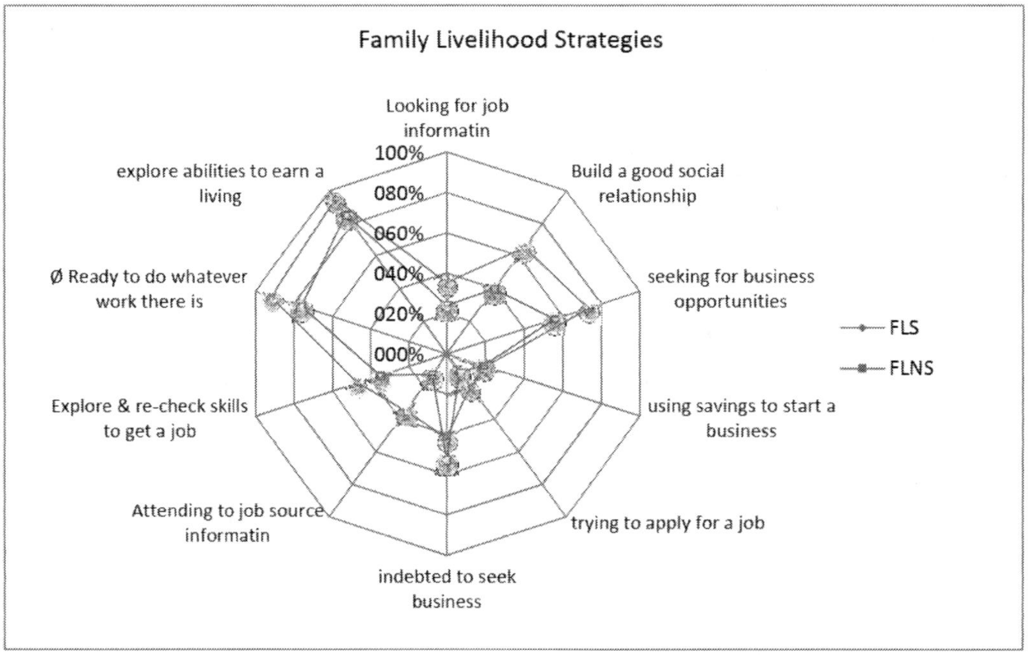

Figure 1. Family livelihood strategy in FLS and FLNS

3.2.2. House health and condition

The condition of houses in permanent housing provided by government assistance to refugees (FLS) is better than houses occupied by families of victims who did not evacuate from the start because they did not meet the criteria for obtaining temporary shelter (FLSN). Permanent housing, which has just been built and occupied after 2 years after the disaster, is considered to have a cleaner condition, an orderly layout of the house, and a yard. Each house faces a fairly wide residential street, although the conditions at the time of the research were still muddy because the construction had not yet been completed. The condition and health of the houses after the disaster of the families living in the shelters were much better than the families living in the non-shelters (Figure 2).

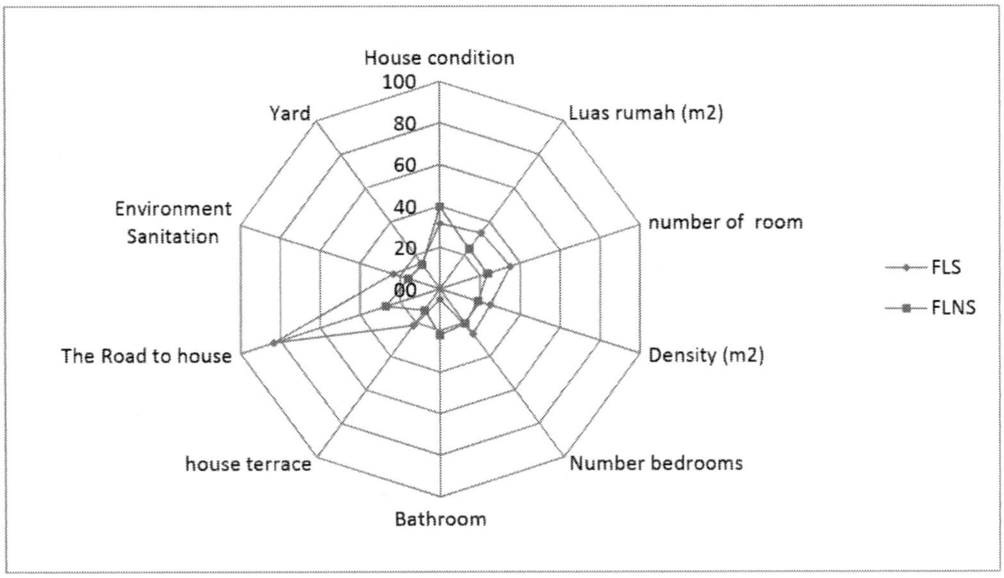

Figure 2. House's health and condition of FLS and FLNS

3.3. *Family Welfare*

Family welfare was measured objectively and subjectively. Subjective welfare is divided into 3 sub-dimensions, namely economic, social, and psychological subjective well-being.

3.3.1. *Objectives welfare*

There was no significant difference between the total objective welfare of FLS and FLNS. The results showed that several objective welfare indicators scored very low, namely "consume fruit at least once a day" and "the condition of house is healthy (no leak, no dirt floor, etc)". The objective welfare scores of temporary residents (FLS) and non-temporary residents (FLSN) are presented in Figure 3. Although there was no significant difference in objective welfare in total, data showed that FLS have better or cleaner toilets. This is because they have just occupied permanent housing, assistance from the government. This is different from FLNS whose houses are crammed and irregular in a beachfront location.

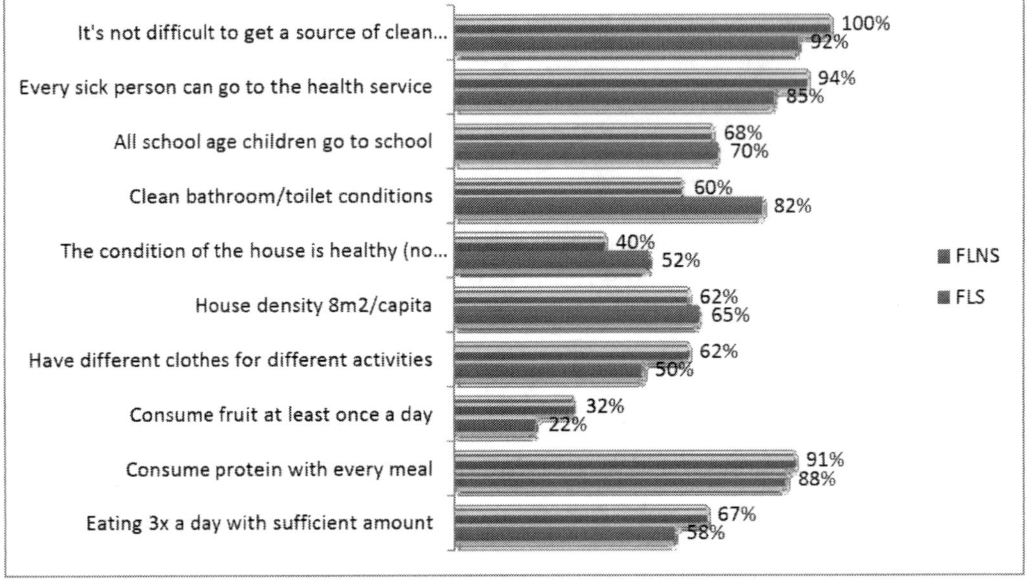

Figure 3. Indicators of Objective welfare of FLS dan FLNS

3.3.2. Economic aspect of subjective well-being

Subjective welfare was measured based on 3 aspects, namely physical-economic, social, and psychological. The results of the different test showed that there was no significant difference between the economic aspect of subjective well-being of FLS and FLNS. In both groups (Table 3), indicators that have low scores were owned savings (score 25), assets or assets owned (score 45,6), and family income (score 58,9).

Table 3. FLS and FLNS scores of economic aspect of subjective welfare

No	Indicators	Families		
		FLS	FLNS	Total
1	Food consumed	69.5	72.8	72.7
2	Owned & clothes used	71.0	68.9	71.4
3	Condition of the house occupied	83.6	65.5	75.8
4	Ability to finance health services	68.1	66.7	68.9
5	Ability to finance child education	69.8	64.4	68.4
6	Family income	62.6	52.6	58.9
7	Assets or property owned	51.9	38.3	45.6
8	Owned Saving	26.2	23.4	25.0
9	Self financial management	69.8	68.0	69.7
10	Financial management by partner	69.8	63.9	68.1

3.3.3. Social aspect of subjective welfare

The data in Table 4 shows that the social aspects of subjective well-being of FLS and FLNS are high. Relationships between family members showed the highest scores in both groups. Both groups had scores of support and assistance from outside the family below 80. The data on subjective economic well-being of FLS and FLNS are presented in graph 6.

Table 4. Scores of subjective well-being of social aspects of FLS and FLNS

No	Indicators	Families		
		FLS	FLNS	Total
1	Family decision process	85.2	83.9	86.2
2	Time management & family togetherness	87.6	85.3	87.3
3	Handling problems in family	76.9	80.0	79.4
4	Relations with neighborhood	86.7	88.0	89.0
5	Relationship among family members	93.8	94.1	95.4
6	Relationship with extended femily	90.0	85.5	89.3
7	Support given to others	82.6	80.0	82.7
8	Support and assisstance from external family	78.6	79.8	80.3
9	Participation in the community	83.6	78.7	82.3
10	Social acceptance to family	91.7	88.7	91.8

3.3.4. Psychological aspect of subjectvie well-being

The results of the analysis showed that the psychological satisfaction scores of FLS and FLNS were high as presented in Table 5. Although there was no significant difference, the subjective psychological well-being indicators of FLS are slightly higher (range 86-94.5) than FLNS (range 78-91). ,4). Data on subjective psychological well-being of FLS and FLNS are presented in graph 7.

Table 5. Scores of subjective well-being of social aspects of FLS and FLNS

No	Indicators	Families		
		FLS	FLNS	Total
1	Comfort family environment	88.6	88.9	90.4
2	Daily worships	89.3	78.5	85.0
3	Family member behavior & development achievement	81.07	78.0	80.7
4	Family healthy condition	91.4	85.5	89.7
5	Family harmonious	93.8	89.1	92.8
6	Communication & family ties	94.5	91.4	94.5
7	Role & task as husband and wife	91.7	89.1	92.0
8	Role & task as parent	90.0	90.5	91.9
9	Role in extended family	86.0	81.6	85.4
10	Role as community members	89.0	81.0	86.3

3.4. Results of test effect

The results of the SEM-PLS test (Table 6) showed that objective economic pressures directly have a negative effect on objective welfare, subjective welfare economic aspects, and subjective welfare psychological aspects. The higher the objective economic pressure felt by FLS and FLNS, the lower the objective welfare, subjective economic welfare, and subjective psychological well-being. Meanwhile, subjective economic pressures, either directly or indirectly, affected the objective welfare, and the economic aspect of subjective welfare. The lower the subjective economic pressure of FLS and FLNS, the higher the economic satisfaction. It is interesting to find that the higher the subjective economic pressure, the higher the objective welfare. This shows that the higher the dissatisfaction with the economy, the higher the fulfillment of the family's objective needs. Family characteristics directly had a significant positive effect on psychological social welfare. This means that the better the family characteristics, the better the psychological satisfaction of the families of disaster victims.

Table 6. Factors directly and indirectly effects on the objective and subjective welfare of families

Variabel	Direct effect	Indirect effect	Total effect
Objective economic pressure → Objective welfare	**-0.534***	-0.010	**-0.534***
Subjective economic pressure → Objective welfare	**0.198***	**0.181***	**0.379***
Objective economic pressure → Economic subjective welfare	**-0.540***	0.011	**-0.528***
Subjective economic pressure → Economic subjective welfare	**-0.497***	**0.191***	**-0.305***
Family characteristic → Psychological welfare	**0.232***	0.022	**0.254***
Objective economic pressure → Psychological welfare	**-0.210***	0.021	-0.190

The R square values of objective welfare, subjective economic well-being, subjective social welfare, and psychological well-being were 0.379, 0.340, 0.061, and 0.150, respectively. This shows that all variables affect the objective welfare of 37.9%, the subjective economic welfare of 34%, the subjective social welfare of 6.1%, and the psychological welfare of 15%. The influence path scheme between variables is presented in Figure 4

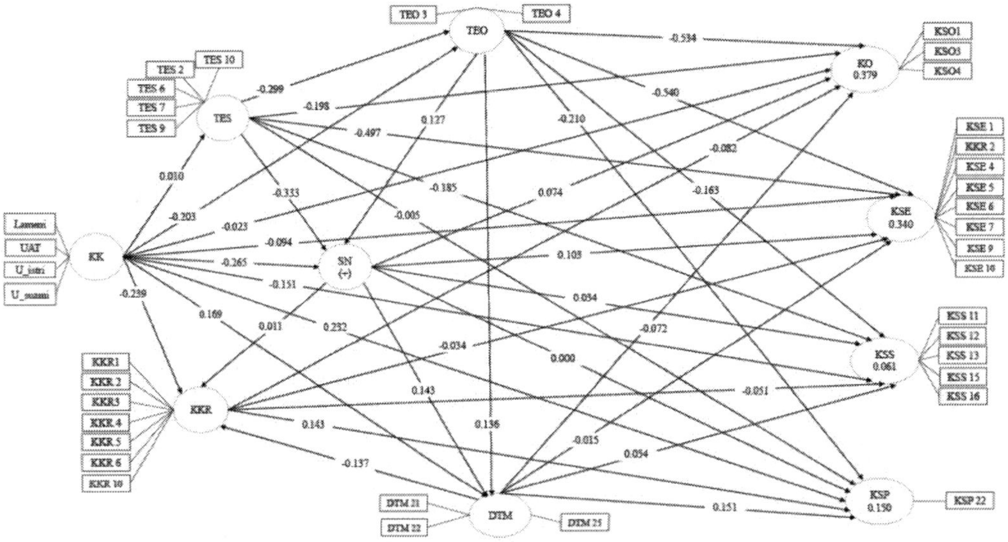

Figure 4. Direct and indirect influence path among research variables

Description:
KO=Objective welfare
KSE= Economic aspect of subjective welfare
KSS=Social aspect of subjective welfare
KSP=psychological aspect of subjective welfare
TEO=Objective economic pressure
TES=subjective economic pressure
KK=Family characteristics
DTM=tsunami impact
KKR=House condition

This study found three factors that affect the welfare of FLS and FLNS, namely family characteristics, objective economic pressure, and subjective economic pressure. These three factors significantly influence both objectively and subjectively (economically, socially, and psychologically) on family welfare. Family welfare is significantly affected by economic pressure due to the lack of recovery in livelihood patterns after the tsunami. Objective economic pressure has a significant negative effect on objective welfare, economic subjective welfare, and psychological subjective welfare. This means that the higher the family economic pressure, the lower the objective welfare, the lower economic family satisfaction, and also the lower psychological family satisfaction. These findings are in accordance with the findings of other studies [11][12]. Economic pressures also make it difficult for families to have business capital. Whereas family welfare is influenced by family capital ownership [13]. The inability to meet economic needs is directly related to household income, asset ownership, and other economic activities that mutually generate strategies and household livelihood outcomes [14]. Lower-middle income families generally have low levels of education and skills so that alternative livelihood strategies available to them are very limited [15][16].

The results of the study showed that up to 27 months after the Sunda Strait Tsunami, the victim's family still has not been able to carry out normal work as before [5]. Apart from the fact that many fishermen have not been able to replace damaged boats, it is also due to changes in the fishing catchment area which is getting further away from the coast and requires boats with a larger capacity [5]. This is where disaster victims need government assistance, which has yet to materialize. Therefore, the effectiveness of post-disaster management is the key to economic recovery for disaster victims, as are similar findings and conclusions found in various disaster studies which state that disaster management and disaster risk reduction have not become a priority for regional development programs, and have not been well integrated in planning and regular development [17]. Measures that support environmental protection, maintenance of social structures and livelihoods should be considered in both the transition and long-term recovery phases, so that post-disaster impacts do not drag on [18], as post-disaster risk reduction requires a holistic understanding of the related issues. complex interactions between interrelated physical, environmental and social factors [19].

Economic pressure is a condition of the family's inability to fulfill economic needs, income uncertainty, job instability, and/or income instability to fulfill desires [20]. Various studies confirmed that disasters increased poverty [9][10]. The families of fishermen affected by the tsunami became poor due to a sharp decline in income, and were potentially trapped by moneylenders' loans [21]. Families who experience economic pressure have difficulty in managing family resources and finances to meet family needs. This is in line with research which stated that economic pressure due to difficult conditions will have an unfavorable impact on meeting daily needs due to limited resources and income [22]. On the other hand, families with physical and financial capital ownership will have a greater chance of being prosperous [23], as well as the ability of the family to manage family resources and finances as a factor in the family's objective welfare [24]. The role of income is important for objective welfare, because higher income increases the capacity to meet the needs of life [25]. On the other hand, one's education and knowledge are also very important and are believed to be an investment for the future to improve the socio-economic status of the family [26].

The results of the analysis show that compared to FLNS, FLS actually does more 4 livelihood strategies, namely (1) building good social relationships in order to get work opportunities, (2) seeing or looking for business opportunities to earn a living, (3) looking for information on job sources, and (4) ready to do whatever work there is. This is seen as a logical consequence of the larger number of FLS who have lost job opportunities compared to FLNS whose houses are still closer to fishing ports and the center of economic activity (markets). The families of the victims of the Sunda Strait tsunami carried out various livelihood strategies in response to the declining income due to the reduced or even loss of income sources after the tsunami. The number of livelihood strategies carried out shows the greater the family business seeking income. This is consistent with the findings of another study [27] which stated that rural communities generally have diverse livelihood strategies to increase income. Prosperity and survival are one of the ultimate goals of livelihood strategies, so families with lower levels of welfare will make every effort to get out of poverty and become more prosperous [28]. Meanwhile, the diversification of livelihoods carried out by non-poor families aims to develop their businesses and bring them to a higher level of welfare [29].

Subjective economic pressure had a significant positive effect on the objective welfare of the family, which means that the higher the family's perception of the condition of economic pressure, the higher the fulfillment of the family's objective needs. This finding seems contradictory at first glance, but several studies [30][31] confirmed that there is not always a linear relationship or influence between the subjective and the objective. This shows that the higher the awareness of the existing economic conditions and the higher the expectations for better economic conditions for families whose basic needs are getting higher. Another explanation for this is that the higher the social class of the family, the more dissatisfied with the existing conditions, because the greater the expectations and demands. This is in contrast to the research findings which state that families with husbands or wives with higher education and higher incomes have better subjective well-being [32]. Another interesting finding of this study is that subjective economic pressure has a significant negative effect on psychological family satisfaction, which means that the higher the subjective economic pressure, the lower the psychological family satisfaction. Further meaning that the perception of economic pressure in the family is thought to also decrease emotional function and increase feelings of depression [33].

This research was conducted when FLS (whose house was lost and heavily damaged), just occupied a permanent residence with assistance from the government. Although the location of the shelter is far from the place of business and the road infrastructure has not been completed so that the roads become muddy and impassable when it rains, in general, FLS had a much higher satisfaction with the condition of their houses, compared to FLNS who still live in their old houses with sanitary conditions, and a shabby and cramped environment. FLS has higher satisfaction than FLNS regarding the condition of the house, house area, density, bathroom, yard, road to the house, environmental sanitation, and yard. Housing is generally the element that most severely damaged or lost, and often accounted for the largest share of losses in the total impact of disasters on the national economy [34]. The level of damage to houses and property during a disaster was significantly related to the severity of psychiatric morbidity [35], so it is crucial to accelerate housing construction as it impacts the quality of daily life [35] [1][2].

The overall effect test showed that objective and subjective economic pressures, livelihood strategies, earning skills, housing and health conditions (as an impact 27 months after the tsunami), and family characteristics had an effect on 37.9% of objective welfare, 34% had an effect on economic aspects of subjective well-being, 6.1% affected the social aspects of subjective well-being, and 15% affected the psychological aspects of subjective well-being. This indicates the need to conduct a more holistic and comprehensive study to find other factors that affect the welfare of the families of disaster victims. The still significant effect of economic stress (caused by the tsunami) on family welfare up to 27 months after the tsunami confirmed the impact of disasters on life disruption, damage and loss, misfortune and deterioration, a decline in welfare [10][36] as well as poverty [5].

Welfare is one of the indicators of the SDGs, the model linking the SDGs and welfare stated that there is a direct influence between the two, and recent research suggests that the SDGs will affect other SDGs [37]. This deserves attention to the welfare of all development activities, namely alleviating poverty in low-income countries, reducing inequality in high-income countries, and adding to the positive effect on all the SDGs, thereby helping to support SDG priorities by region [38], especially at the micro system level, namely the family. Therefore, an adequate, fast and appropriate disaster management mechanism is needed. The international community for disaster risk reduction has long argued that reducing people's vulnerability to hazards is a key component of sustainable development (9). Especially in developing countries, locations that are frequently affected by disasters tend to be inhabited by low-income households (10).

4. Conclusion

An important finding from this research was that even after 27 months after the Sunda Strait Tsunami event, the victim's families still have not recovered from their economic activities so that their income, assets, and savings are still low. They were also still experiencing economic pressures that have a negative impact on their welfare. Objective economic pressure had a negative effect on both the objective and subjective welfare (economic and psychological aspects) of the family. Likewise, the influence of subjective economic pressures. The higher the perception of economic pressure, the lower the perception of welfare. This is further strengthened by other findings, namely the higher the perception of the economic pressure felt by the family when the objective welfare is higher.

Recommendation

Based on the research findings, the researcher recommends to the government, especially the stakeholders of disaster management in Pandeglang Regency, to immediately take action to collaborate with various parties to restore economic activity. Thus, the family will be able to have a decent income which will reduce economic pressure. To the National Disaster Management Agency, it is recommended to encourage rapid and effective post-disaster recovery, both housing assistance and facility recovery assistance and economic infrastructure that is difficult for families of disaster victims to carry out.

Acknowledgment

The authors would like to thank the 2021 World Class University Research Fund of Bogor Agricultural University for National Collaborative Research. We also would like to thank the enumerators and respondents for their helped during data collection.

References

[1] Sunarti, E., Gunawan, E., Widiyantoro, S., Marliyani, G. I., & Ida, R. (2021). Critical point on housing construction, resilience and family subjective welfare after disaster: Notes from the Lombok, Indonesia, earthquake sequence of July-August 2018. *Geomatics, Natural Hazards and Risk, 12*(1), 922-938.

[2] Sunarti, E. (2010).Pemulihan Ekonomi Dan Psikososial Korban Gempa Bumi Tasikmalaya. Bogor

[3] Ahmed, I. (2011). An overview of post-disaster permanent housing reconstruction in developing countries. *International Journal of Disaster Resilience in the Built Environment.*

[4] Sunarti, E. (2015). Ketahanan Keluarga Indonesia: Dari Kebijakan dan Penelitian Menuju Tindakan.

[5] Sunarti, E. Irwan Meliano, Cecep Pratama, Setya Haksana Aliya Faizah Fithriyah, Samsa Wigar Prayitno, and Winny Novyanti. 2021. The Impact of the Sunda Strait Tsunami and The Covid-19 Pandemic on SDGs. A Change in The Poverty Status, Food Security, and The Welfare of the Tsunami Victim's Families. *Article on going submitted to Journal*

[6] Pomeroy, R. S., Ratner, B. D., Hall, S. J., Pimoljinda, J., & Vivekanandan, V. (2006). Coping with disaster: rehabilitating coastal livelihoods and communities. *Marine Policy, 30*(6), 786-793.

[7] Sunarti, E., Praptiwi, R. N., & Muflikhati, I. (2011). Kelentingan Keluarga, Dukungan Sosial, dan Kesejahteraan Keluarga Nelayan Juragan dan Buruh di Daerah Rawan Bencana.[Family Resilience, Social Support, and Family Welfare of Fishermen and Workers in Disaster Prone Areas]/ *Jurnal Ilmu Keluarga & Konsumen, 4*(1), 1-10. http://dx.doi.org/10.24156/jikk.2011.4.1.1

[8] Bang, H. N., & Few, R. (2012). Social risks and challenges in post-disaster resettlement: The case of Lake Nyos, Cameroon. *Journal of Risk Research, 15*(9), 1141-1157.

[9] Fothergill, A., & Peek, L. A. (2004). Poverty and disasters in the United States: A review of recent sociological findings. *Natural hazards, 32*(1), 89-110. http://dx.doi.org/10.1023/B:NHAZ.0000026792.76181.d9

[10] Masozera, M., Bailey, M., & Kerchner, C. (2007). Distribution of impacts of natural disasters across income groups: A case study of New Orleans. *Ecological Economics*, *63*(2-3), 299-306.

[11] Firdaus, F., & Sunarti, E. (2009). Hubungan antara tekanan ekonomi dan mekanisme koping dengan kesejahteraan keluarga wanita pemetik teh. *Jurnal Ilmu Keluarga & Konsumen, 2*(1), 21–31.

[12] Robila, M. (2006). Economic pressure and social exclusion in Europe. *The Social Science Journal*, *43*(1), 85-97.

[13] Iskandar, A. (2008). Analisis praktik manajemen sumberdaya keluarga dan dampaknya terhadap kesejahteraan keluarga di Kabupaten dan Kota Bogor. *Sodality: Jurnal Sosiologi Pedesaan*, *2*(1).

[14] Thorbecke, E. (2013). Multidimensional poverty: conceptual and measurement issues. In *The many dimensions of poverty* (pp. 3-19). Palgrave Macmillan, London.

[15] Hallegatte, S., Vogt-Schilb, A., Bangalore, M., & Rozenberg, J. (2016). *Unbreakable: building the resilience of the poor in the face of natural disasters*. World Bank Publications.

[16] SAMHSA. 2017. Greater impact: How disasters affect people of low socioeconomic status. Rockville, MD: Substance Abuse and Mental Health Support Administration

[17] Sunarti… Integrating …..

[18] Ingram, J. C., Franco, G., Rumbaitis-del Rio, C., & Khazai, B. (2006). Post-disaster recovery dilemmas: challenges in balancing short-term and long-term needs for vulnerability reduction. *Environmental science & policy*, *9*(7-8), 607-613.

[19] Cardona, O., 2003. The need for rethinking the concepts of vulnerability and risk from a holistic perspective: a necessary review and criticism for effective management. In: Bankoff, G., Frerks, G., Hillhorst, D. (Eds.), Mapping Vulnerability: Disasters, Development and People. Earth Scan Publishers, London (Chapter 3).

[20] Lall, S. V., & Deichmann, U. (2012). Density and disasters: economics of urban hazard risk. *The World Bank Research Observer*, *27*(1), 74-105.

[21] Syafrini, D. (2015). NELAYAN VS RENTENIR Studi Ketergantungan Nelayan terhadap Rentenir pada Masyarakat Pesisir. *Jurnal Ilmu Sosial Mamangan*, *3*(2), 67-74

[22] Conger, R. D., Wallace, L. E., Sun, Y., Simons, R. L., McLoyd, V. C., & Brody, G. H. (2002). Economic pressure in African American families: a replication and extension of the family stress model. *Developmental psychology*, *38*(2), 179.

[23] Salimah, N., & Muflikhati, I. (2016). Family Capitals, Livelihood Strategies, and Family Well-Being of Plantation Worker. *Journal of Family Sciences*, *1*(1), 13-23.

[24] Suandi, S. (2010). Analisis Pengaruh Sosial Ekonomi Demografi Rumah Tangga terhadap Kesehatan Reproduksi di Provinsi Jambi (Analysis Data Sdki Tahun 2007). *Jurnal Penelitian Universitas Jambi: Seri Humaniora*, *12*(2), 43406.

[25] Gartaula, H. N., Visser, L. and Niehof, A. (2012) 'Socio-cultural dispositions and wellbeing of the women left behind: A case of migrant households in Nepal'. Social indicators research, 108 (3), pp. 401–420.

[26] Kwon, H. K., Rueter, M. A., Lee, M. S., Koh, S., & Ok, S. W. (2003). Marital relationships following the Korean economic crisis: Applying the family stress model. *Journal of Marriage and Family*, *65*(2), 316-325.

[27] Paudel Khatiwada, S., Deng, W., Paudel, B., Khatiwada, J. R., Zhang, J., & Su, Y. (2017). Household livelihood strategies and implication for poverty reduction in rural areas of central Nepal. *Sustainability*, *9*(4), 612.

[28] Scoones, I.(2009). Livelihoods perspectives and rural development. Journal of Peasant Studies,36(1), 171-196.

[29] Sabania, H., & Hartoyo, H. (2016). Economic Pressure, Livelihood Strategy, and Family Well-being in Cimanuk Watershed, Garut and Indramayu, West Java, Indonesia. *Journal of Family Sciences*, *1*(1), 24-38.

[30] Sunarti, E. Johan, I,J., Haryati, C. (2010). Hubungan fungsi AGIL dengan kesejahteraan keluarga

nelayan yang rawan terkena bencana alam. Jurn. Ilm. Kel. & Kons.,3(1), 11-17

[31] Sunarti, E. Nuryani N, Hernawati N. (2009). Hubungan antara fungsi adaptasi, pencapaian tujuan, integrasi dan pemeliharaan sistem (AGIL) dengan kesejahteraan keluarga.

[32] Islamia, I., Sunarti, E., & Hernawati, N. (2019). Tekanan Psikologis dan Kesejahteraan Subjektif Keluarga di Wilayah Perdesaan dan Perkotaan. *ANFUSINA: Journal of Psychology*, *2*(1), 91-100

[33] Dennis, J. M., Parke, R. D., Coltrane, S., Blacher, J., & Borthwick-Duffy, S. A. (2003). Economic pressure, maternal depression, and child adjustment in Latino families: An exploratory study. *Journal of Family and Economic Issues*, *24*(2), 183-202.

[34] Lyons, M. (2009), "Building back better: the large-scale impact of small-scale approaches to reconstruction", World Development, Vol. 37 No. 2, pp. 385-98

[35] Sharan, P., Chaudhary, G., Kavathekar, S.A. and Saxena, S. (1996), "Preliminary report of psychiatric disorders in survivors of a severe earthquake", The American Journal of Psychiatry, Vol. 153 No. 4, pp. 556-558.

[36] Thornley, L., Ball, J., Signal, L., Lawson-Te Aho, K., & Rawson, E. (2015). Building community resilience: learning from the Canterbury earthquakes. *Kotuitui: New Zealand J Soc Sci Online*, *10*(1), 23–35. doi:10.1080/1177083X.2014.934846

[37] Pradhan, P., Costa, L., Rybski, D., Lucht, W. & Kropp, J. P. A systematic study of sustainable development goal (SDG) interactions. Earths Future 5, 1169–1179 (2017).

[38] Lusseau, D. & Mancini, F. Income-based variation in sustainable development goal interaction networks. Nat. Sustain. 2, 242–247 (2019)

[39] Global Assessment Report on Disaster Risk Reduction 2015 (UN Ofce for Disaster Risk Reduction, 2015

IOP Conf. Series: Earth and Environmental Science **989** (2022) 012022 doi:10.1088/1755-1315/989/1/012022

Community preparedness in facing earthquake in disaster-prone areas of the Lembang fault at Pasirlangu Village Cisarua District, West Bandung Regency

Sutrisno Utomo and Tukino

Bandung Polytechnic of Social Welfare`

*Email: sutrisno.utomo.001@gmail.com, tukinoktn@gmail.com

Abstract. This research aims to determine the community's preparedness in dealing with earthquake disasters in the Lembang Fault disaster-prone area, Pasirlangu Village, Cisarua District, West Bandung Regency, covering six critical aspects including respondent characteristics, knowledge aspects, policy, emergency response planning, early warning system, and system mobilization resources aspects in dealing with earthquake disasters.
This research was using descriptive research method with a quantitative approach. The population in this research was the head of the family of 3,564 people with the Proportional Random Sampling technique and obtained 100 people from 10 communities areas as respondents.
The results showed that the community's preparedness in dealing with earthquake disasters was in the moderate to low category with a score of 6,240 out of a total score of 12,400. In Conclusion, the community at Pasirlangu Village as a Disaster-Prone Areas are barely prepared to face the earthquake disaster. The program proposed to solve the problem is "*Program Pasirlangu Siap Siaga, Cisarua Tanggap Bencana (Pasaga Cinta)*" or the Pasirlangu Ready-Alert Program, Cisarua Disaster-Response which aims to improve community preparedness in dealing with earthquake disasters. The method used is the Community Organization/ Community Developement methods by applying the collaboration techniques, Technology of Participation techniques, and campaigns.

1. Introduction

As a country located between four tectonic plates and consisting of islands located in a series of volcanic belts in the south and east, Indonesia has been named one of the countries that has a high level of seismicity, up to ten times the level of seismicity in the United States. Throughout 2020, there have been twenty-eight earthquakes that hit various regions in Indonesia (dibi.bnpb.go.id). The earthquake disaster occurred in several provinces, including North Sumatra, Central Java, North Sulawesi, Central Sulawesi, West Sulawesi, Maluku and Papua Provinces with the number of intensities from each province as much as once, West Nusa Tenggara Province with an intensity of twice, Nanggroe Aceh Darussalam Province and North Maluku Province with four incident intensity, and West Java Province with the highest earthquake disaster intensity throughout 2020, which was eleven times.

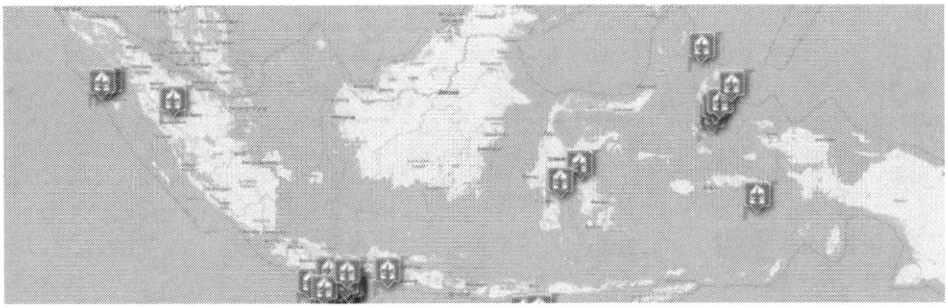

Figure 1. The earthquake disaster occurred in several provinces

West Java Province has a topography of lowlands in the north, highlands and mountains in the middle and hills and beaches in the south. West Java Province is also located on the Circum Pacific and Mediterranean routes, so it has an unstable area, characterized by many active volcanoes and a high frequency of earthquakes. Tangkuban Perahu Active Mountain and Lembang Fault are two of the many potential earthquakes in West Java Province, located in the northern part of Bandung City. The Lembang fault, which extends for twenty-nine kilometers, shifts by 3 millimeters per year and has the potential to cause earthquakes with a magnitude of up to 7 Richter scale.

Indonesia as a country that has a high intensity of earthquakes has a great influence on the social conditions of its people. The earthquake disaster which claimed many lives, created enormous losses, which prompted the Indonesian people to always be ready if at any time they had to deal with an earthquake disaster. Since 2004 after the earthquake and tsunami in Aceh, Indonesian Government has adopted various policies and programs for disaster prevention and management, for example the policy for building and designing earthquake-resistant houses. This policy is very important considering the high percentage of the death toll due to the destruction of houses and buildings, which reached sixty-two percent. In addition to the policy aspect, the knowledge aspect is also an important aspect that must be considered. The rate of self-inflicted safety is 35%, according to a study in Japan [1].

Referring to various information in the previous section, it is known that community preparedness in dealing with disasters, especially earthquakes, is very important. This research was conducted with the aim of knowing the preparedness of the people of Pasirlangu Village, Cisarua District, West Bandung Regency in dealing with earthquake disasters. The research location was chosen based on the geographical location of Pasirlangu Village, which is right on the West Bank of the Lembang Fault. The research was conducted with the aim of measuring the preparedness of the Pasirlangu Village community, with reference to six critical factors, including respondent characteristics, aspects of knowledge, policies, emergency response plans, early warning systems and mobilization of source systems aspects to 100 respondents spread over 10 areas in the village. The research is also carried out based on the concept of Social Work research, in accordance with the field of Research, namely Social Worker in the field of Disaster Management.

2. Literature Review

2.1. Earthquake Disaster

Earthquake is a type of disaster that is caused by sudden ground movements and has a certain duration of time. Indonesian National Board for Disaster Management (BNPB) defines an earthquake as an "energy release event caused by a sudden shift/movement of the earth's interior (crust). Earthquakes are caused by the release of pressure from moving plates, getting bigger and bigger and finally reaching a state where the pressure cannot be resisted by the edges of the plates [2]. The pressures that cannot be resisted by the edges of the slab are released in the form of vibrational energy, causing catastrophic earthquakes to occur. Earthquake disaster is one of the natural disasters that is difficult to predict because earthquakes occur very quickly and have a large impact in a wide area.

2.2. Preparedness
Preparedness is a series of activities carried out to anticipate disasters through appropriate and efficient organization and steps. The Big Indonesian Dictionary defines preparedness as "a state of readiness". Thus, preparedness can also be defined as a state of readiness carried out to anticipate disasters in an effective and efficient manner. Furthermore, according to Judah Carter (in LIPI UNESCO/ISDR 2006: 5) explains, that "preparedness is actions that enable governments, organizations, communities and individuals to be able to respond to a disaster situation appropriately, quickly and effectively. Included in preparedness actions are the preparation of disaster management plans, maintenance of resources and training of personnel".

This research was carried out by adopting the UNESCO-LIPI framework (2006:13-14) which was compiled to determine the level of community preparedness in anticipating the earthquake and tsunami disaster, with the following details:

a. Knowledge, as the beginning of an action and one's awareness, so that with the capacity of knowledge it is hoped that it can be the basis of action of someone.
b. Policies, environmental management, building structures, facilities and infrastructure as well as disaster management systems are factors that directly or indirectly affect the level of preparedness in dealing with disasters as regulated in a policy.
c. Plans for Emergencies, relating to the process of evacuation, aid and rescue efforts in order to minimize casualties during and after a disaster event occurs.
d. Early Warning System, as a series of systems that function to notify an impending disaster
e. Resource Mobilization, aims to facilitate access to basic needs, clothing, food, security and strengthen the sources of life.

2.3. Community
Community is defined as "Unity of human life that interacts according to a certain system of customs that is continuous and bound by a sense of shared identity" [4]. From this definition, it can be explained that the community is a group of people who have a mutually supportive relationship and live in a certain geographical area such as villages, hamlets and urban villages. The term community is defined by Koentjaraningrat (2002:143-144), is a term that comes from the Arabic root "syarakat" which means to participate. So that it can be interpreted that Community is indeed a group of humans who "get along" with each other or in scientific terms, "interact". Thus, Community is the organization of social activities or activities that are able to provide access and ability to its members to get the fulfillment of needs for daily life.

2.4. Social Work in Disaster Settings
Max Siporin (1975:27) defines Social Work as "...an institutional method of helping people to enhance their social functioning and to resolve their social problems". Furthermore, Zastrow H.C (2002:12) defines social work as "...the professional activity of helping individuals, groups, or communities to enhance or restore their capacity for social functioning and to create societal conditions favorable to theirs goal" [7,9]. Referring to the two definitions, it can be concluded that social work is a profession that can provide assistance, both to individuals, groups, and communities in solving their problems and improving and enhancing their social functioning so that they are able to carry out their social functions and roles. Prevention and management of earthquake victims is one of the fields of social work practice. Adjustment and improvement of social conditions are important in efforts to improve and heal for earthquake victims. The definition in principle has 3 main elements, including:

a. Social work is a professional service
b. The practice is based on knowledge and skills
c. Its purpose is to help individuals, groups and communities achieve personal and social satisfaction and freedom.

IOP Conf. Series: Earth and Environmental Science **989** (2022) 012022 doi:10.1088/1755-1315/989/1/012022

Social work is a profession that has a responsibility to improve and develop interactions between people, so that people have the ability to overcome difficulties experienced and realize their aspirations and values. Social workers in carrying out their goals in addition to working with clients, also cooperate with other disciplines as well as with the physical and social environment. In dealing with earthquake victims, social workers work together with communities in the neighborhood of the earthquake victims, as well as related parties who can provide assistance and services to victims of natural disasters in disaster-prone areas.

Referring to Tuti Kartika (2014), regarding the roles of social workers, social workers in the field of disaster have an important role in overcoming problems regarding earthquake disasters. Some of the roles of social workers include [3]:

a. Facilitator, to facilitate or enable individuals or groups to be able to make changes that have been mutually agreed upon.
b. Educators, as educators in providing information, knowledge and understanding to the community about disasters.
c. Broker, connecting individuals or groups who need assistance or services with resources, because the individual or group themselves do not know where the source is and how to reach it.
d. Mediator, helps to mediate when there is a dispute between various groups in order to reach a compromise, reconciliation or reach a mutually beneficial agreement.
e. Counselor (therapist). There is a tendency to view social workers more as therapists than as counselors. The counselor carries out counseling while the therapist does psychotherapy.
f. Case Manager, plays a role in coordinating all assistance activities provided to individual clients that are provided to clients individually or in groups.
g. Consultant, acting as a consultant for aspects of human relations (with training experience in the social psychology of individual behavior).
h. Advocacy, defending the interests of clients to claim their rights and providing input in improving client service programs and policies, especially policies related to disaster to the local government.
i. Planner, planning services needed by clients related to disaster problems based on the results of the assessment, including developing referrals.

3. Methods

The method used in this research uses a quantitative approach with a descriptive survey design. Moh. Nazir (2011: 56) states that the Descriptive Survey Research Method where "Descriptive survey method is an investigation conducted to obtain factors from existing symptoms and seek factual information, whether social, economic or political institutions from all groups or regions [6]." Sugiyono (2016:36) also suggests descriptive research, namely "Descriptive research method is a problem formulation relating to statements about the existence of variables, either only on one or more variables (stand-alone variables) [8]. So in content research the researcher does not make comparisons of that variable to other samples, and looks for the relationship of that variable to other variables.

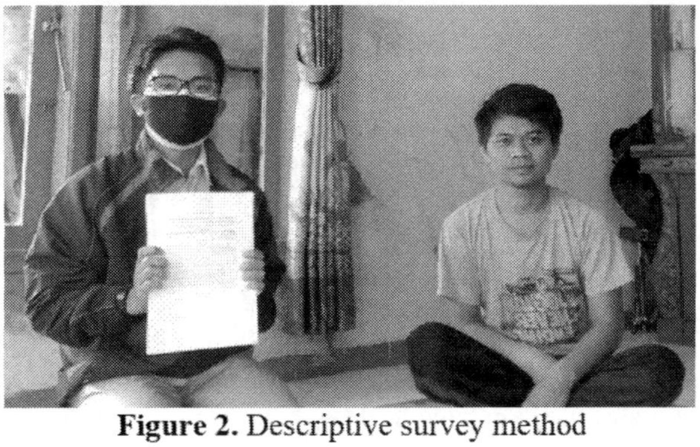

Figure 2. Descriptive survey method

The source of data from this research is a self-administrated questionnaire containing 31 statements regarding aspects of preparedness, namely aspects of knowledge, policies, plans for emergencies, early warning systems, mobilization of sources, and respondent's personal data as complementary data. The population of this study was 3,567 families with a sample of 100 families spread over 10 areas in Pasirlangu Village.

The sampling technique used is proportional random sampling and the Slovin formula is used to determine the sample with an error limit of 10%. Efforts to obtain samples encountered problems where some areas were inaccessible due to the implementation of regional quarantine as an effort to stop the spread of the Covid-19 outbreak.

The measuring instrument used by the researcher is a Rating Scale by dividing the answers to the questionnaire into four answers with graded scores (the highest meaning score is four to the lowest score is one). The validity test in this study uses face validity and the reliability test of the measuring instrument uses the Cronbach Alpha coefficient with the help of the Statistical Product and Service Solutions (SPSS) program. The results of testing the reliability of the measuring instrument show the number 0.965 to the answers of 10 respondents and it can be concluded that the instrument has met the reliable criteria for use in the data collection process.

4. Result and Discussion

The results showed that the preparedness of the people of Pasirlangu Village as one of the villages located in disaster-prone areas was in the medium to low category. Referring to 100 respondents' answers to the 31 statements presented, regarding aspects of knowledge, policies, emergency response plans, early warning systems and aspects of resource system mobilization, a score of 6,240 points out of a maximum total score of 12,400 points was obtained. This is very concerning as it is known that Pasirlangu Village is right above the Lembang Fault which has a high risk of earthquake disasters.

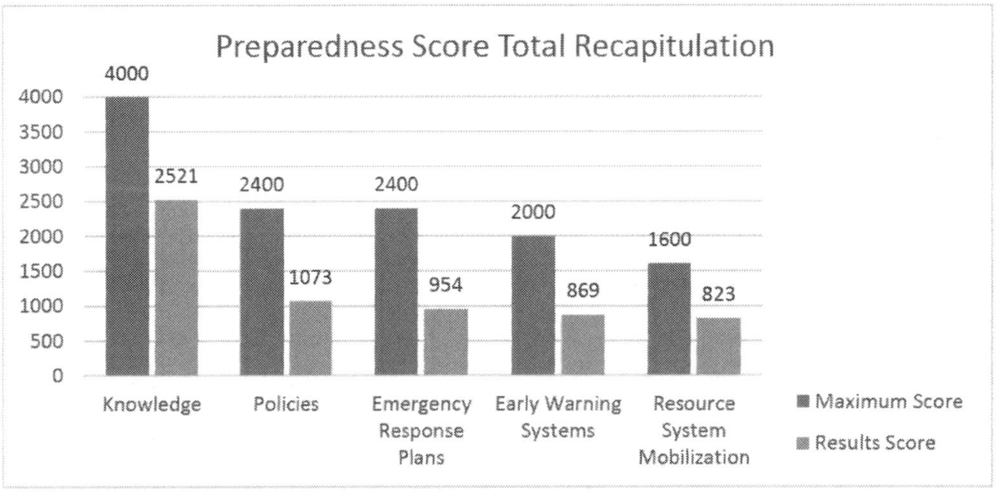

Figure 3. Preparedness Score Total Recapitulation

Based on the diagram above, information is obtained that the knowledge aspect and the resource system mobilization aspect are in the middle category (2,521 points out of 4,000 points and 823 points out of 1,600 points maximum). This was made possible by the basic knowledge possessed by the people of Pasirlangu Village regarding the dangers, risks, signs, and policies regarding earthquake disasters, as well as the mobilization of source systems that have been carried out by some of the people of Pasirlangu Village in dealing with the earthquake disaster. The coordination that has been established, as well as the Karang Taruna which doubles as Disaster Preparedness Cadets have carried out various efforts in solving problems related to disaster issues.

IOP Conf. Series: Earth and Environmental Science **989** (2022) 012022 doi:10.1088/1755-1315/989/1/012022

Based on the same diagram above, it is also known that aspects of policy, emergency response plans, and aspects of the early warning system are in the low category (1,073 and 954 points out of 2,400 points and 869 points out of 2,000 points maximum). Referring to the research results, the three aspects are in the low category, caused by several factors. In the policy aspect, understanding and low implementation of central and regional policies have not been optimal in the Pasirlangu Village area. Policy is one of the important aspects where policy is the legal foundation for infrastructure development, as well as policies in order to regulate emergency response plans, evacuation routes and points, mitigation guidelines and the formation of a disaster preparedness team for the people of Pasirlangu Village.

Most of the people of Pasirlangu Village have not prepared an emergency response plan, contingency plan, evacuation routes and points as well as evacuation locations in the face of an earthquake disaster. Understanding factors that need to be improved, as well as collective action is needed, so that an effective and efficient emergency response plan can be implemented with the hope that the number of fatalities can be minimized. Similarly, aspects of the early warning system as one aspect of preparedness in dealing with disasters. Availability and easy access to an early warning system, as well as routine control over the early symptoms of an earthquake, are still very limited. The available early warning system is still conventional, and has never been used in training conditions or when an earthquake occurs.

4.1. Discussion

Preparedness in dealing with disasters is one of the important factors that must be considered for people living in disaster-prone areas. The results of the research indicate that the level of preparedness of the people of Pasirlangu Village is still and must be improved, so that they can be "prepared" in the face of an earthquake disaster. This can be done in several ways. Increasing public understanding and awareness of the risk of earthquake disasters in the vicinity is one of the important factors in increasing preparedness. Access to correct and easily accessible information is also necessary, considering that in general, humans act according to their knowledge. The Village digitization program also needs to be supported by all parties, considering that internet access in the Pasirlangu Village area, which is located right on the slopes of Mount Burangrang, is still quite limited.

The existence of local policies that regulate various aspects in the field of disaster can be immediately compiled and included in the agenda for discussing village regulations. Thus, the emergency response plan and early warning system established by the community and the Village Government can be realized. Furthermore, good coordination between the community, the Village Government, the Regional Disaster Management Agency (BPBD) of West Bandung Regency, as well as other multidisciplinary in helping the community of Pasirlangu Village to be ready to face the earthquake disaster.

Social Workers with Disasters and Refugees have the capacity to provide assistance, both before and after an earthquake disaster occurs. Social Workers as experts, have various roles to improve community preparedness in dealing with earthquake disasters through various methods and techniques they have. Community Development is a social work intervention method, and can be combined with collaboration techniques (social workers collaborate with local communities in providing assistance), participatory technology techniques as an effort to solve problems and find the best solutions, and campaign techniques.

In addition, several specific roles of Social Workers with Disasters and Refugees that can be applied are the role of Educators, to increase awareness and understanding of the community. Social workers can also act as advocates, by advocating for the necessary policy designs to local governments, as well as planners in formulating the direction of change so that they can achieve the expected results. Thus, social workers comprehensively act as agents of change. These various roles and activities were proposed in the "*Program Pasirlangu Siap Siaga, Cisarua Tanggap Bencana (Pasaga Cinta)*" or the Pasirlangu Ready-Alert Program, Cisarua Disaster-Response Program as a form of research contribution, including community consultations, disaster counseling, village policy formulation and the establishment of an information system based on Whatsapp messages for the people of Pasirlangu Village.

5. Conclusion

Indonesia as a country that has abundant natural resources also has a high risk of disasters, especially earthquakes. "An ounce of prevention is worth a pound of cure" is a saying that is often touted, not least when it comes to dealing with a disaster situation. Although there are no specific efforts to prevent the occurrence of earthquake disasters, there are still various efforts to reduce and even prevent the loss of life through a series of activities within the framework of disaster preparedness.

Various efforts and participation from all parties are needed considering that disaster issues are not the responsibility of only a few parties. Central and Regional Government, Central and Regional Disaster Management Agency, Health Institutions, Education, Security Institutions, business world, and various other parties, including Social Workers with Disasters and Refugees. Reflecting on past catastrophic events, the magnitude of the loss of life and total damage can be greatly reduced by taking appropriate precautions and preparedness.

At the end of this paper, the author hopes that a better synergy will be created from various parties regarding disaster mitigation and management efforts. Togetherness is a unique value that is owned by the Indonesian people, has been passed down from generation to generation, and it is the responsibility of all of us to maintain it. The author also hopes that the results of this research can be useful in various sectors and various scientific fields.

Acknowledgements

I would like to thank the Director and Head of the Research Center of Bandung Polytechnic of Social Welfare, who have allowed this research.

References

[1] BNPB. 2011. *Indeks Rawan Bencana.* Jakarta: Direktorat Pengurangan Risiko Bencana, Deputi Bidang Pencegahan dan Kesiapsiagaan
[2] Ella and Usman. 2008. *Mencerdasi Bencana.* Jakarta: Grasindo
[3] Kartika, Tuti. 2011. *Anak Jalanan dan Model Penanganannya: Studi kualitatif tentang Anak Jalanan yang dibina oleh Yayasan Dian Mitra, YKAI dan Yayasan Amalia di Jakarta.* Jakarta.
[4] Koentjaningrat. 2002. *Pengantar Ilmu Antropologi.* Jakarta: Rineka Cipta
[5] LIPI. 2006. *Kesiapsiagaan Masyarakat dalam Mengantisipasi Bencana Gempa Bumi dan Tsunami.* Jakarta. LIPI-UNESCO/ISDR
[6] Nazir, Mohammad Ph.D. 2011. *Metode Penelitian.* Jakarta: Ghalia Indonesia
[7] Siporin, Max. 1975. *Introduction to Social Work Practice.* New York: Macmillan Publishing Co, inc.
[8] Sugiyono. 2016. *Metode Penelitian Kuantitatif, Kualitatif dan R&D.* Bandung: Alfabeta.
[9] Zastrow, H.C. 2000. *Introduction to Social Work dan Social Welfare.* USA: Thomson Books/Cole

IOP Conf. Series: Earth and Environmental Science **989** (2022) 012023 doi:10.1088/1755-1315/989/1/012023

Optical and Electrical Based Indoor Landslide Early Warning System Based on the Change of the Wall Crack Width

Ahmad Marzuki[1*], F. A. Purnomo[2], N. M. Yoeseph[2], A. M. Imran[3], W. Suryanto[4], and T. Juhana[5]

[1]Department of Physics, Sebelas Maret University, Surakarta, Indonesia
[2]Diploma 3 Informatics Engineering Vocational School, Sebelas Maret University, Surakarta, Indonesia
[3]Department of Geology and Environmental Science, Hasanuddin University, Makassar, Indonesia
[4]Department of Geophysics, Gadjah Mada University, Yogyakarta, Indonesia
[5]Department of Electronics Engineering, Bandung Institute of Technology, Bandung, Indonesia

*Email: amarzuki@mipa.uns.ac.id

Abstract. Wall Cracks in a slope area with high landslide risk is very common. Triggered by small horizontal land movement, relative movement of walls separated by the crack line to create wider crack can occur. In the present work, an early warning system for landslide utilizing such movement is presented. For this purpose, a bar is connected to the displacement sensor. Sensor is fixed to one side of wall at either before or after crack line. The bar from this sensor is fixed to the other side of the wall at the direction perpendicular to the crack. With this configuration, small movement of one part of wall (before or after crack line) activate the sensor. In this paper, two displacement sensors are used, i.e., displacement sensor based on slide resistor and fiber sensor. In the present work, warning is given in the form of sound characterized the increase of frequency as the width of crack increase.

1. Introduction

In a mountainous country like Indonesia, landslide is common disaster that occurs almost every year. Rapid growth population, quality of air and beautiful view that the mountainous area can offer increase the number of people living in such hazardous area. Effort to migrate people from such hazardous area to a saver area is not easy. They feel that they have been there since long time ago and have been used to be in such condition for decades. Housing and agricultural extensification are blamed to be the main source causing landslide is more frequent than before. When it come to a very heavy rain, people living in a slope area are haunted by a possible landslide disaster that can kill them.

To help them reducing such anxiety, various landslide early warning systems have been made. The most common method to detect early land movement is that based on borehole data [1,2] utilizing tiltmeters, extensometer and inclinometer [3,4,5]. In the above literatures, landslide monitoring was carried out by directly looking at land surface movement. In this paper, we present a novel method to monitor land

IOP Conf. Series: Earth and Environmental Science **989** (2022) 012023 doi:10.1088/1755-1315/989/1/012023

movement indirectly, that is by measuring the width of wall crack that several house may have. It is very common that wall crack can be very common in area where a land where a house is built is displaced.

2. Methods

Figure 1 shows a typical wall crack commonly occur in a house located in a slope area. Figure 2 is basic arrangement of landslide sensor developed in this research. When land displacement occurs, one side site of wall separated by the crack is moving relatively to the other site creating the width of crack become wider. Sensing was carried out by placing the unit sensor (Figure 2) such that the sensor box is adhered to one site and the hole of the arm numbered 1 is fixed at the other site. When one site of the wall is displaced, the arm (#2) pulls the variable resistor or optical fibre sensor. Sliding the arm (#2) causes the variable resistor changes its resistance. If the sensing element is fibre coil, sliding the arm (#2) cause the change of the coil shape from circular to elliptical shape and therefore reducing light intensity of light passing through the coil. The working principle of later sensor is described in [6]

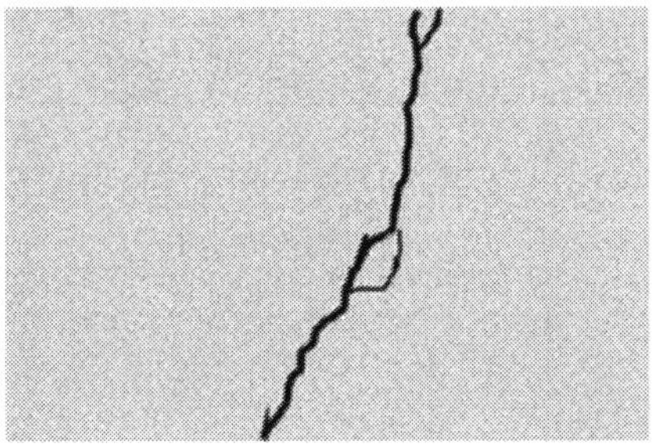

Figure 1. Typical wall crack commonly occurs because of land movement

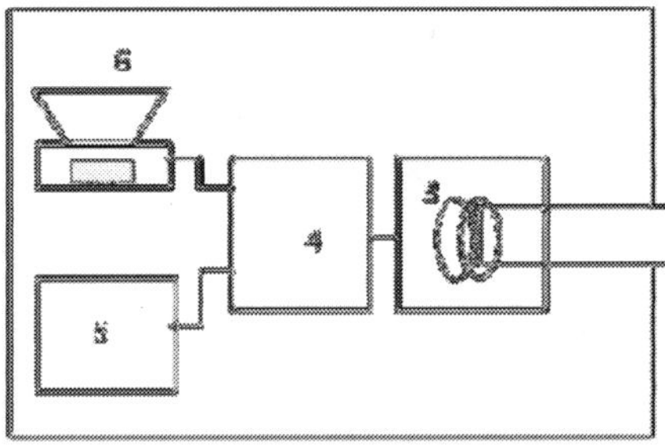

Figure 2. Basic setup of landslide sensor developed in this research. (1) hole, (2) sensor arm, (3) fiber coil or variable resistor, (4) microcontroller, (5) energy source, (6) siren

3. Results and Discussion

Figure 3 give relation between relative movement of slider (distance) and output voltage for the type of sensor as given by Figure 2 at which the sensing element is variable resistor. Figure 4 is relation between light transmitted through fibre coil as the slider is pull by the moving wall. As seen, the transmitted light decreases linearly with the increase of the reducing vertical diameter (minor axis) of coil (indicated by y axis in Figure 5).

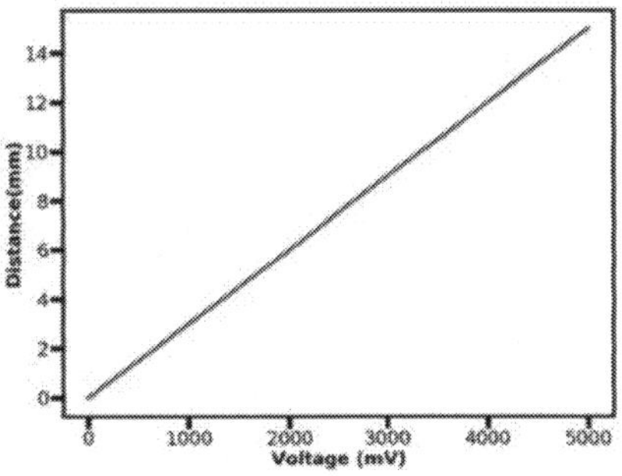

Figure 3. Measured output voltage representing relative distance movement of slider

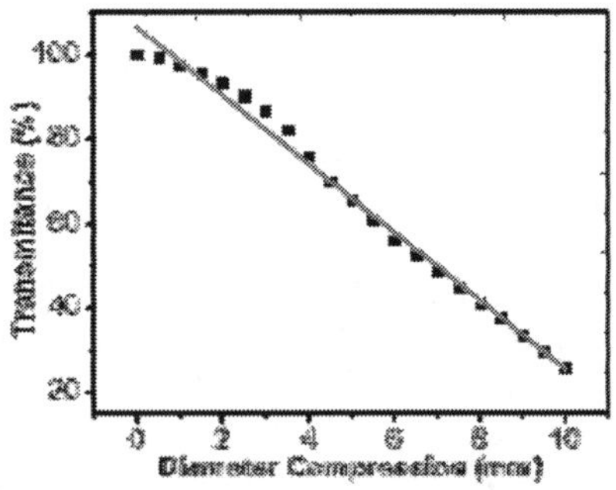

Figure 4. Transmittance change resulted from the change in coil shape from circular shape to elliptical shape

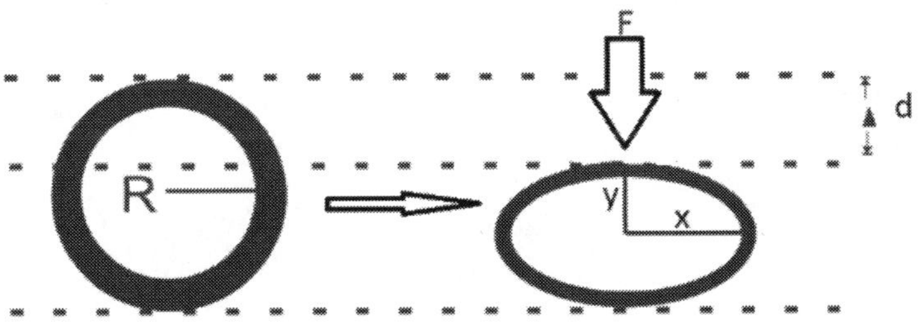

Figure 5. Change of coil shape from circular to elliptical shape that decrease the transmitted light

Decreasing the transmitted light as shown in Figure 4 can be explained using basic optic for light propagation in fibre optic. As shown in Figure 5, increasing F due to increasing displacement d causes increasing curvature of the fibre at the point crossing x axis ($\pm x_{my}$, 0). The crossing points at x axis which represents the length of mayor axis of an ellipse have the absolute values which increase with increasing d. Relation between fiber curvature (κ_{max}) at this crossing points and d is given by [6]:

$$\kappa_{max} = \left(r^2 + rd - \frac{d^2}{4}\right)^{1/2} \left(r^2 - rd + \frac{d^2}{4}\right)^{-1} \quad (1)$$

Where r is radius of the circular shape of the fibre before being displaced. Increasing d results in increasing κ_{max}. The attenuation coefficient (α) contributed by such configurational change therefore can be written as [6]

$$\alpha = A.\,exp\left[-B\left(r^2 + rd - \frac{d^2}{4}\right)^{1/2} \left(r^2 - rd + \frac{d^2}{4}\right)^{-1/2}\right] \quad (2)$$

where A and B are constants. It is clear from equation (2) that attenuation coefficient increases with increasing d, meaning that light transmitted through optical fibre decreases as the width of the wall crack increases. Moreover, equation (2) shows the exponential nature of α and therefore for light transmittance T. In order to have a good sensor having a good linearity, the initial condition for the sensor before being displace is not necessarily in circular shape. Instead, an elliptical shape of fibre coil having a certain length of major axis corresponding to a linear part of the transmittance – displacement curve is chosen.

Unlike outdoor landslide sensor, indoor landslide sensor made using a method developed in this research limited to displacement up to 5 cm but the accuracy can be adjusted to the displacement accuracy 0.01 cm. Having such high accuracy, detecting very small displacement of wall as shown by Figure 1 can be made very easily. Figure 6 is an example siren frequency change to describe how land movement behaves. Example as shown in Figure 6 gives frequency change from initially low frequency which corresponds to small distance of landslide to high frequency siren informing that land is moving continuously and the crack wall become wider and wider.

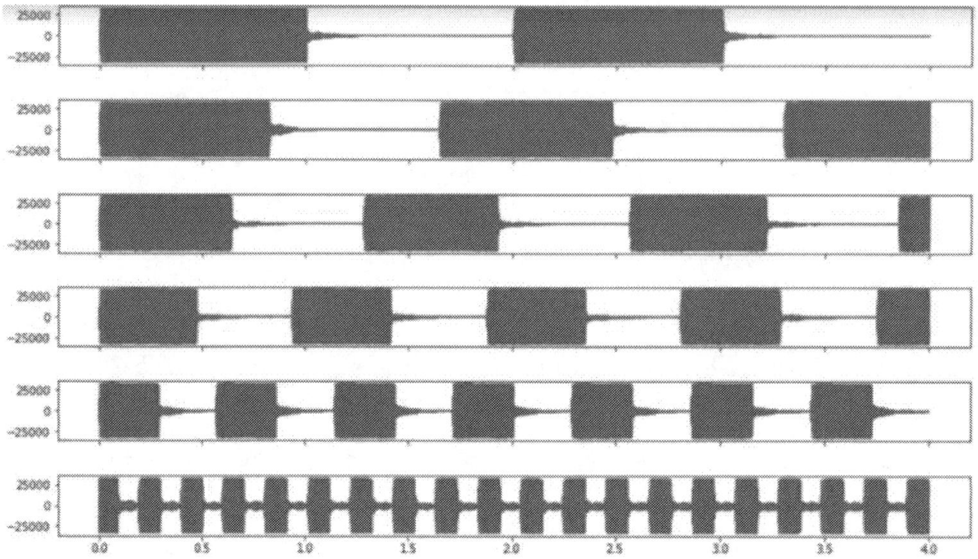

Figure 6. Example of siren frequency change (from top to bottom) made as the wall crack get wider and wider.

4. Conclusions

Land slide sensor made based on variable resistor and fiber sensor have been developed for detecting land movement through an indirect method that is by monitoring the width change of the crack wall occurs because of land movement. According to this method, sensing was made by placing the sensor across the crack such that when one site of the wall make relative movement, physical properties of sensing elements (variable resistor or fiber sensor) are changed.

References

[1] Sun YJ, Shi B, Zhang D, Tong H J, Wei G Q, Xu H Z 2016 *J. Sensors* 2016 1–8.
[2] Zhu Z W, Liu B, Liu P, Zhao B Y, Feng Z Y 2017 *Catena* 155 30–40.
[3] García A, Hördt A, Fabian M, 2010 *Geomorphology* 120 16–25.
[4] Intrieri E, Gigli G, Mugnai F, Fanti R, Casagli N 2012 *Eng. Geol* 147–148 124-136.
[5] Pei H, Jing J, Zhang S 2020 *Measurement* 151 107172.
[6] Marzuki A, Heriyanto M, Setiyadi I D, Koesuma S 2015, *J. Phys. Conf. Ser.* 622 012059

IOP Conf. Series: Earth and Environmental Science **989** (2022) 012024 doi:10.1088/1755-1315/989/1/012024

Level of Vulnerability of Aipiri Village to Climate Change

Rima HS. Siburian[1,*], Julia R. Tapilatu[2,3] and Maryrose E. Tapilatu[2,4]

[1] Faculty of Forestry and Postgraduate Program, University of Papua, Manokwari 98314 – Papua Barat INDONESIA

[2] OurConservaSea Community Group, Manokwari 98314 – Papua Barat INDONESIA

[3] Faculty of Fisheries and Marine Science, University of Papua (UNIPA), Manokwari 98314 – Papua Barat INDONESIA

[4] Marine Science Graduate Program, University of Diponegoro, Semarang 50275 – INDONESIA

*Email: r.siburian@unipa.ac.id

Abstract. Aipiri Village in the eastern part of Manokwari District, West Papua, is one of the villages directly adjacent to the Pacific Ocean. This village is very vulnerable to the threat of sea-level rise. This study aims to determine the level of vulnerability and adaptation of the community at Aipiri village to climate change. Data were collected through a survey method with interview techniques and direct observation to obtain data on the level of vulnerability of the village to climate change. The results of this study indicated that sea-level rise is the biggest threat to the people of Aipiri village. In addition, when referring to the level of resilience, the pattern of community adaptation through community local wisdom and knowledge, the environmental infrastructure aspect is an adaptation solution that needs to be developed. Ecotourism areas are developed as a form of local wisdom in defending the area against climate change.

1. Introduction

The climate change phenomenon that has occurred in the last decade is the result of the implication of increasing greenhouse gases in the earth's atmosphere, which causes various problems of imbalance in the atmospheric system and parts of the earth's surface [1]. Various problems can arise from this phenomenon, such as the increased potential for rainfall, sea level, coastal abrasion, decreased beach quality, turbidity, changes in salinity, increased parasites, and algae [2]. The adverse impact of climate change can increase vulnerability of the environmental, social, economic, and political sectors in various regions, especially in the coastal areas directly adjacent to the sea.

The coastal area is one of the strategic areas that play a vital role in socio-economic factors, with a population range higher than other land areas, which is 60% [3]. As a society that depends on the conditions and the existence of marine resources, climate change has great potential that can affect people's lives from various dimensions, such as the possibility of damage to infrastructure, settlements, and facilities needed to support the lives of coastal communities, decreased fishing income due to waves, high levels of poor education, declining levels of public education, and the emergence of various diseases due to climate change [4], [5].

In this regard, various efforts and adaptations of coastal communities need to be carried out to prevent the impact of vulnerability and increase resilience to the threat of climate change [6]. Furthermore, various climate change adaptation and mitigation efforts need to be supported by various parties such as the government, non-governmental organizations, academics to optimize efforts to prevent the adverse impacts of climate change. This is also in line with the statement of [7] which explains that efforts to reduce the risk of disasters can increase community resilience in facing threats that occur in their environment, such as climate change.

Aipiri Village is one of the residential areas with a stretch of coastal area directly opposite the Pacific Ocean. When viewed based on its geographical location, this village area also has a relatively high level of vulnerability to climate change. The high potential vulnerability of the Aipiri community to climate change needs to be addressed with the right strategies and adaptations in order to be able to maintain their lives in the present and in the future [8]. Based on this, this study was conducted to determine the level of vulnerability and adaptation efforts of Aipiri village to climate change.

2. Methods

2.1 Time and Location
This research was conducted in September 2021 in Aipiri Village, East Manokwari District, Manokwari Regency, West Papua Province. In this case, the Aipiri village area has a significant topographical difference from the coastal area to the mainland with various levels and divisions of different village areas.

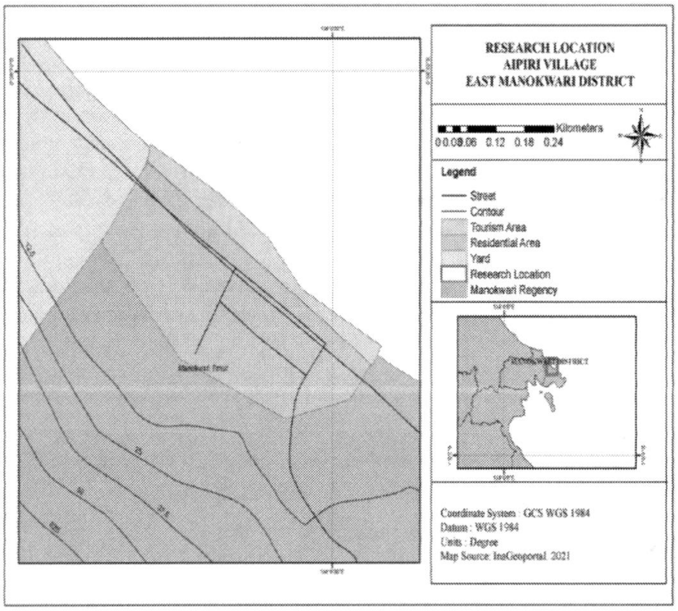

Figure 1. The Location of Aipiri Village, Manokwari Regency, Indonesia

2.2 Data Analysis
The implementation of this research was carried out using a two-way survey method in interviews and observations. In this case, there are five main aspects to be considered, including:

1. The business aspect consists of the type of work, economic level, and entrepreneurship of the village community.
2. Aspects of human resources that are reviewed based on age, gender, level of education, and population density
3. The aspect of surrounding resources is reviewed based on the potential of coastal resources and the development of coastal resource management.

4. Aspects of readiness and adaptation to climate change consist of components: the shape of the building, the distance of the building, the availability of facilities, and regulations that appeal to disaster warnings.
5. Environmental aspects and infrastructure conditions are reviewed based on: the condition of buildings and houses, the availability of clean water, and the availability of health facilities

2.3 Data Analysis
Furthermore, the data obtained from the results of this study will be analyzed and categorized into five groups to determine the level of vulnerability of the aipiri village to climate change conditions.

3. Result and Discussion
The phenomenon of climate change has the potential to impact various aspects of human life, where one area that has the potential to experience danger from climate change is Aipiri Village, Manokwari Regency. In this regard, various adaptation and mitigation efforts need to be taken in order to minimize the impacts that will occur. Based on the results, the readiness and level of vulnerability of Aipiri village to climate change can be reviewed based on two main aspects, including: human aspects, business aspects, resource aspects, environmental and infrastructure aspects, and aspects of disaster preparedness and climate change. The results and discussion of research can be reviewed based on the following points:

3. 1 Human Aspect
Based on the KKP Coastal Village Resilience Assessment Guidelines [9], the human aspect is one of the aspects included in the fixed, variable assessment section, which is reviewed based on the characteristics of the Aipiri village community, such as race and ethnicity, age, occupation, and education. In this case, the population of Aipiri is 365 people, consisting of three age groups, namely the productive age group, children, and the elderly.

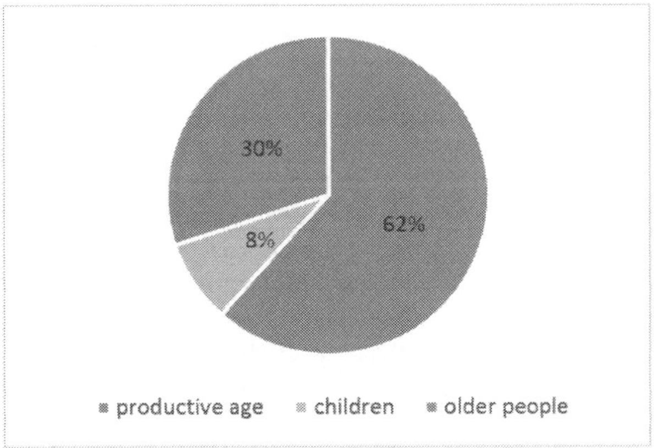

Figure 2. Age group percentage

In this case, the age group affected by the threat of climate change in Aipiri village is the age group of children and the elderly. This is because, the age group has limitations in dealing with disasters and the high risk of a disaster. Children and the elderly are the age group with a high level of vulnerability to a disaster [7]. Based on the data, it can be seen that the age range of the people of Aipiri village is dominated by productive age, which is 62%. Besides age, aspects of education level, occupation, and race and ethnicity are important factors in increasing understanding and adaptation to climate change. Someone who has completed junior high school education generally has good knowledge of reading, writing, and arithmetic. This will facilitate delivering information and efforts to adapt society to the threat of climate change. In this regard, 51% of the people of Aipiri village have completed junior high school education. This then affects the mindset of ethnic and racial acculturation that occurs harmoniously between village communities. Based on the data results, it can be concluded that the support for the human aspect of Aipiri village has a strong potential for resilience in facing the threat of climate change.

3. 2 Business Aspect
The business aspect is one aspect of regional vulnerability assessment, which is reviewed based on the type and diversity of livelihoods of the people of an area. In this case, the results of the survey of the business aspects of Aipiri village have a significant difference compared to the human aspect. This may happen as the level of diversity in the livelihoods of the villagers is included in the low category and is generally dominated by agricultural livelihoods such as fishermen and farmers. This livelihood condition is also very common in various other coastal areas, where people use the area between land and sea to support life and fulfill daily needs.

Figure 3. The agricultural livelihood of Aipiri Coastal Community

However, due to climate change that disrupts soil salinity levels, the emergence of storms, coastal abrasion, and water problems, this has the potential to cause new problems for agricultural production in coastal areas [10], [11], [12]. If this condition continues to be left without alternative livelihoods, then this will become a substantial problem that has an impact on economic growth, education, and even the health of the village community. This was called the gripping condition of climate change occurs and may have an impact on the paralysis of the economic system and people's sources of income [13]. Based on this, the role of the government, organizations, and the community is very much needed to deal with this problem and create alternative jobs that can help the Aipiri village community to adapt and survive when the problem of climate change in the coastal environment is increasingly gripping.

3.3 Resources Aspect
The resource aspect is one aspect of assessing the level of vulnerability reviewed based on the condition of the coastal ecosystem, which is the supporter of the community's life resources. Basically, the tropical coastal ecosystem consists of three complex ecosystems: coral reefs, seagrass beds, and mangroves that are interconnected. In addition to resources, coastal ecosystems also have an important role as a buffer area that protects coastal areas by reducing wave energy and distributing nutrients to nearby ecosystems [14].

Figure 4. Ecosystem in Aipiri Village

The study results indicate that the availability of coastal resources has a strong contribution in supporting the resilience of coastal community life. This is also reinforced by the high value of the use of coastal resources to support the business aspects of the community in the village, which is 85%. If one part of the coastal ecosystem is disturbed, there will be an imbalance and affect other ecosystems [15]. In this regard, sustainable management of coastal resources and ecosystems is needed in order to support community welfare and protect villages from the threat of climate change, without compromising the primary function of coastal ecosystems [16]

3. 4 Environment and infrastructure Aspect

The environmental and infrastructure aspects are reviewed based on the conditions and availability of facilities and infrastructure that support the lives of the people of Kampung Aipiri. Infrastructure conditions and development require the role of the government and the community to assist in realizing and carrying out their functions in a sustainable manner [17].

Figure 5. Tourism infrasturcture in Aipiri Village

In this regard, the survey results on environmental and infrastructure aspects of Aipiri village are categorized in the good category, where the supporting infrastructure conditions such as road access from the village outside the area, the condition of the asphalt road, and the availability of adequate clean water. Furthermore, the entire state of village infrastructure is also supported by the existence of various types of facilities by the government and the private sector such as tourism facilities, gathering facilities, and health facilities. Good facilities and infrastructure need to be managed through an infrastructure management system for the availability of clean water. It is important to pay attention and manage infrastructure development to help reduce business risks, support environmental, social, and economic aspects of the village area [18], and support efforts to minimize the impact of climate change disasters at Aipiri village.

3.5 Disaster Preparation and Climate Change Aspect
Disaster preparation and climate change are essential aspects that need to be considered to reduce the adverse effects of phenomena or disasters on an area. In this case, the people of Aipiri village already have some knowledge and understanding about disaster mitigation and management efforts and the phenomenon of climate change, which is expressed through the manufacture of artificial reclamation from organic materials, construction of semi-permanent houses, raising the base of the house, and the preservation of coastal mangrove ecosystems and coral reefs.

Figure 6. Semi-permanent and raising the base of the building in Aipiri Village

Furthermore, the infrastructure that supports efforts to prepare and manage disasters due to climate change in this region is still inadequate. This can be seen from the observation that there are no facilities or infrastructure for protecting coastal areas and early disaster warnings in this village. However, the government and various private parties are also trying to support for mitigation and efforts to reduce the impact of climate change through outreach activities with the Aipiri village community.

4. Conclusion
The geographical position of Aipiri village directly facing the Pacific Ocean makes this area very vulnerable to the impacts of climate change such as sea-level rise, the emergence of various storms, tsunami events, the emergence of various types of diseases. In this case, the entire impact of climate change can then affect economic inequality in the education of coastal communities.

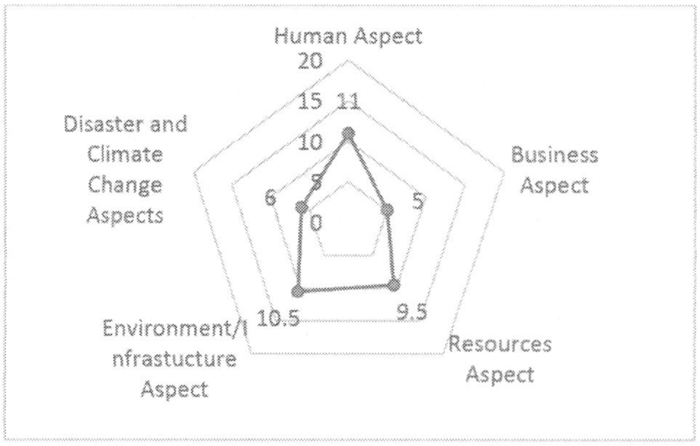

Figure 7. The resiliency of Aipiri Village

In this regard, various prevention efforts need to be carried out to reduce the adverse effects of the climate change phenomenon in Aipiri Village. The government, the private sector, institutions, and the community plays a vital role in developing this effort. In this case, the government has provided various facilities and infrastructure to support community's needs in the form of paved roads, the availability of posyandu, and tourism buildings, as well as the provision of clean water by the institution. In addition to contributions from outsiders, adaptation and prevention efforts by the community play an important role in adapting to the phenomenon of climate change. This can be seen from the results of graphical analysis, where human aspects such as age range, race and ethnicity, type of work, have the highest score of 11. On the other hand, the business aspect is the lowest value with a value of 5. This is because the least number of variations of community alternative work so that this aspect has the lowest value compared to other aspects.

References

[1] Susandi, A., Herlianti, I., Tamamadin, M., & Nurlela, I. (2008). Dampak perubahan iklim terhadap ketinggian muka laut di wilayah Banjarmasin. *Jurnal ekonomi lingkungan*, *12*(2).

[2] FAO. 2015. Coping with climate change–the roles of genetic resources for food and agriculture. Rome.

[3] Doukakis, E. (2005). Coastal vulnerability and risk parameters. *European Water*, *11*(12), 3-7.

[4] Howden, S. M., J. F. Soussana, F. N. Tubiello, et al. 2007. Adapting agriculture to climate change. Proceedings of the National Academy of Sciences 104(50): 19691-96

[5] Kolopaking, L. M., Adiwibowo, S., & Pranowo, M. B. (2014). Adaptasi perubahan iklim komunitas desa: studi kasus di kawasan pesisir utara Pulau Ambon. *KOMUNITAS: International Journal of Indonesian Society and Culture*, *6*(1), 57-69.

[6] Smit, B. and Wandel J. 2006. Adaptation, adaptive ca-pacity and vulnerability. Global Environmental Change, 16: 282-92

[7] Siregar, J. S., & Wibowo, A. (2019). Upaya pengurangan risiko bencana pada kelompok rentan. Jurnal Dialog dan Penanggulangan Bencana, 10(1), 30-38.

[8] Marshall N.A., Marshall P.A., Tamelander J., Obura D., Malleret-King D. and Cinner J.E. 2010. A Framework for Social Adaptation to Climate Change; Sustaining Tropical Coastal Communities and Industries. Gland, Switzerland, IUCN

[9] KKP. Panduan Penilaian Kondisi Ketangguhan Desa Pesisir. 2012.

[10] Gopalakrishnan, T., Hasan, M. K., Haque, A. T. M., Jayasinghe, S. L., & Kumar, L. (2019). Sustainability of coastal agriculture under climate change. *Sustainability*, *11*(24), 7200.

[11] Hoque, M. Z., Cui, S., Xu, L., Islam, I., Tang, J., & Ding, S. (2019). Assessing agricultural livelihood vulnerability to climate change in coastal Bangladesh. *International journal of environmental research and public health*, *16*(22), 4552.

[12] Van Dan, T., Bond, J., & Thuy, N. T. T. (2018). Assessment of Impacts of Climate Change on Agriculture and Fisheries in the Coastal Areas of Thua Thien Hue Province, Vietnam. *Agriculture and Development Notes*, *8*, 1-4.

[13] Kantamaneni, K., Rice, L., Yenneti, K., & Campos, L. C. (2020). Assessing the vulnerability of agriculture systems to climate change in coastal areas: A novel index. *Sustainability*, *12*(11), 4771.

[14] Dahuri, Rokhmin. 2003. Keanekaragaman Hayati Laut: Aset Pembangunan Berkelanjutan Indonesia. PT Gramedia Pustaka Utama. Jakarta.

[15] Kusmana, C. (2007). Konsep pengelolaan mangrove yang rasional. *Makalah pada Kegiatan Sosialisasi Bimbingan Teknis dan Pemantauan Pelaksanaan Rehabilitasi Mangrove. Makassar*, *13*.

[16] Adli, A. (2016). Profil ekosistem lamun sebagai salah satu indikator kesehatan pesisir Perairan Sabang Tende Kabupaten Tolitoli. *JSTT*, *5*(1).

[17] Yuningsih, A. (2017). Local government relations sebagai mitigasi risiko sosial pembangunan. *BOOK CHAPTER PUBLIC RELATIONS AND POLITICS*, 37.
[18] Soemitro, R. A. A., & Suprayitno, H. (2018). Pemikiran Awal tentang Konsep Dasar Manajemen Aset Fasilitas. Jurnal Manajemen Aset Infrastruktur & Fasilitas, 2.

IOP Conf. Series: Earth and Environmental Science **989** (2022) 012025 doi:10.1088/1755-1315/989/1/012025

Identification of Land Use and Land Change due to its impact on Groundwater Level in Surakarta city

Sorja Koesuma[1,*], Sunaryo[2], Ari Handono Ramelan[1], Doddy Sutarno[3]

[1]Department of Physics. Sebelas Maret University.
[2]Department of Geophysical Engineering, Brawijaya University.
[3]Department of Earth Physics and Complex System. Bandung Institute of Technology

[*]Email: sorja@staff.uns.ac.id

Abstract. Surakarta City is the center of economic activity in the great Solo area. As a result, land use changes occur very quickly. This study aims to identify land use and land changes due to their effects on groundwater levels in Surakarta City. The method used in this research is a descriptive-quantitative method. The analysis technique used is time series, by comparing image data for 2003, 2008, and 2019. The image is obtained from Google Earth and then visually interpreted in a GIS application to get the land use area. Land use in Surakarta City includes buildings, agriculture, vacant land, tourist land, waters, settlements, green open spaces, rice fields, shrubs, and dryland agriculture. The results of image processing show that there is a change in land use in Surakarta City. The results show that there is a relationship between land use/change with an increase in population and ground water use. The areas most affected by the decline in groundwater are in the central part of the city of Surakarta, especially in the villages of Sangkrah and Semanggi.

1. Introduction

Indonesia's population growth rate from 2010 to 2019 was 1.31% [1]. The rapid growth rate causes an increase in the need for land, especially for settlements, resulting in changes in space use. Changes in space use are changes in certain land uses into other land uses caused by economic developments and population dynamics which always increase every year. [2]. The increasing demand for urban land has led to diversification of land use patterns (eg, industrial use, housing and infrastructure, and ecological conservation) [3]. Most of the world's urban areas have experienced substantial land cover changes over the years [4]. In the last three decades, urban development in many global cities, such as Shanghai in China, Singapore, and Jakarta, Indonesia, has been marked by the proliferation of high-rise buildings [5], [6]. An urban area can be broadly defined as a city or a metropolitan area, it is the most dynamic land change system in the world [7].

Population growth, agricultural expansion and economic development are the drivers of increasing water demand worldwide [8]. This is often the case in large cities which are dynamic areas. The increasing demand for water in urban areas makes water intake higher. If groundwater abstraction exceeds natural groundwater recharge for a large area in the long term, over-exploitation or continuous depletion of groundwater occurs [9]. Water is becoming increasingly scarce as the human population continues to grow and demands high quality water for household and economic activities [10]. Groundwater is the water source that is most often used by humans to support domestic needs (households), industry, services, and agriculture [11].

One of the big cities in Indonesia is Surakarta City. The city of Surakarta is one of the cities that has considerable potential as a centre of economic activity, in this case the role of the industrial sector is still dominant in the formation of the city budget and the trade sector, so that the industrial sector still exists. economic driver in the Surakarta City. [12]. The development of the city of Surakarta occurs in the economic, industrial, education and tourism sectors which will spur the flow of economic growth in the Surakarta City. [13]. This condition will spur people to move to the Surakarta City or referred to as the phenomenon of urbanization. This phenomenon will increase the demand for residential land. The rapid rate of urbanization with various purposes and limited land availability, allows land conversion from vegetated land to non-vegetated land or settlements [14]. The settlement area is one area that is increasing every year.

Increasing land changes will have an impact on groundwater in Surakarta City. Several studies on land use changes affecting groundwater have been carried out in other areas. As in Guwahati City The urban environment has increased at the expense of vegetation, vacant land and open areas, long-term analysis of groundwater level data shows a steady decline, this downward trend is due to urbanization and population growth [15]. The results of this study indicate that the increase in population and changes in land use greatly affect the availability of ground water. Changes in land use in Surakarta City allow for an influence on groundwater. The previous explanation shows the importance of research with the aim of knowing changes in land use that have an impact on groundwater in Surakarta City. The purpose of this study is first to determine land use changes in Surakarta City and second to determine the impact of land use changes on groundwater levels.

2. Methods

This research was conducted in Surakarta City, Central Java Province, Indonesia. Surakarta City consists of 5 districts which include Jebres District, Banjarsari District, Serengan District, Laweyan District and Pasar Kliwon District. Based on the 2020 population census, the population in Surakarta City has reached 522,364 people. The purpose of this study is first to determine land use changes in Surakarta City and second to determine the impact of land use changes on groundwater levels.

The method used in groundwater monitoring is the geoelectric method. The geoelectrical resistivity technique, involving vertical electrical sounding (VES) surveys, was performed using the Schlumberger electrode configuration [25]. This technique has been utilized for a wide variety of groundwater exploration studies [26]. The tools used in this research are OYO Resistivity meter with 4 cables @400 m, and 4 electrodes.

This research also uses remote sensing method. Remote sensing plays an important role in generating land use/cover information from regional to global scale, not only because of its spatially explicit representation of the earth's surface, but also because of its frequent temporal coverage and relatively low cost of observation [16], [17]. Very little research has been done to use Google Earth imagery as a direct data source for land use/cover mapping [18]. The spatial resolution is very high, it is possible to visually look at images, buildings, roads, bodies of water, etc. and on-screen digitization in GIS can easily be done to prepare land use maps [19]. Analysis of land use change using quantitative description methods and time series analysis techniques. In this study, 3 maps of land use were used for comparison, namely maps for 2003, 2008, 2019. The land use maps were made based on visual interpretation of high-resolution imagery from Google Earth. Image selection is based on images that have a cloud cover level of 0-10%.

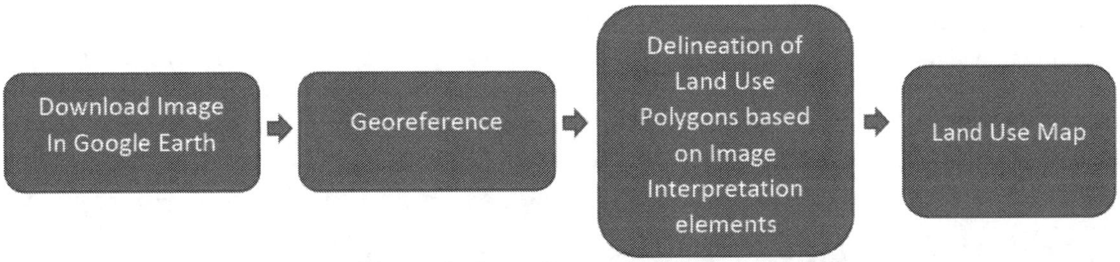

Figure 1. Image Processing Flowchart

Georeference is done in every corner of the image [20]. Georeferenced Google Earth Imagery can be done manually on GIS devices [21]. Georeference will produce a Root Mean Square (RMS) value which shows the Ground Control Point (GCP) value [22]. A good RMS result is less than 0.5. After georeference is done, then manual delineation is carried out with visual interpretation techniques. Delineation based on the characteristics of objects depicted in the image is used to identify objects which is called image interpretation [23]. Elements of Image Interpretation Include hue or colour, size, shape, texture, pattern, shadow, site and association. Image interpretation will produce land use polygons including settlements, green open space, tourism areas, rice fields, shrubs, buildings, empty land, and farms. The results of the delineation need to be checked for accuracy or it is called a validation test. The validation test is carried out by ground checking the interpretation results which are stated with overall accuracy [24]. Accuracy of permitted land use classification is more than 75%. The results of the delineation that have been tested for image accuracy can be analysed for land use area and its changes using the ArcGIS application.

3. Results and Discussion

3.1 Land Use Change Surakarta City

Land use change in Surakarta City was detected based on Google Earth image processing in 2003, 2008 and 2019. The selected image is an image that has a cloud cover of 0 – 10% to facilitate visual delineation of land use. The result of the delineation is in the form of vector data that can be calculated. The area of land use is analysed to detect changes in land use. The area of land use in Surakarta City in 2003, 2008 and 2019 is presented in table 1.

Table 1. Surakarta City Land Use Area in 2003, 2008, and 2019

No	Land Use	Area (Ha)		
		2003	**2008**	**2019**
1	Building	58,93	58,93	65,48
2	Farm	57,69	47,60	22,46
3	Empty Land	236,81	173,56	40,79
4	Tourism Land	95,32	95,32	95,32
5	Waters	59,92	59,92	55,50
6	Settlement	3668,66	3775,03	3995,71
7	Green Open Space	68,18	68,18	68,18
8	Rice Field	183,73	149,55	73,30
9	Shrubs	103,02	104,19	181,39
10	Dryland Farming	92,62	92,62	26,76

Land use in Surakarta City changes every year. Changes in land use are characterized by the increase and decrease in the area of each land use. Land uses that experience continuous increase are settlements, buildings, and shrubs. Settlement land use in 2003 had an area of 3,668.66 Ha, increasing to 3,775.03 Ha in 2008 and 3,995.71 Ha in 2019. Settlements from 2003 to 2008 increased by 2.96% and from 2008 to 2019 increased by 5.69%, especially in suburban areas. Many suburban areas have changed from agricultural land such as rice fields, dryland agriculture, and farms to settlement areas (See Figure 2). One of the outskirts of Surakarta City which has experienced a change from agriculture to settlement is Karangasem Village which is directly adjacent to Colomadu District. It is clear on Google Earth that the use of rice fields (vegetated objects in the plot) has turned into settlements (marked by red square line).

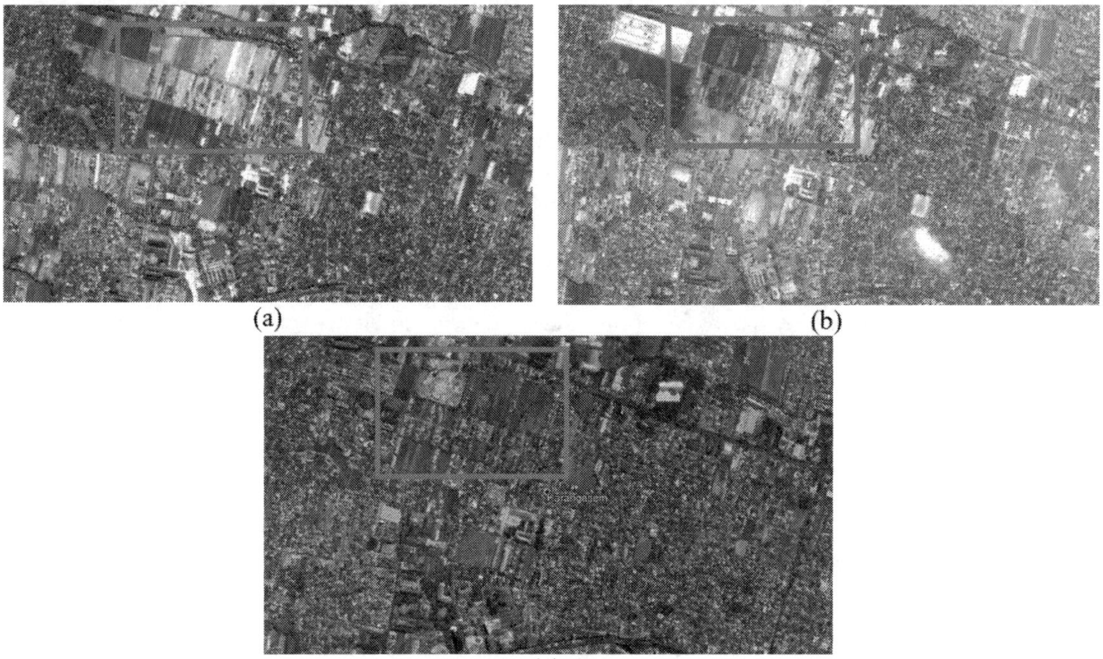

(a) (b) (c)

Figure 2. Identification of land use changes in Karangasem village, Surakarta city viewed from Google Earth (a) 2003 (b) 2008 (c) 2019

The use of building land also experienced an increase in area from 2003 to 2019 by 11.12%. The increasing use of building land shows an increase in large buildings such as hotels, malls, and industries. This is reinforced by the increase in the number of hotels in Surakarta City from 2003 to 2019.

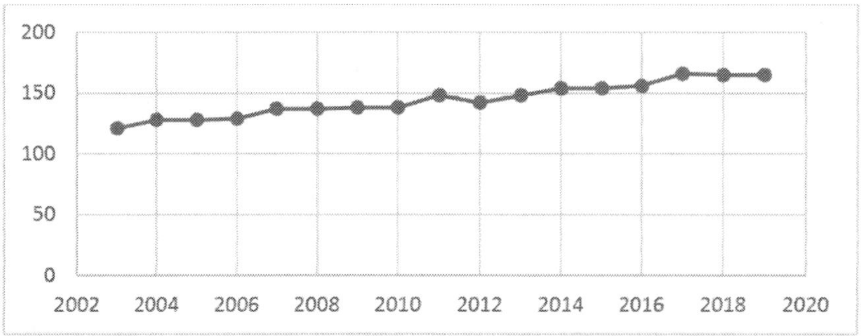

Figure 3. Graph of the number of hotels in Surakarta city in 2003 – 2019 (Source: *Badan Pusat Statistika (BPS)* Surakarta City)

IOP Conf. Series: Earth and Environmental Science **989** (2022) 012025 doi:10.1088/1755-1315/989/1/012025

Based on data from the *Badan Pusat Statistik (BPS)* of the Surakarta City, the increase in the number of hotels from 2003 to 2019 was 36.36%. The number of hotels in 2003 was 121 and increasing every year until 2019 the number of hotels was 165. The city of Surakarta as an economic and tourism centre makes this area suitable for hotel development, so the number of hotels is increasing rapidly.

In 2003 the use of agricultural land was still quite extensive in the suburbs of Surakarta, especially in the districts of Jebres, Banjarsari, and Laweyan (See Figure 4). Meanwhile, in 2008 and 2019 the symbols of agricultural land were replaced with symbols of residential land and buildings. Expansion of agricultural land use towards settlements and buildings due to the increasing need for houses, industrial buildings, and hotels.

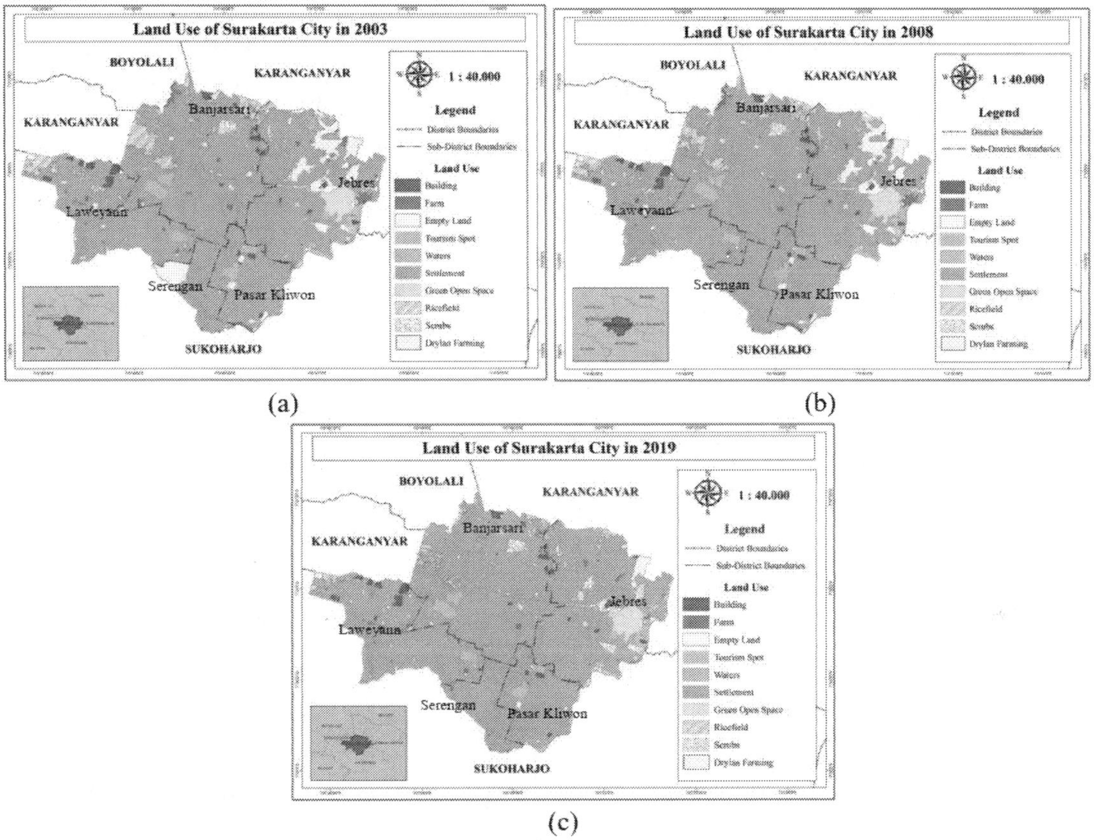

Figure 4. Land Use Map of Surakarta City (a) 2003, (b) 2008, (c) 2019

Meanwhile, land uses that have decreased are farms, dry land agriculture, empty land and rice fields. This is due to the conversion of agricultural land into buildings and settlements. The area of rice fields decreased from 2003 to 2008 by 11.60% and from 2008 to 2019 it decreased to 50.99%. Not only moor, it also decreased by 71.1% from 2003 to 2019. The decline in agricultural land was very clearly visible in the northern area of Surakarta City which borders to Karanganyar Regency.

Surakarta City also has a fixed land use, namely green open space and tourism land use. Green open spaces include parks and the university complex. The widest green open space is in Jebres district because there are 2 major universities, namely Universitas Sebelas Maret and the Indonesian Art Institute. The university complex is included in the green open space because this land use has a dominant vegetation appearance. Tourist land also has a fixed area. Tours in the city of Surakarta include the Keraton Kasunanan Surakarta, Manahan stadium, Jurug zoo, Balekambang park, and many more.

IOP Conf. Series: Earth and Environmental Science **989** (2022) 012025 doi:10.1088/1755-1315/989/1/012025

The narrower green open space in Surakarta City has an impact on increasing the coverage of waterproof surfaces such as settlement areas and buildings. Increased coverage of impermeable surfaces due to urbanization has greatly reduced the area of the potentially high groundwater recharge zone and thereby lowered groundwater levels in wells [15]. The need for groundwater is also increasing along with the increasing number of residents in Surakarta City. Settlements not only cause the amount of groundwater to decrease, but are also affected by the number of industrial buildings and hotels that consume high ground water. Because we know that Indonesia is a country with 15 km^3/year of groundwater absorption, the largest in Southeast Asia, with 93% for household purposes, 5% for industry, and 2% for irrigation. [27].

3.2 Population Density

Population density indicates the level of land demand for settlements that exceeds the availability of land. As we know, in the Surakarta City there is an increase in the area of settlements which indicates an increasing need for settlement land due to the higher population growth rate every year.

Based on data analysis of the *Badan Pusat Statistik (BPS)* of Surakarta City, it shows that there are 12 villages that have increased population density. The villages include Pajang Village, Kerten, Karangasem, Semanggi, Sangkrah, Pucang Sawit, Jebres, Mojosongo, Kadipiro, Manahan, Sumber and Banyuanyar. This is due to the increasing number of residents in limited settlement areas.

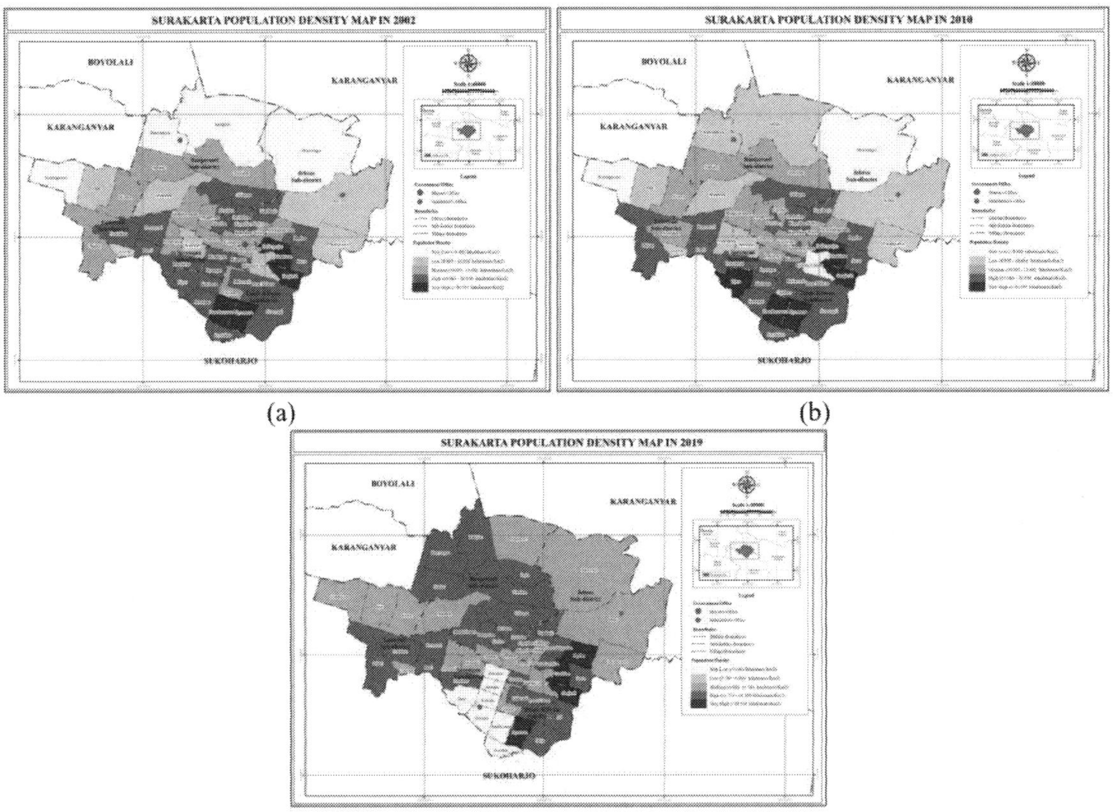

(a) (b)

(c)

Figure 5. Population Density Map of Surakarta City (a) in 2002 (b) in 2010 and (c) in 2019

The highest increase in population density occurred in Semanggi Village, from 2002 to 2010 it increased by 1356 people/km and from 2010 to 2019 it increased to 6258 people/km. Population density shows an increase in population, so that the need for water in this area is increasing. Previous studies have shown that population growth also raises water standards [27]

3.3 Monitoring Groundwater

We conducted geoelectric surveys in 2003 and 2010 in some area of Surakarta city. Based on the result of the geoelectric survey, we obtained the depth of groundwater level. The following are the measurement results for both years. We divide into three parts of area for easy analyse, i.e., northern part, central part and southern part of Surakarta city.

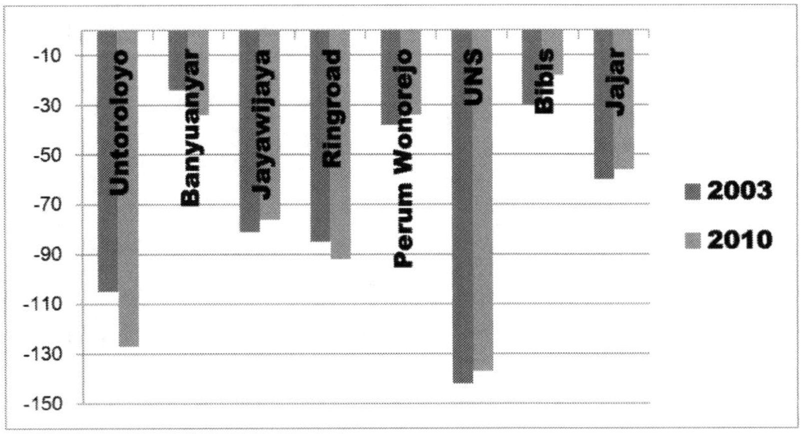

Figure 6. Depth of groundwater in the northern part of Surakarta

The northern part of Surakarta City has a groundwater depth of 6-130 meters. In 2003 the Sebelas Maret University area had the deepest groundwater depth, and in 2010 it was also. While the shallowest groundwater depth in 2003 was Banyuanyar and in 2010 was Bibis.

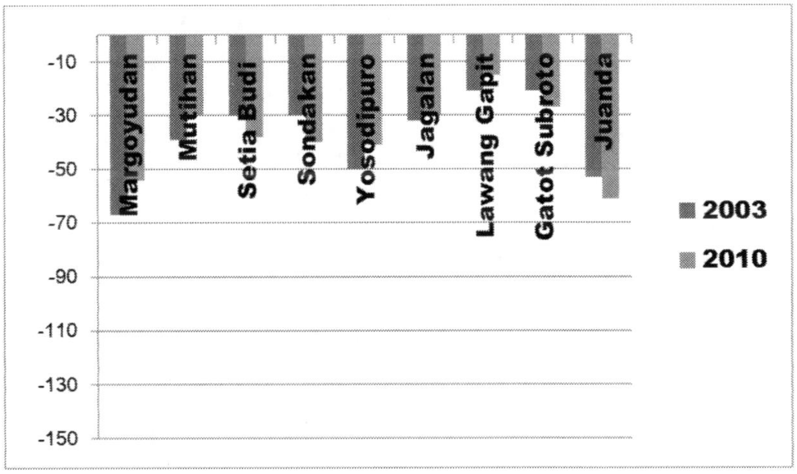

Figure 7. Depth of groundwater in the middle of Surakarta

The central part of Surakarta has a groundwater depth of 20-50 meters. The deepest groundwater depth in 2003 was Margoyudan and in 2010 was Juanda. Meanwhile, the shallowest groundwater depth in 2003 and 2010 was Lawang Gapit.

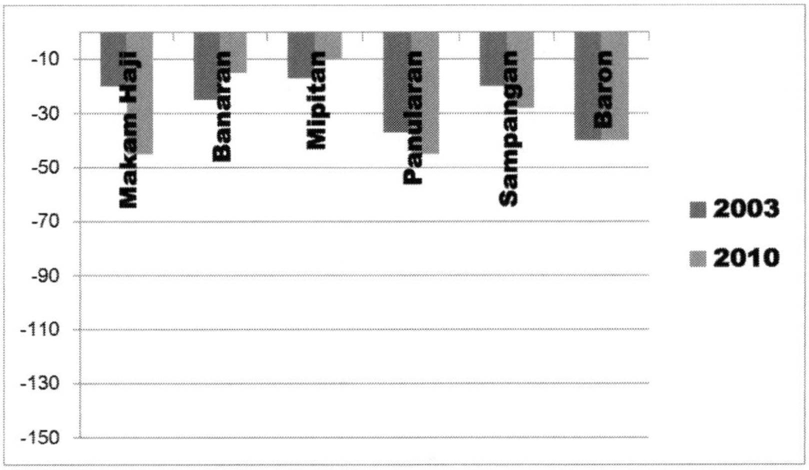

Figure 8. Depth of groundwater in the southern part of Surakarta

The southern part of Surakarta has a groundwater depth of 10-40 meters. The deepest groundwater depth in 2003 was Baron and in 2010 was Makam Haji. Meanwhile, the shallowest groundwater depth in 2003 and 2010 was Mipitan.

Based on land use change and geoelectric data, we found that Sangkrah village and Semanggi village are areas experiencing land changes and increasing population. And this is correlated with the results of the geoelectric survey at the Sampangan survey location, with a decrease in the groundwater level in the area. In general, the southern part of the city of Surakarta experienced a decrease in groundwater level in the period 2003 - 2010 and also experienced land use change. This can be a reference for the local government of the city of Surakarta, for instance, preparing surface water resources for residents or industry from resources other than deep groundwater aquifer, and opening new land for housing so as not to increase the density in Pasar Kliwon district.

4. Conclusion

Changes in land use occur in Surakarta City with a pattern of changes from agriculture to residential land and buildings. Changes in land use are closely related to the increase in population, population density and groundwater. The areas most affected by the decline in groundwater are in the southern part of the city of Surakarta, especially in the villages of Sangkrah and Semanggi.

Acknowledgement

The authors would like to thank to Google Earth for providing images that used in this research. Secondly, The authors also thank all members of the Geophysics Laboratory – Sebelas Maret University, for their support in acquiring geoelectrical data in the city of Surakarta in 2003 and 2010

References

[1] B. P. Statistik, *Statistik Indonesia 2020*. Indonesia: Badan Pusat Statistik, 2020.

[2] Q. Weng, "Land use change analysis in the Zhujiang Delta of China using satellite remote sensing, GIS and stochastic modelling.," *J. Environ. Manage.*, vol. 64, no. 3, pp. 273–284, 2002, doi: 10.1006/jema.2001.0509.

[3] Y. Wu, X. Zhang, and L. Shen, "The impact of urbanization policy on land use change: A scenario analysis," *Cities*, vol. 28, no. 2, pp. 147–159, 2011, doi: 10.1016/j.cities.2010.11.002.

[4] S. Patra, S. Sahoo, P. Mishra, and S. C. Mahapatra, "Impacts of urbanization on land use /cover changes and its probable implications on local climate and groundwater level," *J. Urban Manag.*, vol. 7, no. 2, pp. 70–84, 2018, doi: 10.1016/j.jum.2018.04.006.

[5] R. Cervero and J. Murakami, "Rail and property development in Hong Kong: Experiences and extensions," *Urban Stud.*, vol. 46, no. 10, pp. 2019–2043, 2009, doi:

10.1177/0042098009339431.

[6] Y. Kwon, "Sejong Si (City): Are TOD and TND models effective in planning Korea's new capital?," *Cities*, vol. 42, no. PB, pp. 242–257, 2015, doi: 10.1016/j.cities.2014.10.010.

[7] A. M. Hersperger *et al.*, "Urban land-use change: The role of strategic spatial planning," *Glob. Environ. Chang.*, vol. 51, no. May, pp. 32–42, 2018, doi: 10.1016/j.gloenvcha.2018.05.001.

[8] Y. Wada, L. P. H. Van Beek, C. M. Van Kempen, J. W. T. M. Reckman, S. Vasak, and M. F. P. Bierkens, "Global depletion of groundwater resources," *Geophys. Res. Lett.*, vol. 37, no. 20, pp. 1–5, 2010, doi: 10.1029/2010GL044571.

[9] T. Gleeson *et al.*, "Groundwater sustainability strategies," *Nat. Geosci.*, vol. 3, no. 6, pp. 378–379, 2010, doi: 10.1038/ngeo881.

[10] N. Khatri and S. Tyagi, "Influences of natural and anthropogenic factors on surface and groundwater quality in rural and urban areas," *Front. Life Sci.*, vol. 8, no. 1, pp. 23–39, 2015, doi: 10.1080/21553769.2014.933716.

[11] S. Koesuma, M. L. Asrib, and P. Wijayanti, "Analysis of free groundwater vulnerability level to pollution using 'DRASTIC METHOD' development in Surakarta city," in *Journal of Physics: Conference Series*, 2021, vol. 1825, no. 1, doi: 10.1088/1742-6596/1825/1/012009.

[12] M. Arif and Y. P. Utomo, "Konsentrasi Spasial Industri Industri Unggulan Kota Surakarta," *3rd Univ. Res. Colloq.*, no. February, pp. 35–53, 2016, [Online]. Available: http://publikasiilmiah.ums.ac.id:80/handle/11617/6710.

[13] H. C. Aprilia, J. Jumadi, and A. N. . Mardiah, "Environmental Critical Analysis of Urban Heat Island Phenomenon Using ECI (Environmental Critically Index) Algorithm in Surakarta City and Its Surroundings," *Int. J. Disaster Dev. Interface*, vol. 1, no. 1, pp. 1–15, 2021, doi: 10.53824/ijddi.v1i1.4.

[14] M. Harahap, Erny Poedjirahajoe, Sigit Heru Murti Budi Santosa, and Mariah Ulfa, "Assessment of Five-Year Vegetation Cover Changes to Support Green Open Spaces Monitoring in Surakarta, Central Java, Indonesia," *J. Sylva Indones.*, vol. 4, no. 02, pp. 87–95, 2021, doi: 10.32734/jsi.v4i02.6496.

[15] B. Nath, W. Ni-Meister, and R. Choudhury, "Impact of urbanization on land use and land cover change in Guwahati city, India and its implication on declining groundwater level," *Groundw. Sustain. Dev.*, vol. 12, no. October 2020, p. 100500, 2021, doi: 10.1016/j.gsd.2020.100500.

[16] W. Wu, R. Shibasaki, P. Yang, Q. Zhou, and H. Tang, "Remotely sensed estimation of cropland in China: A comparison of the maps derived from four global land cover datasets," *Can. J. Remote Sens.*, vol. 34, no. 5, pp. 467–479, 2008, doi: 10.5589/m08-059.

[17] D. Bargiel and S. Herrmann, "Multi-temporal land-cover classification of agricultural areas in two European regions with high resolution spotlight TerraSAR-X data," *Remote Sens.*, vol. 3, no. 5, pp. 859–877, 2011, doi: 10.3390/rs3050859.

[18] Q. Hu *et al.*, "Exploring the use of google earth imagery and object-based methods in land use/cover mapping," *Remote Sens.*, vol. 5, no. 11, pp. 6026–6042, 2013, doi: 10.3390/rs5116026.

[19] K. Malarvizhi, S. V. Kumar, and P. Porchelvan, "Use of High Resolution Google Earth Satellite Imagery in Landuse Map Preparation for Urban Related Applications," *Procedia Technol.*, vol. 24, pp. 1835–1842, 2016, doi: 10.1016/j.protcy.2016.05.231.

[20] A. Y. Chang *et al.*, "Combining google earth and GIS mapping technologies in a dengue surveillance system for developing countries," *Int. J. Health Geogr.*, vol. 8, no. 1, pp. 1–11, 2009, doi: 10.1186/1476-072X-8-49.

[21] V. Visser, B. Langdon, A. Pauchard, and D. M. Richardson, "Unlocking the potential of Google Earth as a tool in invasion science," *Biol. Invasions*, vol. 16, no. 3, pp. 513–534, 2014, doi: 10.1007/s10530-013-0604-y.

[22] Y. Lisnawati and A. Wibowo, "Penggunaan Citra Landsat Etm+ Untuk Monitoring Perubahan Penggunaan Lahan Di Kawasan Puncak," *J. Penelit. Hutan Tanam.*, vol. 4, no. 2, pp. 79–90, 2007, doi: 10.20886/jpht.2007.4.2.79-90.

IOP Conf. Series: Earth and Environmental Science **989** (2022) 012025 doi:10.1088/1755-1315/989/1/012025

[23] Sutanto, *Penginderaan Jauh Jilid 1*. Yogyakarta: Gadjah Mada University Press, 1994.
[24] D. P. Prabowo, S. Bachri, and B. S. Wiwoho, "Prediksi Perubahan Penggunaan Lahan Dan Pola Berdasarkan Citra Landsat Multiwaktu Dengan Land Change Modeler (Lcm) Idrisi Selva 17: Studi Kasus Sub-Das Brantas Hulu," *J. Pendidik. Geogr.*, vol. 22, no. 1, pp. 32–48, 2017, doi: 10.17977/um017v22i12017p032.
[25] E. D. Ebong, A. E. Akpan, C. N. Emeka, and J. G. Urang, "Groundwater quality assessment using geoelectrical and geochemical approaches: case study of Abi area, southeastern Nigeria," *Appl. Water Sci.*, vol. 7, no. 5, pp. 2463–2478, 2017, doi: 10.1007/s13201-016-0439-7.
[26] T. H. Abdel Hafeez, H. S. Sabet, A. N. El-Sayed, and M. A. Zayed, "Geoelectrical exploration of groundwater at West Dayrout Area, Assiut Governorate, Egypt," *NRIAG J. Astron. Geophys.*, vol. 7, no. 2, pp. 279–296, 2018, doi: 10.1016/j.nrjag.2018.07.004.
[27] J. Margat and J. Van Der Gun, *Groundwater around the world: a geographic synopsis*. New York: CRC Press, 2013.

IOP Conf. Series: Earth and Environmental Science **989** (2022) 012026 doi:10.1088/1755-1315/989/1/012026

Proposed scenario of low carbon development from waste sector in West Papua

Syafrudin Raharjo[1, 5], Saraswati Prabawardani[2, 5*], Hendri[3, 5], Egi Suarga[4], M, Faruk Rosyaridho[4], Suriany Ida[4], and Julia Kalmirah[4]

[1]Fishery and Marine Science Faculty, University of Papua
[2*]Agriculture Faculty, University of Papua
[3]Forestry Faculty, University of Papua
[4]World Resource Institute, Indonesia
5Postgraduate Program, University of Papua

*Email: s.prabawardani@unipa.ac.id

Abstract. The increasing of population and economic growth, urbanization, development progress and living standard has accelerated the rate and amount of waste generation. Waste is a serious problem of a worldwide concern because of its impact on environmental pollution, health and flood disasters in some drainages and rivers. Proper waste management is needed to reduce or prevent its negative impacts. For this reason, it is necessary to identify the types of waste and their potential to produce greenhouse gas emissions in order to obtain an accurate picture in an effort to formulate policies and strategies for reducing GHG emissions. The method used was thinking system and dynamic system by looking at the BAU scenario, 29%, 41% and 44% emission reduction. Based on the model of the dynamics system, a new scenario is needed through the intervention of waste processing by 3R at TPS and TPA, applying aerobic composting, methane capture and incineration.

1. Introduction

In line with the population growth, increasing income and living standards, urbanization and development will have an impact on increased waste production. An increase in various types of potential waste becomes a serious problem because the accumulation of waste can cause environmental damage and health problems. The trend of increasing global GHG emissions from solid waste management have been widely reported including [6,7]. To avoid or reduce the negative impact of waste, the management or proper handling is required according to the composition and characteristics of the wastes types.

There are four types of waste based on their shape or form, namely solid, liquid, gas and sound waste. Solid waste is the residue resulting from industrial activities or domestic activities in solid form and is known as waste. According to SNI 19-2452-2002, waste is solid scraps consisting of organic and inorganic materials which are considered useless and must be managed to the environment and safeguard development investment. Meanwhile, according to RI Law No. 18 2008, waste is the remains of human daily activities and/or natural processes in solid form. Examples of waste include paper, plastic, iron, wood, cloth, food waste and so on. Solid waste generally produces greenhouse gas emissions in the form of CH_4 and CO_2. The activity of burning waste produces CO_2, while open dumping in the landfill causes the accumulation of organic waste with anaerobic decomposition and produces CH_4 [4]. These gases are included in the greenhouse gases that contribute to global warming. To reduce greenhouse gas emissions, proper waste management is important.

West Papua Province with an area of 567 km^2, had a population of 765,258 people in 2010 but increased to 1,134,068 people in 2020 with a population growth rate of 2.65% and a gross regional domestic product of 5.2% per year [1, 5]. The increase in population and GRDP will have unavoidable consequences for an increased volume of solid waste, which can have serious impacts on the environment, social and economic. Therefore, solid waste management should be prioritized to reduce the impact of GG in the atmosphere.

Until 2021, West Papua province has a total of 8 FDA (final disposal area) with a total area of 49 ha and 475 TD (temporary disposal) units with a capacity of 475 m^3 spread across all districts. The average waste generation in West Papua Province is 0.518 kg/person/day or equivalent to 185.64 kg/person/year. However, the problem of waste in West Papua is that it has not been managed properly. TD in the form of garbage bins (containers) at various points of the area (TD transfer depot system) and the application of FDA systems of open dumping in each district show that waste management in West Papua has not been handled properly and mitigation efforts, especially for reducing carbon emissions. Waste management is aimed at reducing and dealing with waste generation by sorting and reducing and maximizing its utilization to provide positive value for the community and a positive contribution to low-carbon development. Until now, waste management at the TFD in West Papua is still carried out using an open dumping system. Open dumping is a waste management system in which waste material is left open and piled up in a final disposal location so that it can destroy the environmental and health conditions.

Another obstacle in building a waste management system in the province of West Papua is data collection. Data related to various waste variables is very minimal, this makes it quite difficult to calculate the contribution of GHG emissions from the waste sector and its management efforts to avoid environmental and social problems. The landfill extension program to accommodate solid waste can be executed, but without efforts to reduce waste sustainably through proper management or handling, GHG emissions massively increase. This study aims to build a dynamic system model for estimating waste generation and GG emissions, and to propose policies and strategies related to waste that can be managed with the right system to keep GG emissions at a minimum level following the capacity set in West Papua.

2. Methods

The data collected is in the form of the population and the amount of waste generated in West Papua, which are independent variables to obtain GG emissions waste. All data are formulated using a dynamic system model to simulate GG emissions and their interventions for mitigation actions. The system dynamics methodology includes two modeling steps: (1) Qualitative modeling known as a causal loop diagram (CLD) to show how different variables in an interrelated system, and (2) stock and flow diagram modeling (stock-flow diagram/SSWDS) to provide a better understanding of the system behavior by visualizing a well-off representation through simulation.

CLD is built based on these variables and relationships, and then the integrated CLD is converted into a stock-flow diagram (SSWDS), which describes the dynamic relationship between system quantities. The relationship between waste emission variables in this study was simulated using Powersim Studio 10 Professional.

The stock-flow diagram (SSWDS) is based on the principle of system dynamics which includes the state of level (accumulation or stock), flow/rate (input rate and output rate) and variables (Figure 1). System equations can be in the form of auxiliary, level, lookup, initial value, data and constants and are represented by symbols to be used in constructing the model structure and computer operations to perform simulations.

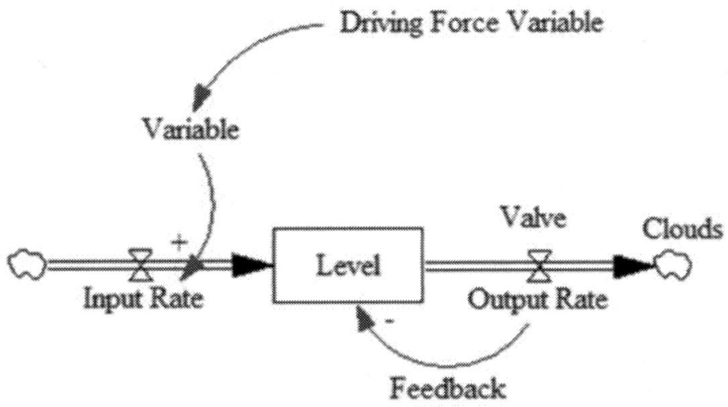

Figure 1. System Dynamics Model Structure

The structure notation of the system dynamics model with Powersim Studio 10 Professional software:

a. The stock or level is a fixed variable represented by a rectangular box;
b. Clouds represent sources or discharges that are outside the system or in the environment;
c. The incoming material flow (input rate) is represented as a pipe or arrow with a double line pointing to the stock. This flow will add material in stock;
d. The outflow of material (output rate) is represented as a pipe or arrow with a double line coming out of stock. This flow will reduce the material in stock;
e. Flow control expressed by a valve (valve);
f. Variables are expressed as variables and can be stated as constants or as driving variables (variables that affect the model but are not affected by the model;
g. The flow of information, represented by a single arrow, can be positive or negative;
h. Feedback indicates a causal relationship, where the arrowhead reveals the cause and effect.

Powersim helps to create stock and flow diagrams to represent larger system structures in detail with many interrelated workspaces and has a highly interactive user interface.

2.1 Model Validation

The statistical test was used to verify the model output with empirical data, namely by comparing the average value and calculating the error percentage of the average value. The average value (E_i) is calculated by the following formula [2]:

$$E_i = \frac{|\bar{S}-\bar{A}|}{\bar{A}} \quad (1)$$

with: $\bar{S} = \frac{1}{N}\sum_{i=1}^{N} S_i$, $\bar{A} = \frac{1}{N}\sum_{i=1}^{N} A_i$

Information : A = actual value, S = simulation value, N = Time

IOP Conf. Series: Earth and Environmental Science **989** (2022) 012026 doi:10.1088/1755-1315/989/1/012026

2.2 Variable determination and data collection

The main variables used in the system dynamics model are listed in Table 1. These variables include exogenous and endogenous variables. Exogenous variables are mostly leverage variables in the system dynamics model. When a small change is made in the variables, it will cause a large impact on the model.

Table 1. Variables for the simulation of the dynamic waste model

No	Name of the Variables	Unit	Source
1	Population	kg	BPS West Papua in 2011-2021, 2021 DLHP Waste Management Report.
2	Waste generation	kg	
3	Waste generation managed by the government (30%)	kg	
4.	Waste composition	%	
5.	Temporary Disposal (TD) Capacity	kg	
6.	Final Disposal/ Landfill Capacity	kg	
7.	Waste Bank	kg	
8.	Composting	kg	
9.	Processing 3R	kg	
10.	TD Capacity	kg	
11.	FD / Landfill capacity	kg	

2.3 BAU and Proposed Model

The BAU (Business as Usual) model provides an interlinked relationship between the total population and waste generation. These factors are vital to represent the growth rate of waste every year. This BAU model is the basis for designing the proposed scenario to achieve low carbon emissions from the waste sector. The proposed model includes several interventions as part of a scenario to reduce carbon emissions. The intervention is described as processing and processing waste at TDA and FDA to reduce the amount of waste and ultimately reduce GHG emissions.

3. Results and Discussion

3.1 Model Conceptualization

The generated waste in West Papua Province is equivalent to the population. The population of West Papua is determined by the birth rate and the imigration rate. While the population reduction is influenced by the death rate and outmigration rate. Population dynamics determine the changes in West Papua's waste generation. This waste generation will determine how much GHG emissions will be generated. This model also observes the amount of waste handled by the West Papua Provincial government and the amount of waste that is not handled. Therefore, in this model, a managed GHG emissions and abandoned waste was calculated. Waste management handled by the government will go through the stages of transporting waste to TDA and FDA. The amount of waste disposed in the landfill produce carbon emissions.Waste management at the TDA will be carried out by reducing waste, as a scenario for reduction policies (baseline, LCD-medium, LCD-High and LCD-Plus) in the form of changes in the level of 3R processing, composting and waste bank. Reduction of waste is also carried out at the household level, with a policy scenario of reducing waste generation per person per day.

3.2 Causal Loop Diagram

A causal loop diagram (CLD) visualizes how different variables in a system are related. CLD provides an overview of the SD model.

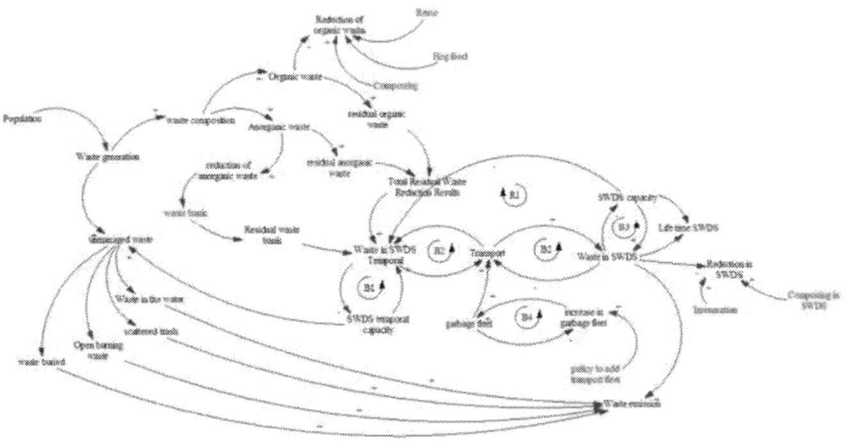

Figure 2. Causal Loop Diagram of Waste

In Figure 2, there is a causal relationship from one variable to another. Each arrow indicates the effect of one variable on the other because it has a positive (+) or negative (-) loop. Loop + shows that the relationship of the variable is directly proportional. If there is an increase in the value of the variable, it will cause an increase in the value of the variable it affects. On the other hand, a negative loop shows an inversely proportional relationship (not unidirectional) so that if there is an increase in the value of the variable, it will cause a decrease in the value of the variable it affects.

3.3 Stock-Flow Diagram

Stock and flow diagrams are the most common first steps in building a simulation model because they help determine the types of variables that are important in causing the behavior of the model. The stock-flow diagram of the pond sub-model is presented in Figure 3.

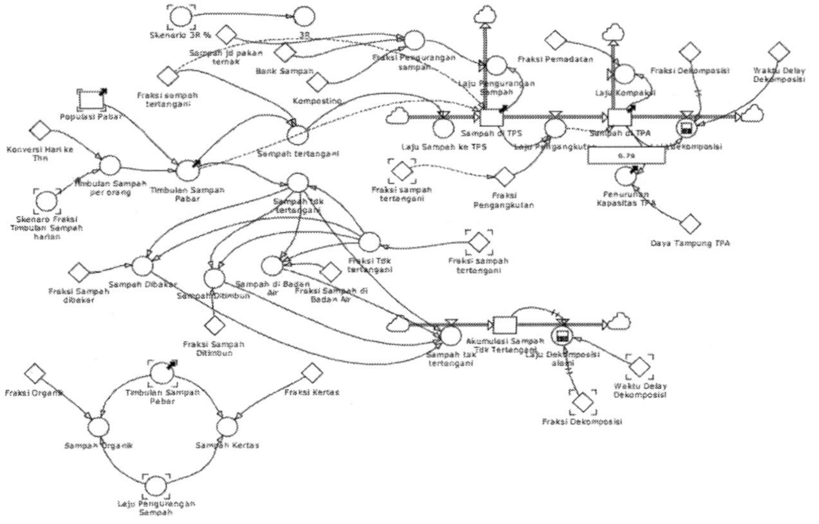

Figure 3. Stock-Flow Diagram (SFD) of Waste Sub-model of West Papua

IOP Conf. Series: Earth and Environmental Science **989** (2022) 012026 doi:10.1088/1755-1315/989/1/012026

3.4 Model Validation
The validation of the solid waste sub-model was carried out on the population and waste generation, by comparing the average value of historical data with simulations.

3.5 Population Sub Model Validation
The historical data on the population in 2010-2020 is used to validate the model (11 years). Validation was carried out for the population of West Papua in the population sub-model. The total population of West Papua varies each year and tends to increase from year to year (overtime). The average population growth rate is 3.08% with variations in the population growth rate between 2.46-3.94%. A very significant increase occurred in 2020, which is assumed to be influenced by the influx of residents from outside West Papua (immigration).

The result of calculating the error rate from the average value (Ei) obtained a validation value of 0.003 %. This means that the system dynamics model that is built can represent the actual condition of population growth over time because the percentage error value is <5% (Figure 4) [3]. stated that the deviation limit that is still acceptable is in the range of 5-10%.

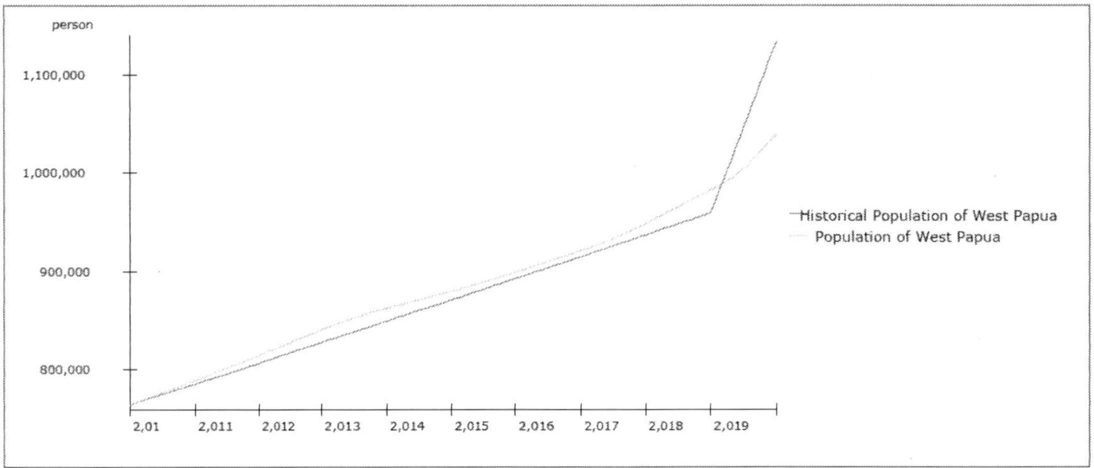

Figure 4. Comparison of Historical Population versus Simulation Data

3.6 Waste Generation Validation
The 2010-2020 historical waste generation data was used to validate the model (11 years). The value of waste generation varies every year and tends to increase from year to year, namely 133,234,195-178,567,625 kg/year (an average of 153,958.061 kg/day).

The result of calculating the error rate from the average value (Ei) obtained a validation value of 5.78%. This mean that the system dynamics model can represent the actual solid waste condition because the error percentage value is <10% (Figure 5) [3]. stated that the deviation limit that is still acceptable is in the range of 5-10%.

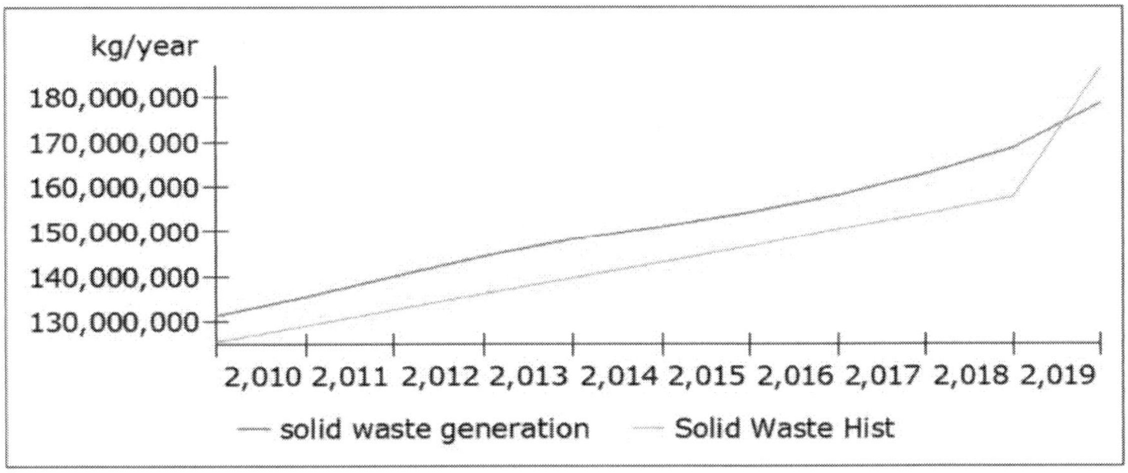

Figure 5. Data Comparison of Historical Waste Generation versus Simulation

3.7 Model Simulation

This model perform simulations using the four scenario approaches as follow:

1) Baseline: No new policy. However, it takes environmental degradation into account—This scenario reflects continuing historical trends. No new policies are introduced in this scenario.

2) Medium-LCD Scenario: Includes a new low carbon policy for 2020-2030; The scenario adopted is reduction at the source, namely reducing household waste generation from 0.51 kg/person/day to 0.4 kg/person/day, doing 3R processing by 5% and composting by 10%.

3) High-LCD Scenario: Covers more ambitious policies than the Medium LCD for 2020–2030; The scenario used is reduction at the source, namely reducing household waste generation from 0.51 kg/person/day to 0.4 kg/person/day, doing 3R processing by 15% and composting by 20%.

4) Plus-LCD Scenario: Includes HIGH LCD for 2020–30 and the implementation of additional, more ambitious policies thereafter. The scenario utilized is a reduction at the source, namely reducing household waste generation from 0.51 kg/person/day to 0.4 kg/person/day, doing 3R processing by 25% and composting by 30%.

3.8 Basic Assumption Scenario (Baseline)

The model simulation on the policy strategy with the baseline scenario has not provided new policies and still describes the continuation of historical behavior until the end of the simulation year.

Table 2. Waste generated at TDA, FDA, waste generation in West Papua, reduction in landfill capacity, total waste CO_2 emissions from 2010 - 2030.

Year	TDA Waste	FDA Waste	Waste Generation, West Papua	Decrease in FDA capacity (%)	Total CO_2 Emissions of Waste (kg*CO_2/year)
2010	42618748.54	12785624.56	142062495.12	98	105473543.94
2011	64089774.40	23921284.97	146649142.87	96	107389812.99
2012	79021117.09	37304945.52	151383875.70	94	109892204.42
2013	89800523.06	50749915.06	156178470.66	92	112620241.54
2014	97890098.34	63116729.26	160336824.60	90	114793788.32
2015	103997118.48	73859705.02	163535539.22	89	116807566.15
2016	108742810.97	82757672.94	167047146.24	87	119289380.29
2017	112744299.06	89850089.43	171058906.09	86	122092425.52
2018	116357375.72	95341331.36	175493692.58	85	125425198.34
2019	119895574.80	99503663.94	180670444.64	85	129904370.20
2020	123800652.61	102648997.93	192060983.22	84	141582409.90
2021	130427240.23	105371131.45	208922627.12	84	150316036.02
2022	137699587.63	108430547.06	216460677.36	83	155549349.12
2023	144310568.79	111715847.66	224270704.86	83	161039934.57
2024	150568813.34	115047366.21	232362522.70	82	166772436.89
2025	156674411.04	118355514.21	240746298.04	82	172739814.62
2026	162758070.84	121632239.78	249432564.90	81	178940448.44
2027	168906188.72	124901397.73	258432237.32	81	185376293.65
2028	175176896.04	128201053.16	267756623.18	80	192051701.58
2029	181610335.20	131573261.62	277417438.32	80	198972668.93
2030	188235240.33	135058484.80	287426821.30	79	203416055.35

Population growth has a positive correlation with the increase in waste generation. Part of this waste generation is not managed, and the composition of the waste generation consists of organic and inorganic waste. To reduce organic waste, three management strategies are needed, namely composting, for pig feed and 3R application, thus the generation of organic waste in the TDA (temporary disposal) can be reduced. In addition to reducing waste that lands in TDA, efforts to reduce waste at TDA are also useful to keep the capacity of waste at TDA from being exceeded. If the capacity of the waste in the TDA is exceeded, there will be scattered waste which will certainly increase GHG emissions. Furthermore, the remaining waste from the reduction will be transported to the FDA. Efforts to reduce both waste at the source and the TDA will also extend the service life of the FDA.

In its management, inorganic waste can be reduced by the availability of a waste bank, and this will minimized inorganic waste transported to TDA and FDA. So far, the handling of waste transportation from TDA to FDA is carried out by waste trucks. Waste transportation will run well if the number of transportation fleets is adequate, so the policy offered is to increase the number of waste transportations according to their capacity. To reduce the load on the capacity of waste in the FDA, it is necessary to handle waste through composting. Thus, waste reduction and management at TDA and FDA will reduce GHG emissions.

For unmanaged waste such as waste which enters water bodies, scattered on the ground surface. burned or enters the ground will have an impact on the increase of GHG emissions. It is hoped that with the addition of capacity and management at TD, unmanaged waste can be reduced.

The generation of waste at TDA is higher than FDA (Figure 6A), this is because several types of waste such as iron, aluminum that are disposed of at TDA are collected by waste collectors. 30% of waste is handled and the remaining 70% is unmanaged waste (Figure 6 B). Waste that is not handled is burned in the open by the community as much as 40%, 17.4% of waste is piled up by the community and 15% of waste is dumped in water bodies, while the rest is scattered waste.

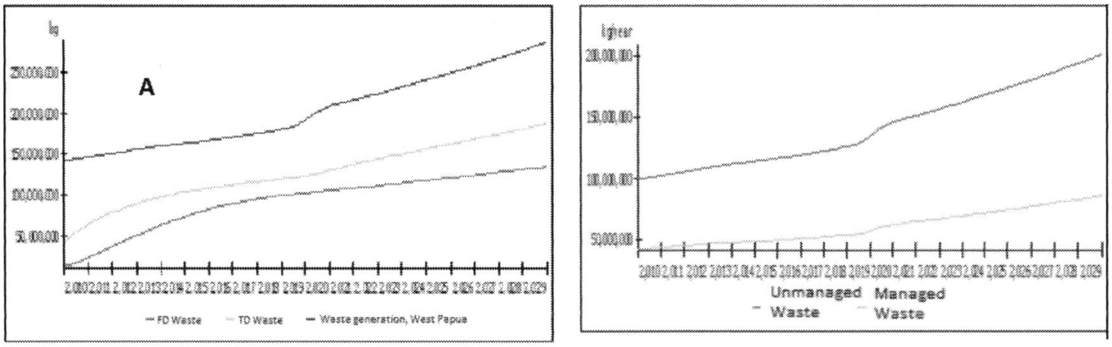

Figure 6. (A) The trend of increasing waste at TDA, FDA and West Papua waste generation in 2010-2030, (B) Amount of managed and unmanaged waste in 2010-2030.

In relation to the condition of the waste, the increase in waste generation causes a growth in CO_2 emissions from 105473543.94 tons CO2e in 2010 and will continue to increase until it reaches 203416055.35 tons CO2e in 2030 (Figure 7). The contribution of CO_2 emissions is dominated by FD which is managed by 78% open dumping, 3.4% waste burning activities, and 18.9% non-FD (scattered waste and marine litter).

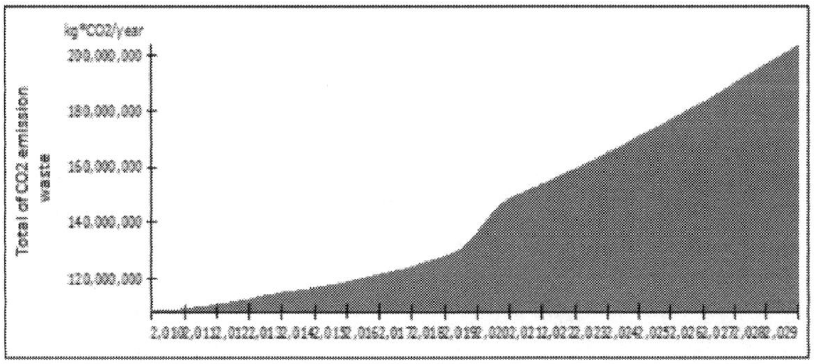

Figure 7. Trend of increasing CO_2 emissions of BAU waste from 2010-2030

3.9 LCD-Medium Scenario (Scenario 1)

This scenario is consistent with Indonesia's efforts to achieve the climate target desired by the West Papua provincial government, namely a 29% reduction in emissions by 2030 compared to the baseline. Scenarios is taken for new low carbon policies for 2020-2030; The scenario adapted is the reduction at the source, namely reducing household waste generation from 0.51 kg/person/day to 0.4 kg/person/day, 3R processing by 5% and composting by 10%.

This new LCD-Medium scenario is taken to intervene in the waste produced to align with the emission targets from the waste sector by proposing several policies in waste management as mentioned above.

Scenario 1, namely reducing household waste generation to 0.4 kg/person/day, 5% 3R processing and 10% composting can reduce waste generation by up to 22% compared to baseline. The level of waste generation reduction can be seen in Figure 8.

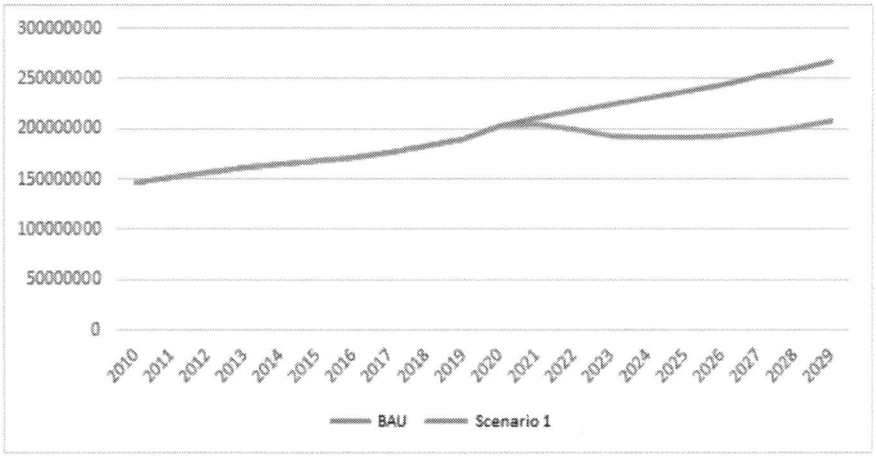

Figure 8. Reduction of Waste Generation in Scenario 1

The capacity of the landfill for waste collection efforts until 2030 in West Papua has not experienced any complications since the carrying capacity of the landfill has not been exceeded (Figure 9).

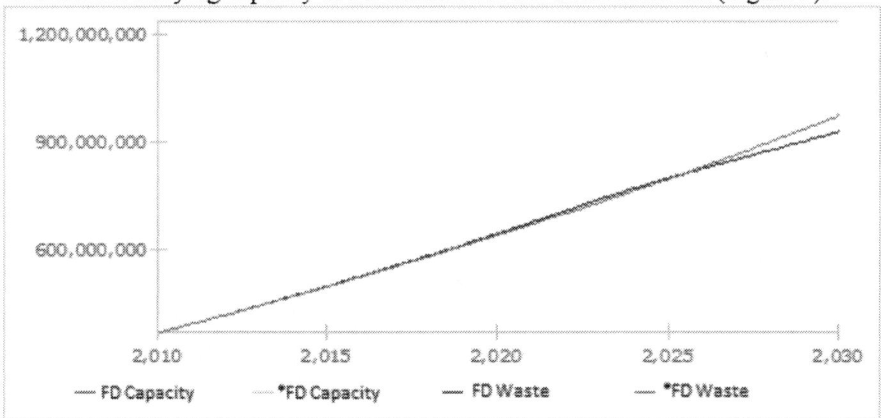

Figure 9. Capacity and Waste in TPA (kg) for Scenario 1

The level of reduction in waste emissions (tons of CO_2/year) in scenario 1 can reach an emission reduction of 18% (28647 tons of CO_2/year) (Figure 10).

IOP Conf. Series: Earth and Environmental Science **989** (2022) 012026 doi:10.1088/1755-1315/989/1/012026

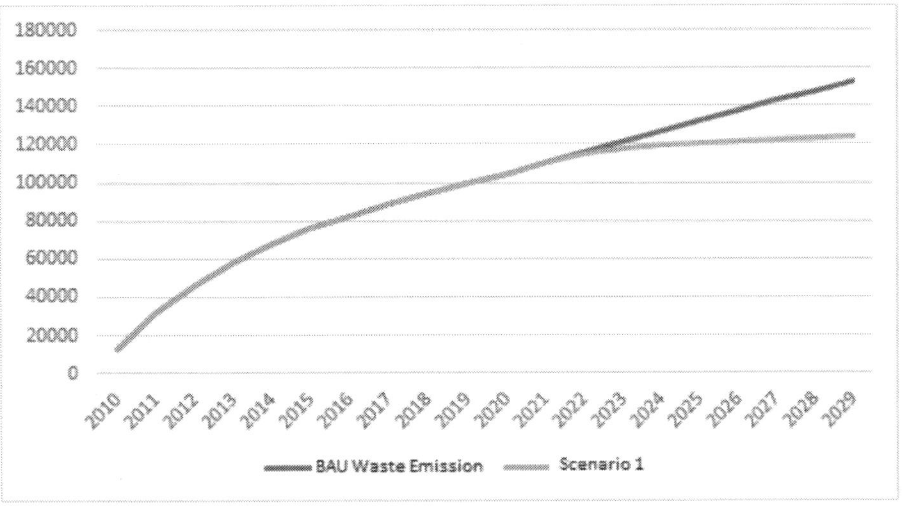

Figure 10. Waste emission reduction (ton CO2/year) for scenario 1

3.10 High LCD Scenario (Scenario 2)

These scenarios include policies that are more ambitious than the Medium LCD for 2020–2030. The scenario used is a decrease at the source, namely household waste generation reduction by 0.4 kg/person/day: 3R processing by 15% and composting by 20%. Reducing waste generation with policy intervention in scenario 2 can reduce waste generation by 22% or 60945183 kg (Figure 11).

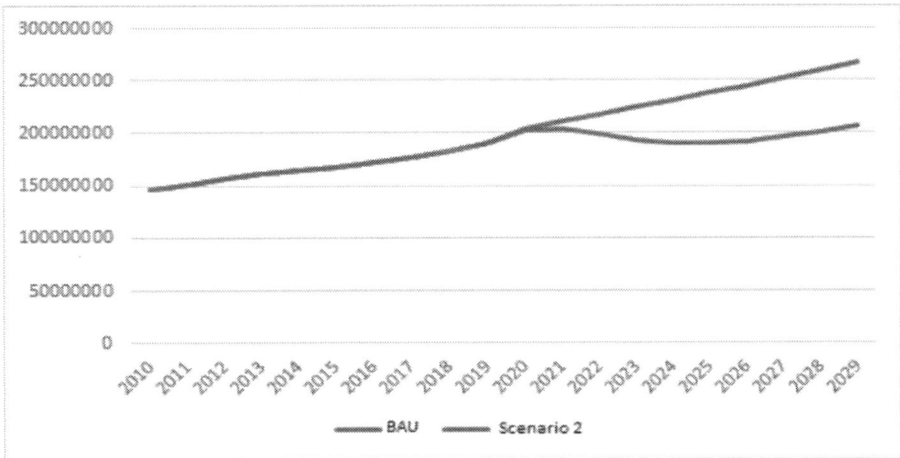

Figure 11. Reduction of Waste Generation in Scenario 2 (kg)

The capacity of the landfill against the efforts of waste collection until 2030 in West Papua has not experienced any difficulties because the carrying capacity of the landfill has not been exceeded (Figure 12).

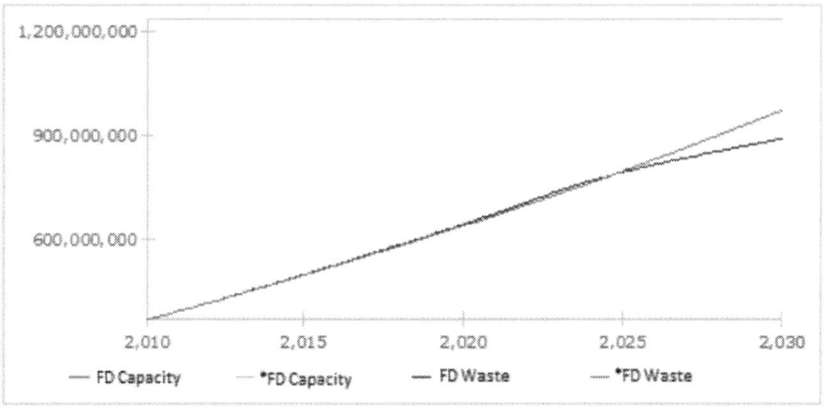

Figure 12. Capacity and Waste in TPA (kg) for Scenario 2

The composition of waste management is very dynamic, where the main composition is scattered waste, marine litter, burned and managed waste. The variation of managed and unmanaged waste is illustrated in Figure 13.

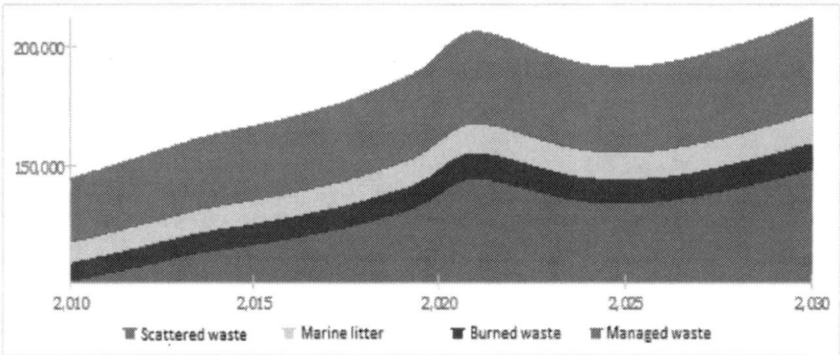

Figure 13. Composition of Waste Management Status (Scenario 2)

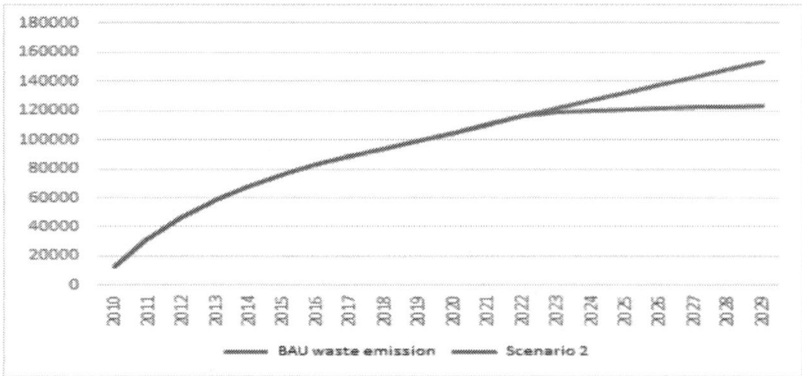

Figure 14. Waste Emission Reduction (tons CO_2/year) for Scenario 2

IOP Conf. Series: Earth and Environmental Science **989** (2022) 012026 doi:10.1088/1755-1315/989/1/012026

The level of reduction in waste emissions (tons CO_2/year) in scenario 2 achieves an emission reduction of 19% (29586 tons CO_2/year) (Figure 14).

3.11 Low Carbon Development (LCD)-Plus Scenario (Scenario 3)

Scenario 3 includes high LCD for 2020–2030 and subsequently, the implementation of additional and more ambitious policies. The scenario operated is a source reduction, namely a reduction in household waste generation by 0.4 kg/person/day, conducting the 3R processing by 25% and composting by 30%. Reducing waste generation with policy intervention in scenario 3 can reduce waste generation by 22% (Figure 15).

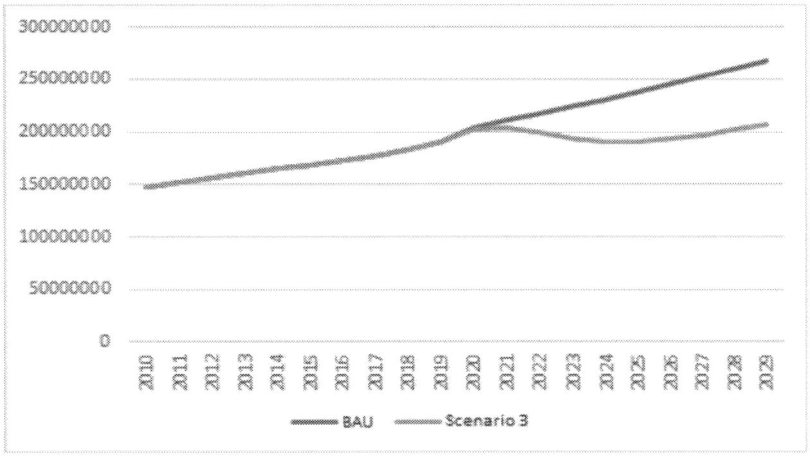

Figure 15. Waste Generation Reduction Scenario 3

The landfill capacity for waste collection efforts until 2030 in West Papua has not experienced problems because the carrying capacity of the landfill has not been exceeded (Figure 16).

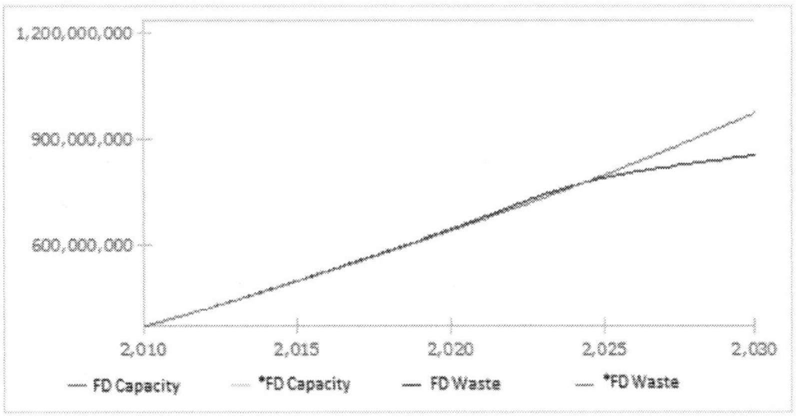

Figure 16. Capacity and Waste in FDA (kg) Scenario 3

The composition of waste management is very dynamic, where the main composition is scattered waste, marine litter, burned waste and managed waste. The variation of managed and unmanaged waste can be seen in Figure 17.

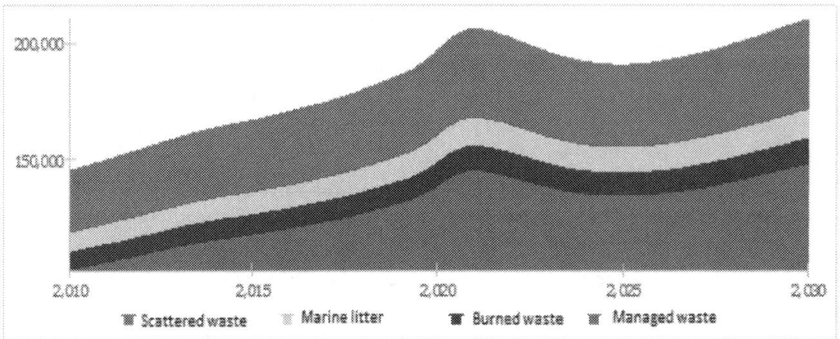

Figure 17. Composition of Waste Management Status Scenario 3

The level of waste emission reduction (tons CO_2/year) in scenario 2 can reach an emission reduction of 19% (29211 tons CO_2/year) (Figure 18).

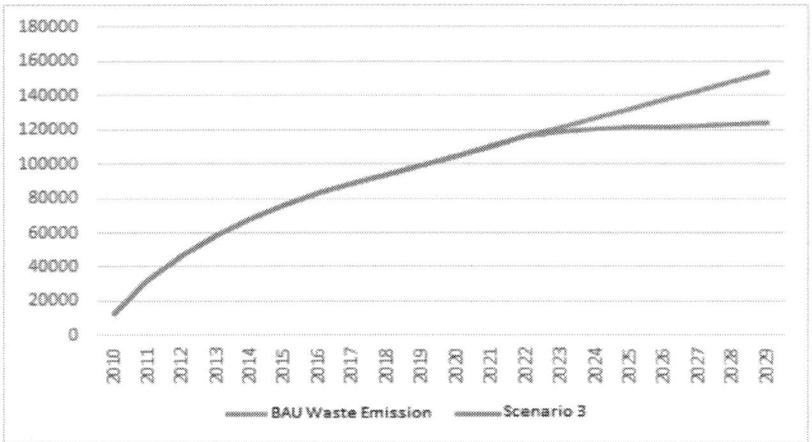

Figure 18. Waste Emission Reduction (tons CO_2/year) for Scenario 3

3.12 Proposed Scenario

Based on the simulation results using four policy interventions, GHG emissions were only reduced by 19%. Therefore, other policy interventions are still required. Other policies that can be adopted to reduce GG emissions are aerobic composting, applying methane capture, and incineration. Waste generation in West Papua is divided into manage and unmanaged waste. An increase in population will upsurge waste generation, thus it is necessary to increase the number of waste bins at TDA and the number of fleets to transport waste from TDA to FDA. To reduce the waste accumulation rate at TDA, waste must be managed through mitigation strategies to improve waste management and reduce GHG emissions. The 3R (Reduce, Reuse and Recycle) strategy and waste banks can be an option in managing managed waste at TDA. It is not only the capacity of the TDA that requires attentiveness in its management to reduce the fraction of waste transported to the FDA to control the concerns that arise on the issue of GHG emissions which continues to rise along with the rate of waste generation in the FDA. Therefore, it is necessary to intervene in waste management at the FDA by composting. The maximum composting technique must be applied to accelerate the decomposition rate; hence it can reduce GHG e

IOP Conf. Series: Earth and Environmental Science **989** (2022) 012026 doi:10.1088/1755-1315/989/1/012026

missions. Compost (decomposed waste) will also have economic value as organic fertilizer. It will have a positive impact on the economic and social aspects.

For unmanaged waste due to burning, stockpiling and entering water bodies will decompose naturally. Nevertheless, the time depends on the type of waste and environmental conditions. For organic waste derives from the rest of living material, the rate of decomposition will be faster compared to other types of inorganic waste such as paper, plastic, cans and others.

4. Conclusion

GHG emissions produced by solid waste or waste in West Papua will continue to increase along with the increase in population and waste generation in both TDA and FDA. The system dynamics model shows that in 2030 waste generation in TDA and FDA will continue to rise along with the population growth so that waste generation in West Papua will increase its GG emissions by up to 123,832 tons of CO_2/year (scenario 3) or equivalent to a 19% reduction in GHG emissions. Therefore, a new scenario is proposed by implementing waste management interventions through 3R at TDA and FDA by aerobic composting, applying methane capture and incineration.

References

[1] Badan Pusat Statistik (BPS). 2020. Papua Barat Dalam Angka, BPS Papua Barat.

[2] Barlas Y. 1996. Formal aspect of model validity and validation in system dynamics. System Dynamics Review. 12(3),183-210.

[3] Muhammadi, Aminullah E, Soesilo B. 2001. Analisis Sistem Dinamis (Lingkungan Hidup, Sosial, Ekonomi, Manajemen). Jakarta (ID): UMJ Press.

[4] Sudarman. 2010. Meminimalkan daya dukung sampah terhadap pemanasan global. *Jurnal Profesional*, 8 (1), 51-59.

[5] Saraswati P, Hendri, Karuniasa K, Syamsudin K, Maharani C, Munichputranto F. 2021. Towards a low carbon solid waste management in West Papua. IOP Conf. Ser.: Earth Environ. Sci. 716 01201. doi:10.1088/1755-1315/716/1/012016.

[6] Rahman MM, Rahman SM, Rahman MS, Hasan MA, Shoaib SA, Rushd S. 2021 Greenhouse Gas Emissions from Solid Waste Management in Saudi Arabia—Analysis of Growth Dynamics and Mitigation Opportunities. Appl. Sci. 11, 1737. https://doi.org/ 10.3390/app11041737.

[7] Xin C , Zhang T , Tsai SB, Zhai YM and Wang J. 2020. An Empirical Study on Greenhouse Gas Emission Calculations Under Different Municipal Solid Waste Management Strategies. Appl. Sci. 2020, 10, 1673; doi:10.3390/app10051673

Localizing disaster risk reduction: a case study of community-based disaster preparedness in Bima and Palangkaraya Indonesia

Rahmawati Husein*

Universitas Muhammadiyah Yogyakarta

*Email: Amahusein@yahoo.com

Abstract Community- Based Disaster Preparedness (CBDP) approaches are important elements to reduce vulnerability and increase resiliency in disaster risk areas. These are part of a new trend in localizing disaster risk reduction (DRR) which values the knowledge and capacities of local people and builds on local resources, including social capital. The CBDP in Indonesia has been initiated both by government and various actors such as NGOs, academia, and privates. This paper focuses on the CBDP initiatives called "MASTANA" that are carried out by Muhammadiyah (MDMC), one of the largest Faith-Based Organization (FBO) in Indonesia in collaboration with local disaster management agency. This paper will describe how local communities through MASTANA program succeeded in formulating disaster preparedness strategies and situating them within wider local development planning. It also depicts how communities are able to integrate contingency and operational plans into planning document that was formalized and signed by the village head to guarantee the sustainability of DRR program. In addition, it will describe the formation of DRR forum that consist of various stakeholders including women and people with disabilities. Last, it will discuss challenges in using CBDP approaches at the village level.

1. Introduction

Indonesia is one of the world's most disaster-prone countries. The natural geographic and physical setting of Indonesia between the Indian Ocean, South China Sea, and Pacific Ocean place the country at the centre of typhoon, tectonic and volcanic belts. It has significant levels of exposure to earthquakes, tsunamis, volcanic eruptions, droughts, floods, and cyclones. From 2007-2017 Indonesia experienced 18,651 natural disasters, small and large, with the total lost up to US $ 20 billion [3, 6]. One of the largest disaster was Tsunami that devastated Aceh province on December 26, 2004 that caused hundreds thousand losses of life and enormous destruction of infrastructure and property. The impact of this disaster led the development of Law Number 24 of 2007 on Disaster Management.

The law has carried out the new perspective on disaster management by having inserted disaster management not only in an emergency response context, but also as pre-disaster and post-disaster. Before that, the disaster management in Indonesia remained emphasizing only emergency response or relief of disaster. In addition, based on the law, there was an establishment of Indonesian Disaster Management Authority abbreviated as BNPB. This agency is directly responsible to the President of Indonesia and the chairman is directly appointed by the President.

Since the launch of the law, the agency has been very active in taking a lead to coordinate the entire disaster management system in Indonesia, from preparedness to response and recovery. BNPB has actively launch many regulations, including regulations that emphasizing disaster risk reduction. In

2010, the chairman of BNPB launched Regulation Number 5 of 2010 on the National Action Plan for Disaster Risk Reduction (RAN PRB). This action plan aimed to serve as a guide and reference of DRR policies at the national level within the period of three financial years and provided a comprehensive picture of the action plans of all relevant stakeholders, including government, non-governmental organizations, international community, as well as the business world.

Following the action plan at the national level, the BNPB launched a Regulation No. 1 of 2012 on Guidelines for Disaster Resilient Village (Destana). This policy aimed at localizing preparedness and risk reduction at the community level. This is based on experiences in responding numerous disaster that the local community is the most affected and vulnerable parties in time of disaster, yet they are the first responder during the event. Therefore, local community preparedness and ability to reduce any risk are needed to build a strong basis for the establishment of a disaster resilient community. Since the launch of the program, hundreds Destana have been initiated throughout regencies and cities in Indonesia. Destana is one of the program priority of Indonesian National Development Plan and between 2015-2019, five thousand (5,000) Destana were targeted to be formed [3]. However, based on the data from BNPB until the end on 2017, there are only 528 Destana were initiated by the national and provincial Disaster Management Agency in all 34 provinces in Indonesia, and around 1,500 disaster resilient communities were carried out by various stakeholders in disaster preparedness, such as universities through community services program, businesses through their CSR (corporate social responsibilities), and NGOs [3, 10].

One of the NGOs that promote community resilient program is Muhammadiyah. Muhammadiyah, the oldest and one of the largest faith-based organization in Indonesia, was founded in 1912 and its member were estimated 35 million. Muhammadiyah has responded disaster since 1919 and worked extensively during one of the largest disaster in history, Tsunami Aceh in 2004 and various disasters that hit Indonesia in the past decades [8]. Later, in 2010 Muhammadiyah established a special unit for disaster management and humanitarian assistance namely MDMC (Muhammadiyah Disaster Management Centre) [9].

From 2015-2018 MDMC Muhammadiyah, in partnership with DFAT Australia completed project of Muhammadiyah Hospital Preparedness and Community Readiness (HPCRED) in various parts of Indonesia. This program was initiated in 2008 namely Hospital and Communities Preparedness for Disaster Management (HCPDM). Starting 2015, it was invigorated through a program called Hospital Preparedness and Community Readiness for Emergency and Disaster (HPCRED), where the author was acted as a National Program Coordinator of the project, beside her current position as a vice chair of MDMC. HPCRED #1 focused more on increasing capacity of the safe hospital trainer and EMT, developing materials, standards and guidelines. Yet, 2016-2018, HPCRED #2 introduced a comprehensive safe hospital, where the focus not only aimed at safety and function of hospitals during disaster but also aimed at increasing capacity of community surrounding hospital to create resilient community which is expected to be able to reduce disaster risks and assist hospital during emergency and disaster [12]. As it stated in WHO (2015), the comprehensive safe hospitals integrate community around hospital to participate in helping to reduce the risk by taking care a minor injured, providing mental support and assisting the health providers [21].

The HPCRED #2 (2016-2018) program was held in Bima West Nusa Tenggara (Sumbawa Island) and Palangkaraya, Central Kalimantan (Borneo Island). These two areas become the target for the implementation of community-based disaster preparedness program because Bima is a high-risk area that in the past suffered from floods and flash flood, as well as social conflicts. In 2016, almost half of the city was affected by the flash flood. Meanwhile, Palangkaraya experienced forest fire and haze disaster frequently. In 2015 for more than two months Palangkaraya were dense with haze produced by forest fires. According to Indonesian Disaster Risk Index these two cities were in high risk [4]. Thus, it is interesting to study how the community preparedness and readiness can reduce disaster risks and meet the objective to create resilient community.

2. Conceptual Understanding of Community-Based Disaster Preparedness (CBDP)

Disaster risk is on the rise throughout the world and over the past two or three decades, the economic, social and physical losses have increased exponentially [14, 16]. To minimize the losses, various efforts have been taken by government and international communities including donors. Yet, despite of growing number of program established, there has remained failed to be sustainable at the local level after the project ended [14]. They further said that sustainability is crucial to preserve the risk reduction and this needs to consider 4 elements of community elements including partnership, participation, empowerment and ownership by the local people. Therefore, community-based disaster preparedness become important because at the community level the effects of disasters are felt the most and where the physical, social and economic risks can be most adequately assessed and managed. Besides, nobody is more interested in understanding and improving local affairs and conditions than the community residents themselves. They also understand local opportunities and constraints better and often to be the first and quickest response to a disaster in any community [2].

There are various concepts on Community based disaster preparedness (CBDP). It is defined as a process of bringing people together within the same community to enable them to collectively address a common disaster risk and to collectively pursue common disaster preparedness and a process that mobilizes a group of people in a systematic way towards achieving a safe and resilient community/group [2]. CBDP helps communities prepare for, respond to and recover from a disaster and it is an approach in which community support systems and sustainable coping mechanisms are reinforced and collective knowledge and capacities are applied to reduce the adverse impacts of recurring disasters (CRS, 2009). CBDP is also an approach that focus on community as the key actor who plans, designs, acts, monitors, and evaluates the disaster risk reduction activities.

Community often the first affected by disaster and in the same time community is also the first responder and forefront in any difficult situation. Assam DMA (2013) also mentioned that disaster impact is felt at the community level although it may hit one or several communities at once. Thus, CBDP emphasizes collective community action and the internal mobilization of resources to support the community's effort to be self-reliant to the extent possible and for a more systematic and effective mobilization and use of external resources needed to implement plans and projects identified by the community members to be most relevant.

Meanwhile, CBDP often considered as a community-based disaster risk reduction (CBDRR) since it constitutes preparatory actions taken before a natural hazard strikes and encourages grass-roots communities in conducting their own interpretation on disaster risks, reducing vulnerability of their communities, setting up the priority based on the risks as well as monitoring and evaluating their efforts in disaster risk reduction [15]. In addition, they state that CBDRR can also be defined as the empowerment of the community to be able to manage disaster risk with the level of involvement of the parties/community groups in the planning and utilization of local resources in implementation activities by the communities themselves.

Furthermore, CBDP also often be categorized as a community-based disaster risk management (CBDRM) which is defined as a process that leads to a locally appropriate and locally owned strategy for disaster preparedness and risk reduction [20]. CBDRM promotes an integration of bottom-up and top-down efforts with a starting-point in the local actors in order to strengthen community's capacity to handle and mitigate disasters, increase their risk awareness and reduce vulnerability. The goal is to reduce risk in areas of concern by empowering individuals. It is a shifting from a reactive approach to hazard events, to one of being proactive in reducing disaster risks (DRR). There is a shift in the understanding that in any disaster occurrence, people were the most affected by the impact of disaster. Therefore, active involvement of local communities in disaster risk reduction to reduce their vulnerability and strengthen resilience is very important.

Izumi & Shaw (2012) that were cited by Sjöstedt and Sturegård, stated that CBDRM was presented as an alternative to traditional DRR in the late 1990's and was initially implemented by NGOs in the developing world. CBDRM is considered successful in raising risk awareness, building on local institutions, resources and capacities and in addressing existing vulnerabilities it quickly gained ground among international organisations and local government. They also say that the approach of CBDRM was further strengthened in the Hyogo framework for action from 2005 and in the succeeding Sendai framework for disaster risk reduction from 2015. It has in other words been mainstreamed to the point where it is now difficult to find Disaster Risk Management (DRM) programmes without CBDRM components [18].

3. Methods

This paper was based on the HPCRED (Community Readiness for Emergency and Disaster) project report including area project report, quarter report, annual and final report. It will also be based on the baseline and endline survey report that were carried out by an independent consultant. Additionally, the data were also collected through the in-depth interviewed with the project officers both at the national and the two areas, as well as from documentation, personal field work, observation and engagement through Focus Group Discussion (FGD) and site visit where the author was actively involved as the National Program Coordinator and member of National Advisory Board. The qualitative data were then categorized based on indicators sets in the NDMA Regulation.

The study focuses on how the five communities in Bima and Palangkaraya were planning disaster resiliency using community-based approach to identify and map risks, develop disaster plan including contingency and action plans and also to form DRR forum and emergency response team. It also depicts on how the local community made the disaster plans legalized by the local village leader and authorization to guarantee the implementation and sustainability of the program.

4. Mastana: A CBDP project in Bima and Palangkaraya, Indonesia

Mastana is abbreviated from Masyarakat Tangguh Bencana or a disaster resilient community. The aim is building community capacity for disaster resilience. As Ginsberg and Hunt (2016, p.5) state that "disaster risk reduction is the foundation of community resilience", where designing program on DRR is building community resilience or community resilience can be considered an outcome of DRR.

Mastana is in line with the Indonesian government program namely Desa Tangguh Bencana (Destana), a Disaster Resilient Village - which was urged by the BNPB through the regulation No. 1/2012. In the regulation, Destana is defined as a village that has an independent ability to adapt and deal with disaster threats, and immediate recovery from the adverse effects of disasters. This similar what Twigg (2009) defines that community resilience is a community that has capacities "(1). to absorb stress or destructive forces through resistance or adaptation, (2). to manage, or maintain certain basic functions and structures, during disastrous events and (3). to recover or 'bounce back' after an event" [19].

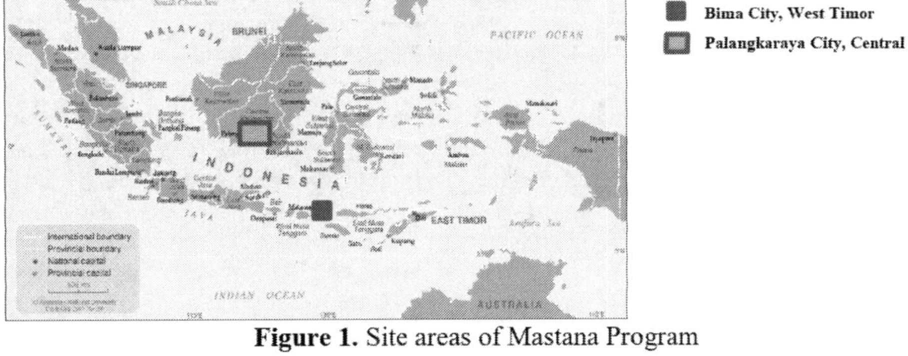

Figure 1. Site areas of Mastana Program

Destana is a self-support, meaning a series of self-empowered and efforts to mobilize resources that villagers have and to be able to recognize disaster risks, village vulnerability and capacity. It is also the ability to monitor and evaluate the resiliency activities. Being at the forefronts, communities need to have capacity to respond to threats themselves. It is for this reason that communities should be involved in managing the risks that may threaten their well-being. There are at least 20 indicators to illustrate the village resilience (BNPB Regulation No. 1/2012).

Between January 2017–January 2018, MDMC Muhammadiyah initiated and completed Mastana in 5 urban communities in the city of Palangkaraya, Kalimantan and Bima, NTB. Two (2) urban villages/communities in Palangkaraya namely Kelurahan Menteng and Langkai and three (3) urban villages/Communities in Bima namely Kelurahan Nae, Pane and Monggono participated in the Mastana projects aimed at forming disaster resilient community where they able to identify and mapping their own risk, creating disaster plans and taking action to implement strategies.

The different between destana, which initiated by government, and mastana that initiated by MDMC Muhammadiyah are the strategy used. Mastana was facilitated by Muhammadiyah members in the grass root, not the village administrative. The process was not only creating plan and standard operating procedure, but also examining plans and procedures through a series of exercises and incorporating disaster plans into the village development plan.

In addition, the different approach was the initiative of the planning which was focus on the collaboration and cooperation between various stakeholders at the district and village level. This collaboration also included local government actors such as the district disaster management agency and district health office, the hospital and clinics, Muhammadiyah board and member from district to community level, the universities/higher institution (public and Muhammadiyah), Red Cross, and Community groups. These local actors may provide local knowledge about disaster issues and formulate plan and action according their priorities as well as agree on mechanism and procedure that needed when disaster hits their community. Besides, in this approach various stakeholders at the national level were also involved to advocate the multi-stakeholder approach and to provide expertise, technical assistance and financial support particularly used for drills and exercises.

At the national level, Mastana was prepared and planned by the division of disaster risk reduction, MDMC, and national facilitator was already well-trained. The module and material used for the training were modified from Destana program and MDMC training material. Meanwhile, at the district level, the collaboration between Muhammadiyah at city level and the city disaster management agency helped in supporting and pushing the village/community leader to work with and support Muhammadiyah at the grassroot level particularly in facilitating the meeting, assisting in data collection and analysis, and disseminating information.

5. Implementation Process of Mastana: stages, steps and activities

To build disaster resilient communities, it first requires empowering community members so that they can cope with the adverse impact of disaster. Thus, Mastana entailed a long process to guarantee the sustainability in dealing with disaster risks. The process of the project implementation including its activities consists of three stages ie: program preparation, program implementation activities, policy advocacy and assessing capacity (HPCRED Report, 2018). The figure below, shows the overall process of Mastana.

The first stage in the process is preparing resources. This is part of the individual and organizational capacity building. In this stage MDMC Muhammadiyah at the national level take a lead role by developing training and workshop material for local facilitators. Along with the local Muhammadiyah member, they also disseminate information of Mastana program and disaster risk reduction education and awareness for individuals and group leaders in the community.

IOP Conf. Series: Earth and Environmental Science **989** (2022) 012027 doi:10.1088/1755-1315/989/1/012027

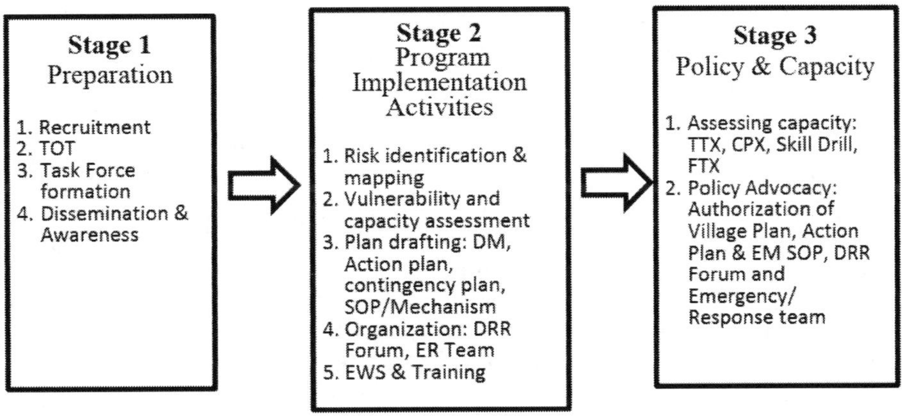

Figure 2. Mastana Process and Activities

Following the dissemination of knowledge and the enhancement of awareness, the local MDMC then recruit and selected 30 persons from each community (village) as Mastana Task Force. In addition, local MDMC also selected 3 representatives from each village/urban community to join Training of Training (TOT) for facilitator of Mastana program. As the program is part of safe hospital program, one of the three persons selected must be represented from the hospital who has the responsibility to ensure the integration of Mastana in the safe hospital program. The other two are from the community (village) surrounding hospital. Some of the requirements to be Mastana facilitators are willingness and capability to be volunteers, have time and ability to organize community meeting and have trust from the community. These selected candidates then must participate in the 5 days TOT for facilitator that was held in the Muhammadiyah headquarter, Yogyakarta, consisting disaster management, disaster risk reduction knowledge and facilitation skills.

The successful of the first stage was influenced by the ability to recruit the right representatives who have networks of supporters in the early-on of the planning process. The recruitment used Muhammadiyah networks in the grassroot level which has lots of contacts, lots of trust, and person-to person relationship with the key community leader including the head of Kelurahan (Village leader) and the city officials. The recruitment strategy was multi-pronged with potential participants identified early-on during meetings and consultation, and invitations were then issued by telephone/WhatsApp, personal contact, followed by face-to-face contact. All member of Mastana task force and facilitators were very engaged and respected in the community and recognized by the village/community leader. In addition, various local actors in disaster management were actively involved since the initiation of the program including the staff of district disaster management agency, health office, red cross, universities and several community-based organizations.

Stage two of Mastana focused on program implementation activities consisting area preparedness and risk reduction activities which was started with a). participatory assessment and appraisal covering risk, vulnerability and capacity assessment, risk identification and risk mapping; b). drafting village disaster management plan (3 years plan); c) drafting annual disaster management action plan that is incorporated in the village development plan; d) Establishment of DRR Forum (FPRB), in Bima FPRB's name as Preparedness Team (TSBK), while in Palangkaraya remain FPRB; e) developing early warning system based on local need and culture. In Bima, it uses wooden/bamboo slit-drum – traditional telecommunication tools that are used to warn people against danger or to call people for meeting, meanwhile, in Palangkaraya, it uses siren; f). drafting contingency plan, and g) training on sector capacity – CPR/emergency health, kitchen soup, EOC management, communication.

The stage two was started once the facilitator and task force member committed to engage the program and understanding the scope and the timeline of the projects. During the process, facilitator and task force received technical assistant from MDMC/HPCRED facilitators and team at national level.

This team assured that activities were carried out respectively and all member of task force participated in the whole process from assessment, from planned and drafted various village plans as well as advocated policy and assessed capacity at the village level. All facilitators and task force member worked collaboratively and attentively for almost one and a half year (HPCRED Report, 2017)

In this stage, the task force and facilitators succeeded in pushing the team to think and work more inclusively to identify risk and to create risk maps, disaster management plan, action plan to reduce impact of disaster as well as to set up the mechanism and procedure once a disaster hit. The team in each village/Kelurahan also succeeded in establishing DRR forum which are representatives from various stakeholders and member of the community. The forum met regularly to draft the contingency plan and work on integrating all DRR documents in the development plan. They were bringing the key stake holders and duty bearers and developed all plans for their villages/communities as next steps toward building disaster resilient communities for themselves and their neighbors. This can also be done as the village leader and their administrative staffs involved since the early-on.

The process in Bima was faster because the two communities have ever implemented the disaster resilient village that enacted by district disaster management agency. Yet, there was no document produced during the program and their preparedness had not been examined through table top exercise and full-scale exercise (HPCRED quarterly report, 2017). Therefore, the focus of communities in Bima was more on drafting disaster plan documents and increasing technical skills of the response team/disaster task force. Meanwhile, in Palangkaraya, the processes started from the very beginning as they have not had any experience yet.

Stage three of Mastana program focused on policy advocacy and assessment of community capacity through drill and exercise. Communities in Palangkaraya and Bima held Table-Top Exercise (TTX), Command Post Exercise (CPX) and Full-scale Exercise (FTX) respectively. Through the drills and simulation, not only skills were exercised but also mechanism and coordination that written in SOP were examined. The scenarios of all exercises were planned and drafted using participatory planning to assess the ability of community in preparing and responding any disaster. In all exercises, various stakeholders including village administrative, health clinics, community groups, individuals, and red cross involved and engaged based on their respective roles and responsibilities.

These exercises in the city of Bima and Palangkaraya succeeded in involving hundreds of people and examining the knowledges and skills they received throughout the program as well as examining the procedure they agree on. Mayors of the two cities even supported the exercises by demanding the city official involved and endorsed financially. During the exercise, expert from various stakeholders at national level participated and acted as evaluator or observer.

In this stage, the task force of Mastana program worked closely with village administrative to ensure that all documents drafted during the project would be signed by the head of the village. This is very important to advocate the policy change at the village/community level to guarantee the implementation of risk reduction program and to reach the goal in creating the disaster resilient community. The documents that were signed by the village leaders of Bima and Palangkaraya included Risk Assessment Report and Mapping, Disaster Management Plan, Contingency Plan, Action Plan, SOP for all sectors such as health, search and rescue, and shelter/protection, and DRR Forum certificate. The signing was also crucial to guarantee the sustainability of disaster risk reduction program as they are integrated in the village development plan which implies the village support on budget, facilities and expertise. In Indonesia, in the Medium-term Development Plan (5 years plan) does address the issue of DRR, however, not all local plan specifically incorporated risk reduction in their plans. In addition, not many local governments have put DRR at the annual work plan (RKP) which is the break-down of the medium-term development plan and DRR constitutes one of the nine national development priorities. Therefore, signing documents particularly the annual work plan will ensure the ability of each community to continue set the priority and provide funding.

IOP Conf. Series: Earth and Environmental Science **989** (2022) 012027 doi:10.1088/1755-1315/989/1/012027

6. Results and Discussion

The community-based disaster preparedness through mastana program was assessed using indicators of Destana which composed in the BNPB Regulation No. 1/2012. Following are six main elements and components of disaster resilient community which have been achieved through Mastana project in five communities in Bima and Palangkaraya. The table shows that almost all elements of disaster resilient village have been met by all five communities in Bima and Palangkaraya, except funding component. This can be explained as during the mastana porject, the funding was supported by the external resources. However, in the plan, communities have commitment to allocate budget for preparedness program in village plans and to use all resources available.

Table 1. Elements of disaster resilient community based on BNPB regulation No 1/2012

No.	Thematic area of Destana	Component of resilient	Palangkaraya		Bima		
			Menteng	Langkai	Monggonao	Nae	Pane
1.	Legislation	•Village regulations governing:	✓	✓	✓	✓	✓
		- Disaster management plan	✓	✓	✓	✓	✓
		- Contingency plan (SOP for all sectors)	✓	✓	✓	✓	✓
		- Adoption of Risk Identification and mapping	✓	✓	✓	✓	✓
		- DRR Action Plan	✓	✓	✓	✓	✓
		•DRR and disaster management - All the above document/plan	✓	✓	✓	✓	✓
2.	Planning	•Contingency plans	✓	✓	✓	✓	✓
		•Community action plan for disaster risk reduction	✓	✓	✓	✓	✓
		•Risk Identification and mapping	✓	✓	✓	✓	✓
		•Mainstreaming DRR in dev. plan					
3.	Institutional	•DRR village forum	✓	✓	✓	✓	✓
		•Cooperation among sectors and stakeholders	✓	✓	✓	✓	✓
4	Funding	•Resource mobilization					
		•Community and private fund					
5.	Capacity building	•Training,	✓	✓	✓	✓	✓
		•Education public awareness	✓	✓	✓	✓	✓
		•Gender mainstreaming	✓	✓	✓	✓	✓
6.	Implementation of disaster management	• Mitigation activities	✓	✓	✓	✓	✓
		• Early warning systems	✓	✓	✓	✓	✓
		• Preparedness for emergency response	✓	✓	✓	✓	✓
		• Recovery			✓	✓	✓

Source: Adapted and modified from Ririn T.N. (2014), BNPB (2012), & HPCRED (2018)

Based on the endline survey report, following will be described the score that were calculated based on indicators set in the NDMA regulation where disaster resilient community fall into three categories: high resilient (score 51-60), medium (score 36-50), and preliminary (score 20-35). In Palangkaraya city, Kelurahan Menteng fall into the high level of resilient with score 51 (2018) compare to the baseline result with score 20. There was significant increase in policy, disaster plan, DRR forum, emergency response team and multi stakeholder's collaboration. Meanwhile Kelurahan Langkai increased from 27 (preliminary) in 2016 to medium resilient with score 50, which is very close to the high level of resilient. Based on the interviews reported in the endline survey, the factors were similar to Kelurahan Menteng [11].

Whereas in Bima, the change was not very significant particularly at Kelurahan Nae and Pane as they have exposed to the program. In Nae the level of resiliency remains in the medium level, it was only increase the score from 38 to 48. Meanwhile in Kelurahan Pane, it increased from preliminary level with the score 35, it moved up to medium resilient with the score of 41. Similar to Palangkaraya, both Kalurahan Nae and Pane increased in in policy, disaster plan, DRR forum, emergency response team and multi stakeholder's collaboration and also early warning system. However, in Kelurahan Monggonao which had not been exposed to the program there was a significant increase from pre-beginning with the score of 2 in the baseline, it increased to 43 in the endline survey (Leman et.al, 2018). The role of the city and neighbouring community which has been exposed to the program help in influencing the adoption of programs and activities at Kelurahan Monggonao.

From the endline survey, it found that community feels that the preparedness program that carry out by the local actor brought some changes: 1) disaster knowledge from not knowing to understanding the risks, 2) skills on preparedness and emergency response have been expanded, 3) the risk behaviour from less care on the risks and impact of disaster to be more thoughtful, 4) networking become more extensive, 5) systematic in disaster management not reactive and unplanned, but more mitigative. There has been a change in mindset from reactive to mitigative. There have been systematic efforts in reducing risk including economic activity and capacity building for better response and recovery. The increasing knowledge, skills and efforts were not only carried out by the task force but community in general. The increasing capacity is very important as Pandey and Okazaki mentioned that "in order to build disaster-resilient community, they first need to be empowered so that community members can cope with the adverse effects of natural hazards (2012).

7. Conclusions: challenges and lessons

Mastana project in Kelurahan Langkai and Menteng, Palangkaraya and Kelurahan Nae, Pane and Monggonao, Bima is one of community-based program that were designed to catalyse and build community to be prepared for, respond to and recover from disaster. The program is considered effective in promoting disaster risk reduction and shaping disaster resilient community. The actively involvement of disaster management task force member and DRR forum as well as village administrative and district stakeholders in the early stages of the project development has created greater awareness about disaster preparedness and risk reduction among the member of community and decision makers.

Some of activities that are considered effective include the integration of disaster plans into the village development plan which were then signed by the village leaders. This process is considered a bottom-up approach and using Arnstein ladder of participation (1969), this reaches higher degree of citizen power where citizens obtain the majority of decision-making [1]. In addition, the project shows the ability of community to examine the knowledge and skills through various exercises such as table-top exercise, skill drills, command post exercise and full-scale exercise.

Yet, community-based disaster preparedness requires a long process and is influenced by several factors such as the ability to select the right local leaders and actors who can mobilize preparedness activities including providing knowledge and skills, identifying local resources to increase capacity, and advocating policy at local level. Also, it is the important to maintain collaboration and network where members of community know with whom the can work together to cope disaster impacts. Last, it is also vital that mastana project specify the action plans and the nature and extend of contribution from internal community as well as external community to meet a need in increasing capacity. Thus, communities need to share agreed criteria for the priority of activities to be funded which remain a challenge for almost all five communities in Bima and Palangkaraya. Other challenges are having regular reviews and revision of disaster plans as they are living documents and performing regular drills and exercises which require a lot of funds too. The cooperation between government and non- government organization become crucial in promoting community-based disaster preparedness for community resilient.

Reference

[1] Arnstein, S (1969). A ladder of citizen participation. *Journal of the American Planning Association,* 35: 4, 216 — 224.
[2] Assam State Disaster Management Authority (Assam DMA) and Doctors for You (2013). *Community Based Disaster Preparedness: Course Book.* Assam State DMA, India
[3] BNPB. (2017). 524 Desa Tangguh Bencana (524 Disaster Resilient Village). Published in August 2, 2017. Jakarta: BNPB
[4] BNPB (2013). Indeks Risiko Bencana Indonesia (Indonesian Disaster Risk Index
[5] Catholic Relief Services (CRS). (2009) Community Based Disaster Preparedness: A How-to Guide.
[6] DIBI (Data Infromasi Bencana Indonesia) BNPB (2017). Downloaded from dibi.bnpb.go.id.
[7] Clark-Ginsberg, A. and Hunt, D. (2016). Disaster Risk Reduction for Community Resilience: A 2015 synthesis of lessons from more than a decade of Disaster Risk Reduction Programming. Concern Worldwide.
[8] Husein, R (2012). Extending transnational network: a case study of Muhammadiyah' collaboration with various actors in major disaster response. Paper presented in the International Conference on Muhammadiyah, Malang 29-3 November 2012
[9] Husein, R. (2017). Muhammadiyah, Indonesia: experiences of a national and local NGO. Power point presentation at conference on "One year on – the grand Bargain and localisation.Wilton Park, UK, 5-7 June 2017.
[10] Kompas (2016). BNPB Targetkan ada 5.000 Desa Tangguh Bencana (in 2019, BNPB targets 5,000 Disaster Resilient Village). Published on October 27, 2016.
[11] Leman, HI, Roy T., & Nurhanita. (2018). Laporan Endline Survey Program HPCRED (endline survey report). Remdec Swaprakarsa Resource Management and Development Consultant.
[12] MDMC (2018). *HPCRED Final Report.* Yogyakarta, May 2018.
[13] MDMC (2017). *HPCRED First Year Report.* Yogyakarta, Dec 2017
[14] Pandey, B., & and Okazaki, K. (2012) Community Based Disaster Management: Empowering Communities to Cope with Disaster Risks. United Nations Centre for Regional Development, Japan
[15] ADPC (2005). *Community Based Disaster Risk Management Guidelines.* The Asian Disaster Preparedness Center, February 2005.
[16] Peacock, W.G, Kang, J.E., Husein, R., Burn, G.R, Prater, C., Brody, S.D., and Kennedy, T. (2009) *"An Assessment of Coastal Zone Hazard Mitigation Plans in Texas.* Unpublished report prepared for the Texas General Land Office and The National Oceanic and Atmospheric Administration, Hazard Reduction and Recovery Centre.
[17] Ririn T.N. (2014). Disaster resilient village as a strong modality for building disaster resilient community in Indonesia. Research Proposal. Downloaded from http://www.respect.osaka-u.ac.jp/satellite-gadjahmada-en/wp-content/uploads/sites/7/2014/12/
[18] Sjöstedt, F., & Sturegård, V. (2015). *Implementation of Community Based Disaster Risk Management in the Mekong Delta, Vietnam.* Division of Risk Management and Societal SafetyLund University, Sweden. Report 5022, Lund 2015.
[19] Twigg, J. (2009). Characteristics of a disaster-resilient community: a guidance note (version 2)
[20] UNISDR (2012). Community-based disaster risk management. Downloaded

IOP Conf. Series: Earth and Environmental Science **989** (2022) 012027 doi:10.1088/1755-1315/989/1/012027

from https://www.unisdr.org/we/inform/events/28132
[21] World Health Organization (WHO) (2015). Hospital safety index: guide for evaluators. 2nd ed.
[22] NDMA (BNPB) Regulation No. 1 of 2012 on Guidelines for Disaster ResilientVillage/Kampong
[23] NDMA (BNPB) Regulation No. 3 of 2012 on Guidelines for Local Capacity Assessment in Disaster Management
[24] PAsTI (Preparedness Assessment Tools for Indonesia) - an assessment tools for community preparedness.

Baseflow separation using graphic method of recursive digital filter on Wae Batu Gajah Watershed, Ambon City, Maluku

Bokiraiya Latuamury *, **Rafael M Osok, Ferad Puturuhu, and Wilma N. Imlabla**

Research and Training Center of Disaster and Climate Change Adaptation, Pattimura University, Ambon-Indonesia 97233.

*Email: bokiraiya.latuamury@faperta.unpatti.ac.id

Abstract. The most commonly used baseflow separation graphics method is the recursive digital filter. Choosing an RDF graphic filter method to estimate the base stream requires a sure accuracy. The application of the RDF graphics method on small island watersheds must take into account the unique hydrological conditions of small island watersheds with a small area of rain catch so that the RDF method approach can be applied. This research aims to analyze graphic methods for the hydrological characteristics of the small island watershed of the Ambon City of Maluku Province. The research method uses five RDF methods: one parameter algorithm, two-parameter algorithm, RDF Lyne & Holick, RDF Chapman, and RDF Eckhard. The results showed that the appropriate filter parameter value for small island watersheds was based on optimal parameters for assessing accuracy. The five RDFs showed that Eckhardt's filters performed better than one parameter algorithm, two parameters algorithm, Lyne and Hollick algorithm, and Champman algorithm, on a more extensive range of conditions. The optimal value of the filter parameters is relatively uniform for all five filter methods depending on the characteristics of the watershed and somewhat similar hydrological inputs.

1. Introduction

According to the hydrometeorology disaster risk study, Ambon Island is on a score of 156, which belongs to the high-risk category. In the last five years, the potential for widespread flood hazards for Ambon Island has experienced flooding and landslides with the impact of infrastructure damage, fatalities, and deaths relatively high. The possibility of residents being exposed to flooding in the five watersheds of Ambon City is relatively high due to the condition of the river border protection area crowded with settlements, and the number of vulnerable groups is high [1]. The study of hydrometeorology of watersheds such as densely populated cities becomes a need for watershed-based regional development.

Disaster-based hydrometeorological studies require accurate and applicable hydrograph separation techniques. The technical application of hydrograph separation has to identify the components of different streams (direct runoff(overland flow), unsaturated flow (throughflow), and saturated flow (groundwater) of the total flow [2]–[4]. The flow separation method generally divides the flow into fast flow components and delayed flow. Using the flow separation method, baseflow separation is undergoing rapid development in hydrological modeling [5], [6]. The delayed flow component is assumed to represent the proportion of discharge coming from the groundwater and other delayed flow sources, which by [7], [8] defined as a base flow.

The base flow of the time series measures the dynamic behavior of the water flow in a catchment area [9]–[11]. The baseflow proportion of the total flow has an index of the catchment area's ability to

IOP Conf. Series: Earth and Environmental Science **989** (2022) 012028 doi:10.1088/1755-1315/989/1/012028

store and release streams during the dry season [12]. A high base flow index implies that the catchment area has a more stable flow regime and can thus maintain river flow during the dry season [13], [14]. The baseflow index indicates the performance of a catchment area in describing the nature of storage [14].

Measuring baseflow components is relatively tricky. In principle distinguishing base flow and quick flow from measurable discharge data in a river requires an accurate and valid flow separation Technique. This is because the measurable discharge in the river combines the quick flow component and the base flow. During the dry season, which has no rain, most rivers in Ambon City still flow continuously. This flow comes from rainwater that seeps into the soil into groundwater reserves and will merge into a stream [15]–[17]. It, therefore, requires a variety of development and testing methods of baseflow separation that are analytical and empirical [10], [18], [19].

Baseflow separation using the RDF method generally considers the usefulness and ease of obtaining more accurate research results and high sensitivity. Testing and development of many RDF methods underwent significant effect. Known RDF methods in hydro office programs include one-parameter algorithm, two-parameter algorithm, IHACRES, BFLOW Lyne and Holick algorithm, Chapman algorithm, Furey and Gupta algorithm, Eckhardt algorithm, and EWMA [9], [10]. Many applications of baseflow hydrograph separation methods are undergoing development and testing throughout the world. [20] rated the performance of three Lyne and Holick RDF models, Boughton and Eckhardt filters in 66 catchment areas. The results showed that Lyne and Holick's filters performed better than Boughton and Eckhardt filters, with optimal filter parameter values varying depending on the characteristics of catchment regions and hydrological inputs. The separation of baseflow in the small island watershed of Ambon City becomes essential to research need to strengthen the study of hydrometeorological disaster risk research in Ambon City.

2. Methods

2.1 Research site

The research location description of the Wae Batu Gajah watershed, Ambon City, is presented in Table 1 and Figure 1. The selection of the research location considers the availability of daily debit data for the year 2007-2017 from SPAS Maluku Provincial River Region Hall, which has geographic coordinates X 408291.73 and Y 9591503.334.

Table 1. Morphometric characteristics of the Wae Batu Gajah watershed

Area (km^2)	Average Slope (m)	River Gradient (m)	Length of main river (km)	Drainage Density (km / km^2)	Circulation Ratio (CR)	Bifurcation Ratio (BR)	Flow patterm
6.35	0.49	324.98	6.58	2.91	12.56	0.95	Pinnate

Source: Image analysis, 2019

Wae Batu Gajah watershed area 7.04 km^2, average slope 0.49 m, river gradient 324.98 m, main river length 6.58 km, drainage density 2.91 km/km^2, Circulation Ratio (CR) 12.56 and Bifurcation Ratio (BR) 0.95 and flow pattern in pinnate shape. Drainage density (dd) Wae Batu Gajah watershed 2.91 includes the range of Dd with good triggering, which is between 1-5, also supported by the ratio of river branching indicates abnormal watershed conditions with very high flood peaks and a short recession time. The river border protection area filled with river dike construction on the left-right of the river shows that the Wae Batu Gajah watershed area has a high flood vulnerability characterized by high surface flow but low permeability and infiltration. The watershed bifurcation ratio of 0.95 indicates that the watershed form tends to round, showing a very high peak discharge (Qp). Rapid rising time and recession time also tend to be fast.

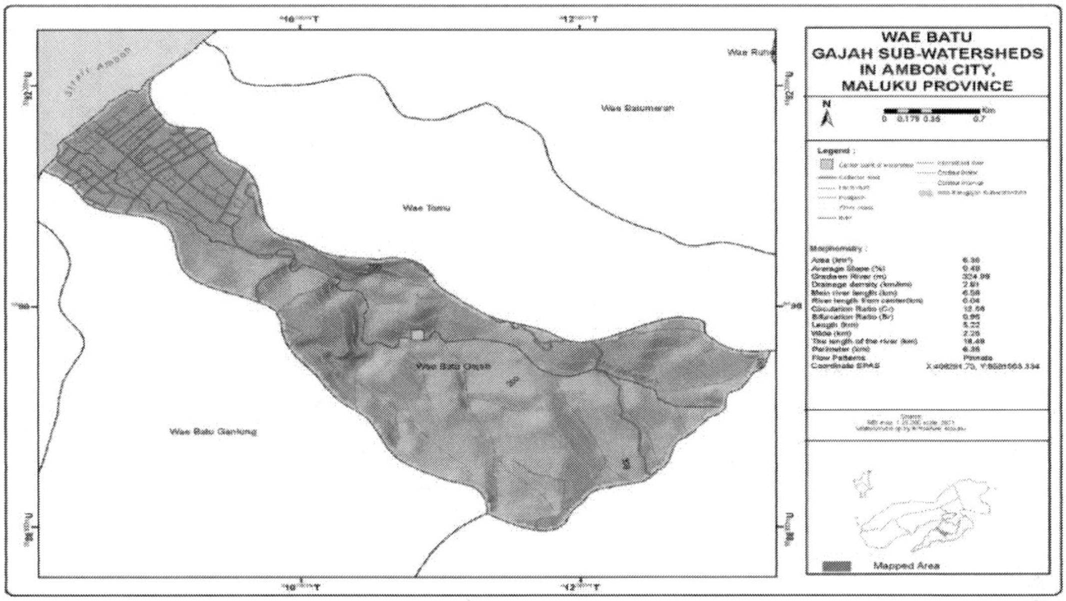

Figure 1. Morphometric map of the Wae Batu Gajah watershed in Ambon City

2.2 Calibration and validation of recursive digital filter methods

The baseflow separation technique uses five recursive digital filter methods in the BFI+ hydro office 12.0 program, each calibrated following the filter parameters presented in **Table 2.**

Table 2. Recursive digital filter methods in the BFI+ 3.0 program

Filter Name	Filter equation	Reference
One-parameter algorithm	$q_{b(i)} = \frac{k}{2-k} q_{b(i-1)} + \frac{1-k}{2-k} q_{(i)}$	(Chapman and Maxweel, 1996)
Boughton two-parameter algorithm	$q_{b(i)} = \frac{k}{1+C} q_{b(i-1)} + \frac{C}{1+C} q_{(i)}$	(Boughton, 1993; Chapman and Maxweel, 1996)
Lynie & Hollick algorithm	$q_{f(i)} = \alpha q_{f(i-1)} + q(i) + q(i-1)\frac{1+\alpha}{2}$	(Lyne & Hollick, 1979; Nathan & McMahon, 1990)
Chapman algorithm	$q_{f(i)} = \frac{3\alpha-1}{3-\alpha} q_{f(i-1)} + \frac{2}{3-\alpha}(q(i) - \alpha q(i-1)$	(Chapman, 1991; Mau & Winter, 1997)
Eckhardt algorithm	$q_{b(i)} = \frac{k(1-BFImax)\alpha b(i-1) + (1-\alpha)BFImaxxq}{1-\alpha BFImax}$	Eckhardth, 2005)
RMSE	$RMSE = \frac{\sqrt{\sum(Qc-Q0)^2}}{n}$	Qc: calculated baseflow (m^3/s), Q0: measured total flow on the river (m^3/s), n: number of sample.

Note:
q (i): total flow (observed flow) at day i, qb(i: calculated baseflow at day I, qf(i): calculated quick flow/direct run off at day I, q(i-1): total flow at day (i - 1), qb(i-1): calculated baseflow at day (i - 1), qf(i-1): calculated quick flow/direct run off at day (i - 1), k: filter parameter ~ recession Constant, α: filter parameter, C: filter parameter. BFImax: maximum baseflow index (constant).

3. Result and Discussion

River discharge is part of the hydrological conditions that take place in watersheds. Discharges often fluctuate promptly. The sustainability of discharge in the river is not constant but changes from year to year. Changes can be an increase or decrease in discharge between the rainy season and the dry season.

The results of static analysis of discharge descriptive during the period 2007-2014 from the Maluku Provincial River Region Hall showed that the minimum value (0.01-0.39), maximum (2.19-4.87), mean (0.40-1.1.0 16), standard deviation (0.32-0.60), variance (0.10–0.37), skewness (0.65-3.96), and kurtosis (-0.13-20.39). Long discharge data recording facilitates the process of calibration and validation of flow separation methods using BFI+ 3.0 hydro office 12.0 software, as presented in **Table 3**.

Table 3. Static discharge (m3/sec) during the period 2007 - 2014

Parameter statistics	Discharge							
	2007	**2008**	**2009**	**2010**	**2011**	**2012**	**2013**	**2014**
Minimum	0.01	0.10	0.39	0.02	0.23	0.20	0.35	0.28
Maximum	3.08	4.87	2.45	3.08	3.37	4.18	3.23	2.19
Mean	0.79	0.86	0.56	0.40	1.02	1.08	1.16	0.98
Std. Deviation	0.35	0.60	0.33	0.32	0.46	0.47	0.44	0.41
Variance	0.12	0.37	0.11	0.10	0.21	0.22	0.19	0.17
Skewness	1.81	2.87	3.18	3.96	1.59	3.24	1.20	0.65
Kurtosis	8.67	13.32	11.42	20.99	4.19	17.25	1.58	-0.13

Source: Primary discharge data 2007-2014

Calculation of the fifth filter parameter of the RDF method

Calculation of the fifth filter parameter of Calibration the RDF method on small island watershed Ambon City using five filter methods, namely one-parameter algorithm (k 0.85-0.90), two-parameter algorithm (k 0.85-0.90; C 0.35-0.50), BFLOW Lyne and Holick algorithm (α 0.905-0.956; BFmax 0.900-0.955), Chapman algorithm (α.900-0.953), and Eckhardt algorithm (α.927-0.962; BFmax 0.900-0.950). The range of filter parameter values for the five filter methods for the Wae Batu Gajah watershed is included in good baseflow hydrological conditions, presented in **Table 4**.

Table 4. Optimization of recursive digital filter parameters

Metode RDF	Parameter filter			
	k	**C**	**α**	**BFmax**
One-parameter	0,85-0,90			
Two-parameter	0,85-0,90	0,35-0,50		
Lyne-Hollick			0,905-0,956	0,900-0,955
Chapman			0,900-0,953	
Eckhardth			0,927-0,962	0,900-0,950

Source: Primary discharge data 2007-2014

Calibration Result from 5 methods RDF

Calibration and validation procedures are performed under different hydrological conditions. The result of the separation of baseflow from the total discharge in the river in daily time intervals using the five RDF separation methods. According to [15], discharge data availability is quite complete, and it is assumed that the morphometry characteristics of the watershed do not change. Hence calibration and validation procedures can be applied in hydrological modeling. The results of the fifth calibration of RDF methods for watershed wae Batu Gajah Ambon City are presented in the following sub-chapter.

3.1. RDF- One-parameter algorithm

The calibration results of the one-parameter algorithm method show that the value of the k filter parameter ranges from 0.85-0.90. Optimization of parameter k in baseflow separation obtains a range of minimum baseflow values between (0.01- 0.28 m3/s), maximum (0.81-1.27 m3/s), mean (0.20-0.58), standard deviation (0.11-0.18), and variance (0.01-0.03). as the security is presented in **Table 5**.

Table 5. Statistical value of baseflow RDF one-parameter algorithm

Baseflow	RDF One-parameter algorithm							
	2007	**2008**	**2009**	**2010**	**2011**	**2012**	**2013**	**2014**
Minimum	0.01	0.09	0.20	0.02	0.12	0.15	0.28	0.22
Maximum	0.84	1.27	0.81	0.81	0.97	1.05	0.93	0.87
Mean	0.39	0.42	0.28	0.20	0.50	0.54	0.58	0.49
Std. Deviation	0.12	0.18	0.11	0.12	0.14	0.11	0.13	0.12
Variance	0.01	0.03	0.01	0.01	0.02	0.01	0.02	0.01
Skewness	-0.43	0.53	1.91	2.02	0.19	1.37	0.40	-0.75
Kurtosis	3.32	1.76	3.33	4.94	1.52	3.04	-0.25	-0.16

Source: SPSS data from 2007 to 2014

The RDF method visualization of the one-parameter algorithm describes fluctuations in both flow components, rapid flow discharge, and baseflow over the study period. These fluctuations are caused by the hydrological conditions of inter-seasonal flows that change significantly, as presented in **Figure 2.**

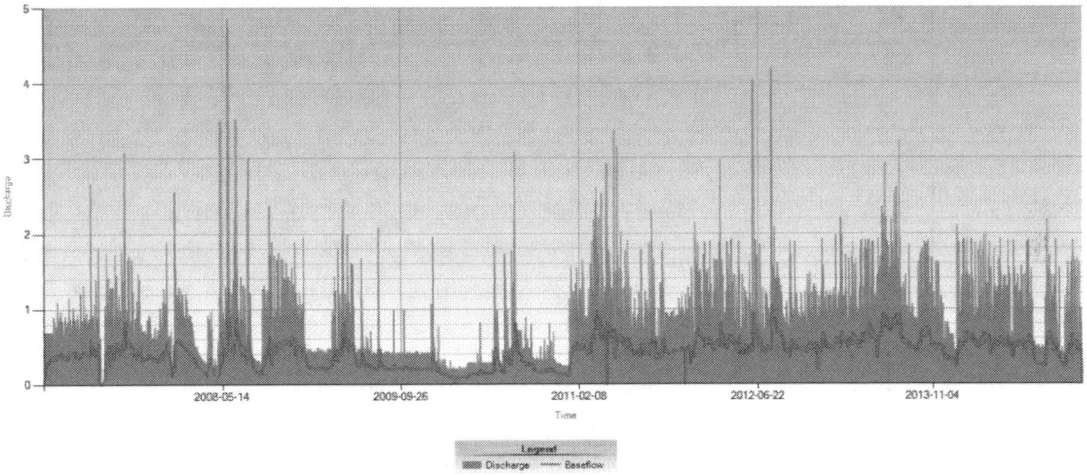

Figure 2. Baseflow separation using RDF One-parameter algorithm

3.2. RDF Two- parameter algorithm

Filter parameter range values for the Two-parameter algorithm.k method 0.85-0.90 and C 0.35-0.50, and baseflow statistical values are minimum value (0.00- 0.33 m^3/sec), maximum (1.41-3.12 m^3/sec), mean (0.32-0.94), standard deviation (0.20-0.36), and variance (0.04-0.13) presented in **Table 6** and **Figure 3**.

Table 6. RDF Baseflow Statistics Two-parameter algorithm

Baseflow	RDF Two-parameter algorithm							
	2007	**2008**	**2009**	**2010**	**2011**	**2012**	**2013**	**2014**
Minimum	0.00	0.06	0.33	0.02	0.19	0.19	0.31	0.25
Maximum	1.78	3.12	1.73	1.77	1.83	2.58	1.75	1.41
Mean	0.63	0.66	0.45	0.32	0.82	0.87	0.94	0.79
Std. Deviation	0.22	0.36	0.20	0.21	0.28	0.25	0.26	0.25
Variance	0.05	0.13	0.04	0.04	0.08	0.06	0.07	0.06
Skewness	0.52	1.73	2.61	3.09	0.64	1.92	0.76	-0.25
Kurtosis	5.00	8.18	8.14	13.40	1.42	8.60	0.20	-0.72

Source: SPSS data from 2007 to 2014

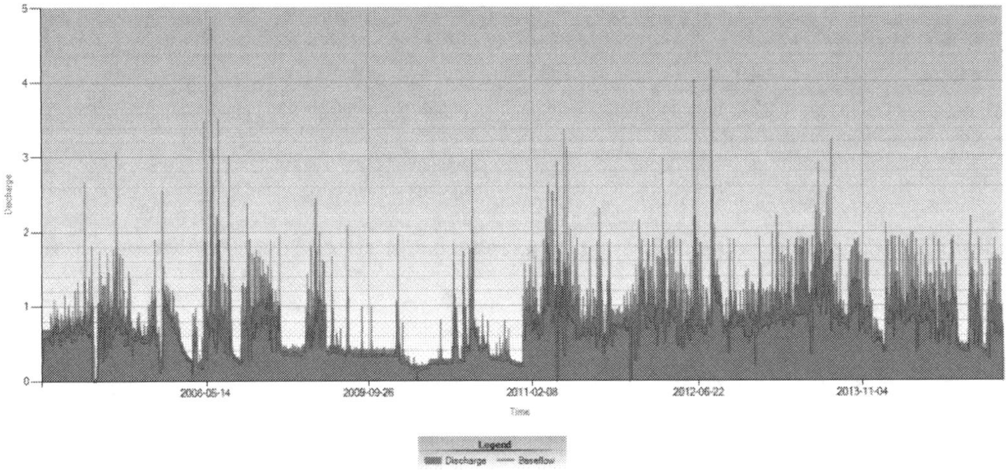

Figure 3. Baseflow separation using RDF Two-parameter algorithm

3.3. BFLOW Lyne and Holick algorithm

The filter parameter range values for the RDF Lyne and Hollick methods are α 0.905-0.956 and BFmax 0.900-0.955, respectively. The results of descriptive statistical analysis of baseflow separation are minimum value (0.00-0.32 m3/sec), maximum (1.22-1.74 m3/s), mean (0.35-1.00), standard deviation (0.17-0.31), and variance (0.03-0.10) presented in **Table 7** and **Figure 4**.

Table 7. RDF baseflow stats Lyne and Hollick

Baseflow	RDF Lyne and Hollick							
	2007	**2008**	**2009**	**2010**	**2011**	**2012**	**2013**	**2014**
Minimum	0.00	0,09	0,29	0.02	0.18	0.20	0.32	0.28
Maximum	1.41	1.71	1.25	1.22	1.65	1.62	1.74	1.26
Mean	0.66	0.67	0.49	0.35	0.85	0.91	1.00	0.83
Std. Deviation	0.22	0.31	0.17	0.17	0.26	0.22	0.27	0.24
Variance	0.05	0.10	0.03	0.03	0.07	0.05	0.07	0.06
Skewness	-0.79	0.15	2.28	1.68	0.18	0.31	0.57	-0.37
Kurtosis	1.72	-0.67	4.43	3.29	0.82	072	-0.07	-1.04

Source: SPSS data from 2007 to 2014

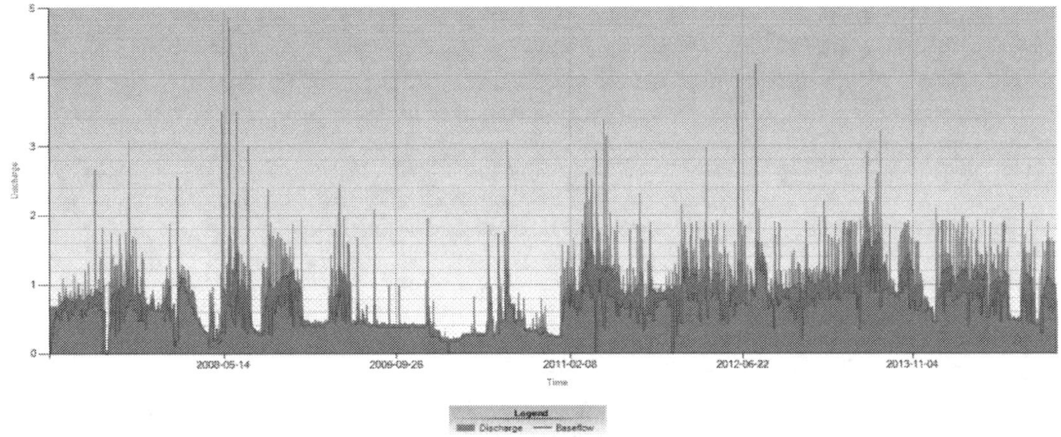

Figure 4. Baseflow separation using RDF Lyne and Hollick algorithm

3.4. *Chapman algorithm*

Calibration of Chapman's RDF method in baseflow separation obtains an alpha parameter value of 0.900-0.953, and base low descriptive statistical values during the period 2007-2014 are the minimum values (0.00-0.33 m3/sec), maximum (0.79-1.25 m3/sec), mean (0.20-0.58), standard deviation (0.11-0.17), and variance (0.01-0.03) presented in **Table 8** and **Figure 5**.

Table 8. RDF Chapman baseflow statistics

Baseflow	RDF Chapmann							
	2007	**2008**	**2009**	**2010**	**2011**	**2012**	**2013**	**2014**
Minimum	0.00	0.10	0.20	0.02	0.12	0.20	0.33	0.22
Maximum	0.85	1.25	0.79	0.79	0.97	1.05	0.93	0.87
Mean	0.39	0.42	0.28	0.20	0.50	0.54	0.58	0.49
Std. Deviation	0.12	0.17	0.11	0.12	0.14	0.11	0.13	0.12
Variance	0.01	0.03	0.01	0.01	0.02	0.01	0.02	0.01
Skewness	-0.62	0.47	1.91	2.02	0.22	1.42	0.40	-0.78
Kurtosis	3.40	1.63	3.38	4.90	1.64	2.91	-033	-0.09

Source: SPSS data from 2007 to 2014

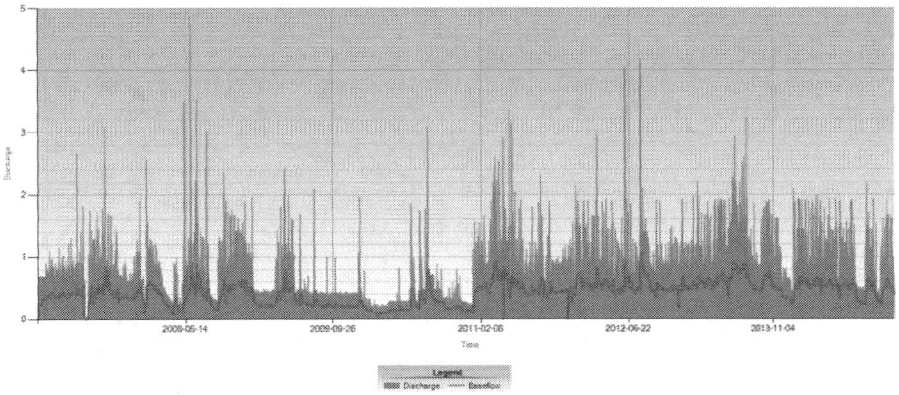

Figure 5. Baseflow separation using RDF Chapmann algorithm

3.5. Eckhardt algorithm

Calculation of filter parameter values using the RDF Eckhradt method is the value of α 0.927-0.962 and BFmax 0.900-9.950; and descriptive baseflow statistical values include minimum values (0.00-0.35 m3/s), maximum (1.68-3.75 m3/s), mean (0.36-1.02), standard deviation (0.24-0.42), and variance (0.06-0.18) presented in **Table 9** and **Figure 6**.

Table 9. Eckhardt RDF baseflow statistics

Baseflow	**RDF Eckhardt algorithm**							
	2007	**2008**	**2009**	**2010**	**2011**	**2012**	**2013**	**2014**
Minimum	0.00	0.10	0.35	0.02	0.21	0.20	0.35	0.28
Maximum	2.11	3.75	1.99	2.11	2.06	3.16	2.06	1.68
Mean	0.69	0.74	0.49	0.36	0.90	0.94	1.02	0.86
Std. Deviation	0.26	0.42	0.24	0.24	0.32	0.30	0.30	0.29
Variance	0.07	0.18	0.06	0.06	0.10	0.09	0.09	0.08
Skewness	0.85	2.24	2.78	3.38	0.85	2.51	0.86	0.00
Kurtosis	6.03	10.95	9.22	15.91	1.66	12.82	0.38	-0.54

Source: SPSS data from 2007 to 2014

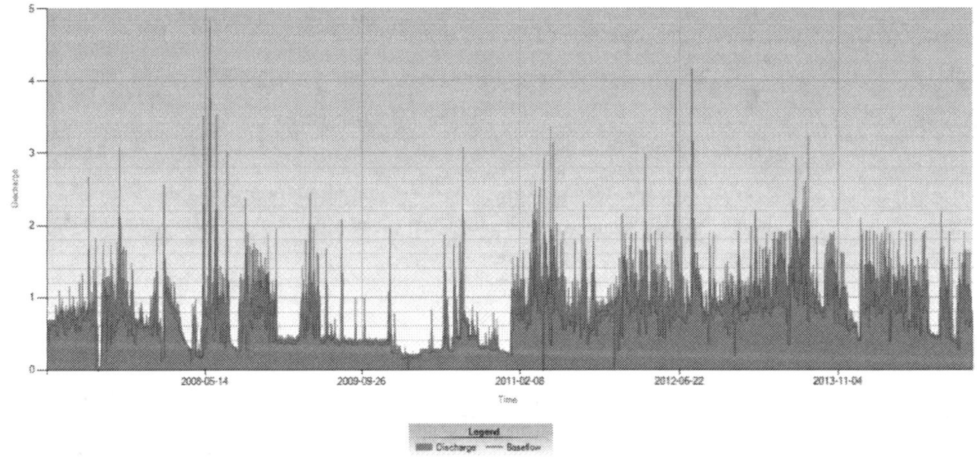

Figure 6. Baseflow separation using RDF Eckhardt algorithm

RDF method Validation Results

The calculation of RSME values and coefficients of determination for the five RDF methods show that the five RDF methods have optimal model conformity ranging from 0.0038 to 0.011. The coefficients of determination for the five RDF methods having the highest to lowest values in a row are Eckhardt, Lyne-Hollick, Two-parameter, One-parameter, and lastly Chapman, as presented in **Table 10**.

Table 10. RSME values and coefficients of determination for all five RDF methods

RDF method	**One-parameter**	**Two-parameter**	**Lyne-Hollick**	**Chapman**	**Eckhardth**
RSME	0.0106	0.0052	0.0059	0.0107	0.0038
Coefficient of determination	0.240	0.255	0.255	0.139	0.424

Source: SPSS data from 2007 to 2014

Calculation of the performance of the five RDF methods based on RSME values and determination coefficients shows that the RDF method of Eckhardt has more optimal performance compared to the other four methods. The research result is evident from the RSME values and coefficients of determination of 0.0038 and 0.424, respectively, which means that the suitability of the Eckhardt method has relatively optimal with relatively small RSME. The performance of the RDF method for the Wae Batu Gajah small island watershed has the usefulness of Eckhardt's RDF method compared to the other four methods.

4. Consclution

The calibration results of the five RDF methods obtain the values of filter parameters k, C, α, BFmax are relatively similar. Optimization of filter parameter values getting description statistical values including minimum, maximum, mean, standard deviation, and variance appropriate for small island watersheds. The fifth baseflow separation of RDF indicates that the Eckhardt filter has the more optimal performance of all four other RDF methods, one parameter algorithm, two-parameter algorithm, Lyne algorithm, and Hollick and Champman. The optimal value of the filter parameters for all five filter methods is relatively uniform, depending on the reasonably similar characteristics of the watershed and hydrological inputs. The relatively small island morphometry provides widespread opportunities to capture limited rainfall inputs and somewhat unique and complex hydrological conditions. Therefore, Eckhardt's baseflow separation method has the optimal suitability of filter parameters to be applied to the watershed conditions of the small island of Ambon City.

Acknowledgment

The authors would like to thank the head and staff of the River Basin Station at Maluku Province for providing the authors with the opportunity to utilize the data recorded at the stream gauge stations installed at the researched watersheds.

References

[1] Latuamury, B, Marasabessy,H, Talaohu,M and Imlabla,W N 2021 Small island watershed morphometric and hydrological characteristics in Ambon Region, Maluku Province *IOP Conf. Ser. Earth Environ. Sci.*, vol. 800, no. 1, doi: 10.1088/1755-1315/800/1/012047.

[2] Tallaksen,L M 1995 A review of baseflow recession analysis *J. Hydrol.*, vol. 165, no. 1–4, pp. 349–370, Feb. doi: 10.1016/0022-1694(94)02540-R.

[3] Jing, Z *et al.*, 2016 An Empirical Method to Investigate the Spatial and Temporal Distribution of Annual Average Groundwater Recharge Intensity-a Case Study in Grand River, Michigan, USA *Water Resour. Manag.*, vol. 30, no. 1, pp. 195–206, doi: 10.1007/s11269-015-1155-z.

[4] He,S, Li,S, Xie, R and Lu, J 2016 Baseflow separation based on a meteorology-corrected nonlinear reservoir algorithm in a typical rainy agricultural watershed *J. Hydrol.*, vol. 535, pp. 418–428, doi: 10.1016/j.jhydrol.2016.02.010.

[5] Szilagyi, J and Parlange, M B 1998 Baseflow separation based on analytical solutions of the Boussinesq equation *J. Hydrol.*, vol. 204, no. 1–4, pp. 251–260, doi: 10.1016/S0022-1694(97)00132-7.

[6] Cimen, M and Saplioglu,K 2004 A Procedure For Separation Of Baseflow Discharge Baseflow," *Time*, no. May, pp. 25–29.

[7] Whiting, P J and Pomeranets,M 1997 A numerical study of bank storage and its contribution to streamflow *J. Hydrol.*, vol. 202, no. 1–4, pp. 121–136, doi: 10.1016/S0022-1694(97)00064-4.

[8] Sujono,J, Shikasho, K and Hiramatsu,K 2004 A comparison of techniques for hydrograph recession analysis *Hydrol. Process.*, vol. 18, no. 3, pp. 403–413, doi: 10.1002/hyp.1247.

[9] Smakhtin,V U 2001 Low flow hydrology: a review *J. Hydrol.*, vol. 240, no. 3–4, pp. 147–186, Jan. doi: 10.1016/S0022-1694(00)00340-1.

[10] Brodie, R S and Hostetler, S 2005 A review of techniques for analysing baseflow from stream hydrographs.

[11] Longobardi,A and Villani, P 2008 Baseflow index regionalization analysis in a mediterranean area and data scarcity context: Role of the catchment permeability index *J. Hydrol.*, vol. 355, no. 1–4, pp. 63–75, doi: 10.1016/j.jhydrol.2008.03.011.

[12] Gregor,M 2010 User's Manual: BFI+ 3.0," *HydrOffice Softw. Packag. Water Sci.*, [Online]. Available: http://hydrooffice.org.

[13] Gregor,M 2012 User Manual for BFI+ 1.

[14] Eckhardt,K 2008 A comparison of baseflow indices, which were calculated with seven different baseflow separation methods *J. Hydrol.*, vol. 352, no. 1–2, pp. 168–173, Apr. doi: 10.1016/J.JHYDROL.2008.01.005.

[15] Indarto,I, Ratnaningsih, A and Wahyuningsih,S 2017 Calibration of six recursive digital filters for baseflow separation in east java *ARPN J. Eng. Appl. Sci.*, vol. 12, no. 12, pp. 3772–3778.

[16] Puspitasari, N D, Wahyuningsih, S, 2001 Studi Pendahuluan Pemisahan Aliran Dasar Menggunakan Metode Grafis dan Filter di Wilayah Administratif UPT PSDA Bondowoso Preliminary Study on Baseflow Separation Using Graphical and Filter Methods in area of UPT PSDA Bondowoso (Eastern Part of East ," no. 1, pp. 2–7.

[17] Tjahyo,A N 2011 Pemisahan Aliran Dasar Bagian Hulu Sungai Bribin pada Aliran Gua Gilap, di Kars Gunung Sewu , Gunung Kidul , Yogyakarta *Indones. J. Geosci.*, vol. 6, no. 3, pp. 165–175, [Online]. Available: http://ijog.bgl.esdm.go.id.

[18] Eckhardt,K 2012 Technical note: Analytical sensitivity analysis of a two parameter recursive digital baseflow separation filter *Hydrol. Earth Syst. Sci.*, vol. 16, no. 2, pp. 451–455, doi: 10.5194/hess-16-451-2012.

[19] Rimmer,A and Hartmann, A 2014 Optimal hydrograph separation filter to evaluate transport routines of hydrological models *J. Hydrol.*, vol. 514, pp. 249–257, doi: 10.1016/j.jhydrol.2014.04.033.

[20] Eckhardt,K 2005 How to construct recursive digital filters for baseflow separation *Hydrol. Process.*, vol. 19, no. 2, pp. 507–515, doi: 10.1002/hyp.5675.

IOP Conf. Series: Earth and Environmental Science **989** (2022) 012029 doi:10.1088/1755-1315/989/1/012029

Tirilolo village community vulnerability and perceptions of the cement industry planning in Baucau District, Timor Leste

J Pereira, G Mahojwala, ET Paripurno*, and MI Effendi

Master of Disaster Management, Faculty of Mineral Technology, University of Pembangunan Nasional Veteran Yogyakarta, Indonesia.

*Email: paripurno@upnyk.ac.id

Abstract. The cement industry investment in Baucau Regency is agreed to be the key to economic growth, creating new jobs, and reducing poverty. To support this investment in the cement industry sector, the government of Timor Leste issued Government Resolution no. 14/2016 regarding the Special Investment Agreement submitted by the Company TL Cement, Lda. Baucau Regency area has a good stretch of karst for the supply of the main raw material for cement production, thus chosen as an investment location. Tirilolo village as well as the surrounding community, where they plan to establish a limestone factory and mine is planned – will receive both benefits and risks from the investment planning. This study uses a mixed-method with a survey approach, interviews, and questionnaires, with socio-economic risk assessment at the local level in the cement industry investment plan sector. This study uses secondary data analysis and field observations. The data obtained will be analyzed descriptively and using a scoring technique on each indicator. Some of the research results obtained indicate a basic vulnerability, that the population density in Tirilolo Village is 52 people/ha, with female household heads 5%, people with disabilities 7%, vulnerable age population 26%. In the education sector, 68.75% of the population did not go to school, while only 17% graduated from elementary school, 5.2% graduated from junior high school, and 8.3% graduated from high school. The average community income is less than $100 per month. The results of structured interviews and focus group discussions show that even though factories and mines carry risks, people look forward to jobs that will emerge to make changes in their lives.

1. Introduction

Investment is agreed to be the key to economic growth, creating new jobs, and reducing poverty. The cement investment in Timor Leste expected by the government to be able to improve the regional economic development process, create jobs and increase people's income. To support this investment in the cement industry sector, the government of Timor Leste, issued Government Resolution number 14/2016, dated 5 May 2016 regarding the approval of the Special Investment Agreement, which was submitted by the TL Cement Company, Lda through Government Resolution No. 14/2016 with an investment cost of 400 million US Dollars, for investment in the cement industry located in Baucau district, the second-largest city after the capital city of Dili. The selection of Baucau Regency as an investment location because the area has a good karst stretch for the supply of the main raw material for cement production, namely limestone, which is based on the initial feasibility study by the postal company in 2014-2015. Following the master plan, this cement industry investment will produce around 1.65 million tons per year of Portland cement clinker with cement refining facilities with a capacity of 5,000 tons per day (TPD) and 100 tons of cement per hour [1]. The main objective of this cement industry investment is to reform the Government's economy, increase private investment, promote economic diversification, sustainable employment, and reduce dependence on cement imports. In the investment plan for the cement industry in the pre-construction and construction stages, the company will employ 3000 people in the pre-construction and construction phase and 1000 people in the production phase, in addition to providing education and training to local communities around the company's location [2]. However, the investment in the cement industry in Baucau Regency, Tirilolo Village, has a potential negative impact on the surrounding community. The researcher conduct a research to find out Tirilolo village community social vulnerability and their perceptions on the risks of the cement industry. Research want to know how the community perceive and their awareness on risk of cement industry.

1.1. Negative Impact of Cement Industry

The negative impact of the cement industry is seen from the social aspect according to the social and economic consequences arising from a development activity or the implementation of a policy and program [3]. Changes in land use, where what used to be agricultural land turned into an industrial area, changed community structures, and social conflicts [4][5][6]. In the aspect of public health, it causes several types of diseases in the respiratory tract such as ARI and lungs [7][8]. On the environmental aspect, including environmental degradation, air, soil, and water pollution, and contributors to emissions to global warming [9][10][11].

1.2. Positive Impact of Cement Industry

The positive impact of the cement industry investment plan brings a million hopes, where the presence of this cement factory will be able to increase the standard of living through employment and business opportunities, increase income, support the emergence of multiplier effects from other sectors [3][12]. Several studies have shown that the presence of the cement industry can improve the socio-economic life of the community around the industrial location are better than those who work in the non-mining sector [12].

1.3. Disaster Risk

Disaster risk is the potential loss caused by a disaster in an area/region within a certain period which can be in the form of death, injury, illness, threatened life, loss of sense of security, evacuation, damage or loss of property, and disturbance activities [13]. Risk is an unavoidable part of life and affects human life everywhere, rich or poor, male or female, adults or small children [14]. Villagran defines risk as a probability of a certain loss which is supported by a certain set of elements as a result of a certain level of danger which can consist of a predictable number of victims, damage, and losses [15]. Risk is the "probability of harmful consequences, or expected losses (death, injury, property, livelihoods, disrupted economic activity and/or environmental damage) resulting from interactions between natural or human-induced hazards and vulnerable conditions.

1.4. Social Vulnerability

Social vulnerability is a multidimensional concept that helps identify the characteristics and experiences of individuals that enable them to respond and recover from harm [16]. Social vulnerability is the potential loss/loss in social groups, for example, groups of the poor, pregnant and lactating women, children, and the elderly further that social vulnerability includes: perceptions of risk and people's views of life related to culture, religion, ethnicity, social interaction, age, gender, and poverty [17][18]. Social vulnerability is related to the level of welfare of individuals, communities, and society, education level, access to human rights, security, traditional values, customs and beliefs, organizational structure, ideology, and social cohesion [19]. The vulnerability of social groups include; position, ethnicity and race, gender, age, education level, disability, health, and household and family. Furthermore, the composition of households with dependents, aged less than 18 years and over 65 years and their single-family head is also vulnerable to disaster hazards [20]. The risk of social vulnerability to the impact of natural hazards is simplified into 4 namely Household level (related to personal attributes), Community (related to how we interact with the people around us), Regional / Geographic (related to how far from the service center)) and Administration/Institutional (related to disaster funding and mitigation studies) [21].

1.5. Economic Vulnerability

Economic vulnerability is the potential loss of assets and economic activities, for example, the disruption of buying and selling activities due to a damaged market, rice fields that cannot be planted due to changes in land function, Economic vulnerability (economic vulnerability) describes a condition of the level of economic fragility in facing hazards (hazards) which include: direct losses such as income, investment, potential loss of goods/related [22][23]. Economic Vulnerability is the economic ability of an individual or a community, which greatly determines the level of vulnerability to hazard threats. In general, poor or underprivileged communities or areas are more vulnerable to hazards, because they do not have the adequate financial capacity to carry out disaster prevention or mitigation efforts [24]. Vulnerability is highly dependent on the status of the condition of individuals, communities, and nations. such as the poor, women, age, and income groups. economic vulnerability also includes the economic level of individuals, communities, access to resources such as loans/credits, insurance, and economic support facilities; transportation and road market [18]. Economic status such as income, poverty, employment, and education where the poor will be very vulnerable to disaster hazards as well as low levels of education and uncertain income [20].

2. Methods

The method that will be used in this study is the mixed method with a survey approach, interviews, and filling out questionnaires. The population in this study is the entire/total community with a population of 7769 and 1699 heads of families (based on village office records, 2016 to September 2019) and the number of hamlets in the research location is 4 hamlets out of a total of 6 hamlets. This study will focus on four hamlets in the location of the industrial investment plan (factory) with a population as shown in table 1.

The sample is part of the population and the type of sample taken must reflect the population. The findings of the information obtained in the sample are findings also in the population so that conclusions drawn from the generalized sample observations will also lead to the population. This means that the sample taken for research can represent the existing population [25]. The sample that will be used in this study is the population at the location of the Cement Industrial Investment Plan (Factory). To get the sample size in this study, the researcher will use the Solving formula [25][26]. In this study, the researcher will use an error tolerance of 10% (0.1) with a confidence level of 90%, so there are 96 respondents.

Table 1. Total population of four hamlets in the study site [27].

No	Hamlet	Male	Female	Total	Household
1	Caisido	389	388	777	188
2	Parlemnto	256	223	479	132
3	Lialaileso	325	275	590	146
4	Osso-Ua	293	306	599	149
Total		1263	1192	2445	615

3. Results and Discussions

The results of the study using a survey showed that almost all hamlets in the Tirilolo Village have persons with disabilities and female-headed households, as detailed by Table 2.

Table 2. Community Population Vulnerability

No.	Hamlet	Total Respondent	Gender Male	Gender Female	Vulnerable Age	Disability	Female-headed Household
1.	Parlamento	19	12	7	4	1	2
2.	Caisido	30	17	13	9	2	3
3.	Lialailesu	23	15	8	7	1	2
4	Osso-ua	23	16	7	5	-	-
Total		**96**	**60**	**35**	**25**	**5**	**7**

All community members in Tirilolo Village have a monthly income of below $100 per month, which means below the Timor Leste minimum wage of $120 per month based on Timor Leste Law No. 4 of 2012. Majority of the person in Tirilolo Village are not going to school. In this case, researcher decide to use term of 'not going to school' as a term to differentiate with 'uneducated', because the role of the education is on their own family. In Caisido and Lialailesu Hamlet, the number of person that not going to school is twenty times greater than person that take elementary school. In The rank followed by elementary school, junior high school, and senior high school, respectively.

Table 3. Community Education and Income

No.	Hamlet	Total Respondent	Income/Household	Education				Economy Asset
				A	B	C	D	
1	Parlamento	19		6	3	2	8	9
2	Caisido	30	< $100/months	2	1	-	27	13
3	Lialailesu	23		1	-	2	20	3
4	Osso-ua	23		8	-	4	11	5
total		**96**		**17**	**4**	**8**	**66**	**30**

A. Elementary School
B. Junior High School
C. Senior High School
D. Not Going to School

Table 4. Land and Cattle Ownership

No.	Hamlet	Total Respondent	Land	Cattle
1.	parlamento	19	Average on 1 ha/household	Cow, Sheep, Goat, Buffalo, Chicken, and Pig
2	Caisisdo	30		
3	Lialaileso	23		
4	Osso-Ua	23		

IOP Conf. Series: Earth and Environmental Science **989** (2022) 012029 doi:10.1088/1755-1315/989/1/012029

Tabel 5. Diversity of Livelihood

No.	Hamlet	Total Respondent	Livelihood		
			Civil Servant	Military/Police	Farmer & Fisherman
1.	Parlemento	19	2	-	17
2	Caisido	30	3	5	22
3	Lialaileso	23	"	2	21
4	Osso-ua	23	4	1	16

As coastal village with dry lands, the profession in Tirilolo Village dominated by farmer and fisherman, with a small number of person woking as civil servant, military, and police. The majority of the community stated that they accepted and agreed to the cement investment in their village. In general, the community has the perception that this investment will create employment opportunities for the younger generation in the village, increasing their income, by which they can open kiosks and stalls around the factory site. Farmers in the village are residents who hope for a mine, to improve family livelihoods.

"Farmers in this village are dryland farmers, planting in the rainy season. The dry season here is approximately six months (April-November) and during the dry season, we have nothing to rely on except selling goats, each of which costs 40 dollars. Our income from agriculture is less than $100/month, and sometimes when it's tough we can only get 15 dollars, which is only enough to buy 25 kg of rice. So, we hope that the cement factory can increase the income of residents in the village." (Farmer, 56 years old, Causido Hamlet)

For fishermen, the situation is better than for farmers but it tends to be based on fish season and weather.

"The income of fishermen depends on the weather. If the weather is good at sea, we can earn 30-40 dollars per day. If the weather is not good they are less than a hundred dollars a day. We agreed with the construction of the mine because later there would be a dock built, our village would automatically be crowded and we could sell grilled fish. Selling fish directly is more profitable because we can make 40% more profit than selling to collectors." (Fisherman, 40 years old, Osso-Ua Hamlet.

Indigenous leaders and veterans as community leaders also agreed with the existence of this mine. The reason is that the economic improvement will be comprehensive, and residents can open stalls and shops around the highway. From conversations with several respondents that the income per person per day is approximately $ 0.50 cent - $ 1.00. with an expense of more than US$100. This is in line with the cement industry's investment plan, which will create jobs for 3000 people in the construction phase and 1000 people in the operation phase.

"There must be a commitment from entrepreneurs to take risk reduction actions. What if after the mine, what will be done to the former excavation. This has to be agreed with the government and the company." (Head of Disaster Management Office of Timor Leste)

Contrary to public perception, local NGOs feel that factories will pose environmental risks. Bridging this situation, the government of Timor Leste stated that development conditions must be supported by efforts to reduce risks so that they do not have an impact on society and the environment.

4. Conclusion

From the statement of the community in Tirilolo Village, it could be seen that the community more consider the economical benefit from the cement industry, and not considering the risk that could arise. The condition of the economic vulnerability of the people in Tirilolo Village is below the Timor Leste minimum wage, and the level of formal education is still low, with the majority not attending formal schools. This condition of vulnerability is the reason why the community's perception supports the existence of a cement factory, solely for economic reasons, and to create an increase in income for residents in Tirilolo Village. Full support for the cement business must be a special concern to keep in mind the risks that may arise outside of the economic and social aspects. Further research could be conducted to assess the ecological vulnerability and emerging risks to springs and post-mining recovery strategies.

Acknowledgments
Authors wishing to acknowledge assistance or encouragement from colleagues, special work by technical staff, or financial support from organizations should do so in an unnumbered Acknowledgments section immediately following the last numbered section of the paper.

References

[1] C. Belt and A. Infrastructure, "Baucau Cement Project Environmental Impact Statement - Cement Plant , Jetty , Conveyor Belt and Associated Infrastructure," vol. I, 2017.
[2] Timor Leste Government, *Timor Leste Government Resolution No. 14*. 2016.
[3] S. P.Hadi, "Aspek Sosial AMDAL Sejarah, Teori dan Metode." Gajah Mada University Press, 2009.
[4] Y. Rahayuningsih, "The Impact of The Cement Industry to The Livelihood System of Bayah Fisherman," *J. Sos. Ekon. Kelaut. dan Perikan.*, vol. 12, no. 2, 2017.
[5] Anang Suhartadi, "Ev aluasi pengelolaan lingkungan kegiatan penambangan batu kapur pt. sinar asia fortuna (saf) di rembang," 2009.
[6] A. Bachtiar, S. Wibisono, and S. Komariyah, "PABRIK SEMEN PUGER JAYA RAYA SENTOSA DI KABUPATEN JEMBER (Studi Empirik Kecamatan Puger) Social and Economic Analysis of Regional Communities in Jember Abstrak Pendahuluan Metode Penelitian," 2014.
[7] E. NKHAMA, "RELATIONSHIP BETWEEN CEMENT DUST EXPOSURE AND ADVERSE RESPIRATORY HEALTH IN A SETTLEMENT RESIDING NEAR A CEMENT FACTORY IN CHILANGA , ZAMBIA . BY UNIVERSITY OF PRETORIA PROMOTOR : PROFESSOR KUKU VOYI CO-PROMOTOR : PROFESSOR SETER SIZIYA," 2017.
[8] S. S. Mehraj, G. A. Bhat, H. M. Balkhi, and T. Gul, "Health risks for population living in the neighborhood of a cement factory," *African J. Environ. Sci. Technol.*, vol. 7, no. 12, pp. 1044-1052–1052, 2013, doi: 10.5897/AJEST2013.1576.
[9] K. S. Devi, V. V. Lakshmi, and A. Alakanandana, "IMPACTS OF CEMENT INDUSTRY ON ENVIRONMENT – AN OVERVIEW," no. Lvii, pp. 156–161, 2017.
[10] B. Anoop Singh Gaharwar, Naveen Gaurav, AP Singh, Hira Singh Gariya, "A Review Article on Manufacturing Process of Cement, Environmental Attributes, Topography and Climatological Data Station: IMD, Sidhi M.P," *~ 47 ~ J. Med. Plants Stud.*, vol. 4, no. 4, pp. 47–53, 2016.
[11] S. Kumar, S. Gopal, and P. Gautam, "Assessment of Reduction in Forced Vital Capacity of School Children Exposed to Air Pollution from Cement Manufacturing Plants," vol. 12, no. 6, pp. 1215–1222, 2017.
[12] Sosmiarti, "Pengaruh Keberadaan PT. Semen Padang Terhadap Sosial Ekonomi Masyarakat (Studi Kasus Kecamatan Lubuk Kilangan)," *J. Bisnis Manaj.*, vol. Vol. 4, 2008.
[13] BNPB, *KEPALA BADAN NASIONAL PENANGGULANGAN BENCANA TENTANG DAFTAR*

ISI KEPALA BADAN NASIONAL PENANGGULANGAN BENCANA TAHUN 2012 TENTANG PEDOMAN UMUM PENGKAJIAN RISIKO BENCANA. 2012.
[14] P. Wingnyo Adiyoso. S.Sos., M.A., "Manajemen Bencana Pengantar & Isu-Isu Strategis." Bumi Aksara, 2018.
[15] Juan Carlos Villagrán De León, "Vulnerability A Conceptual and Methodological Review," 2006. doi: 10.1192/bjp.112.483.211-a.
[16] S. L. Cutter, "Societal Vulnerability to Environmental Hazards Social Sensing and Big Data Computing for Disaster Management View project Exploring twitter reaction to hurricane Matthew: uncovering temporal and spatial response. View project," no. December 1996, 2014, doi: 10.1177/030913259602000407.
[17] D. Hardy *et al.*, "S Ocial V Ulnerability :," no. April, pp. 1–38, 2018.
[18] UN/ISDR, *Living with risk*, vol. 1, no. 4. UNITED NATIONS New York and Geneva, 2004, 2004.
[19] E. a. Phillips, *Social Vulnerability to Disasters; SectIon II Socially Vulnerable Groups*. 2010.
[20] B. E. Flanagan, E. W. Gregory, E. J. Hallisey, J. L. Heitgerd, and B. Lewis, "A Social Vulnerability Index for Disaster Management," *J. Homel. Secur. Emerg. Manag.*, vol. 8, no. 1, 2011, doi: 10.2202/1547-7355.1792.
[21] a Dwyer, C. Zoppou, O. Nielsen, S. Day, and S. Roberts, "Quantifying Social Vulnerability : A methodology for identifying those at risk to natural hazards," *Geosci. Aust.*, p. 101, 2004, doi: ISBN: 1 920871 09 8.
[22] A. Yasir, *The Political Economy of Disaster Vulnerability: A Case Study of Pakistan Earthquake 2005*, no. 20762. 2009.
[23] R. A. Davidson and S. C. Haresh, "AN URBAN EARTHQUAKE DISASTER RISK INDEX Blume Earthquake Engineering Center," no. 121, 1997.
[24] BNPB, *Pedoman penyusunan rencana penanggulangan bencana*. 2008.
[25] V. W. Sujarweni, *Metodologi Penelitian Lengkap, Praktis dan Mudah dipahami*. 2018.
[26] M. S. Dr. Beni Ahmad Saebani and M. A. H. Yana Sutisna, "Metode Penelitian." Pustaka Setia Bandung, 2018.
[27] Tirilolo Village Government, *Tirilolo Village Population Data*. 2019.

IOP Conf. Series: Earth and Environmental Science **989** (2022) 012030 doi:10.1088/1755-1315/989/1/012030

Coral Reef Damage Caused by The Indinurmatalia07 Grounding in Cross-over Reef, Lemon Island of Manokwari - Papua Barat Province

Alex R. Sitanala[1], Hance Krey[1], Marthen Rumfabe[1], Nelman Rumbekwan[1], Melkias Sanyar[1], Rimer Hein Biloro[1], Muda Mofu[1], Vecky Krey[1], Nur Alzair[2], and Ricardo F. Tapilatu[1,3,*]

[1]Ketapang Dive Centre (KDC), Manokwari 98314 – Papua Barat INDONESIA
[2]Geological Engineering Dept, University of Papua, Manokwari 98314 – Papua Barat INDONESIA
[3]Research Centre of Pacific Marine Resources, University of Papua (UNIPA), Manokwari 98314 – Papua Barat INDONESIA

[*]Email: rf.tapilatu@unipa.ac.id

Abstract. The cargo ship, Indi Nurmatalia07, ran aground on 07 August, 2021 at 23.00 at 'Cross Over' reef in the Dorery Bay – Manokwari. The interim report revealed that the ship's officer was not able to control the ship when it left Manokwari port due to bad weather condition when the ship left at 19.00. However, report on the cause of bad weather was doubtful because the distance between the port of Manokwari and the incident site is only about 500m. The vessel was later freed from the reef on 10 August 2021 at 19.00 after disembarking shipload. An initial quantitative assessment of damages was conducted by a local team within the grounding site. Four line transect surveys were conducted, each 25m in length and 2m in width, positioned in the left and right of the damaged and undamaged area. Our initial results demonstrated that a total area of 717,02m^2 has been damaged by the cargo ship. The type of coral reef formation is patch reef with medium diversity of coral species (mainly genus *Acropora*, *Pocillopora*, *Coeloseris*, *Millepora*, and *Goniopora*) and coral cover. Cross over reef has a shallow reef top (between 3-10m depth), slopes down to at least 25m. The reef is subject to high velocity current that might be unsuitable for coral planula to settle permanently. Coral structures were partly crushed and fractured at the edge of the damaged area but totally damaged in central part. The central damaged site has lost remarkable bottom coral structure and diversity from at least 5 different coral genera. By using a comparison study conducted on ship groundings, the Provincial government would sue the cargo ship and insurance company to pay environmental losses due to environmental based on a calculation which considered damage in relation to the ecosystem, economy, society and ecosystem rehabilitation referring to PerMen LH No. 7/2014.

1. Introduction

Coral reef ecosystem is one of the marine ecosystems that has a variety of associated biota and high primary productivity. Coral reefs have several functions such as high physical, ecological and economic functions. Physically, coral reefs act as beach protectors from waves and abrasion [1]. Ecologically, coral reefs are also a place to find food (feeding ground), a place for care and grow up (nursery ground), as well as a spawning ground (spawning ground) for various other marine biotas [2]. Economically, coral reefs are a habitat for various types of marine biota, ornamental fish, construction materials, jewelry, pharmaceutical raw materials that have high economic value [3]. In addition, coral reef ecosystems can be used as marine tourism objects that are very attractive to tourists. However, currently the condition of the coral reef ecosystem is being damaged.

It has been globally known that the incident of vessel grounding would cause physical environmental disruption to coral reefs and benthic habitats that may lead to considerable disturbance and death of benthic biotas. The damages were caused by the direct impacts of the hull, paints, propellers and anchors of the vessel on the benthic communities. The movements of dislodged coral structures, boulders, rubbles, sands and relevant particles also indirectly influenced the condition adjacent to the impacted area. Moreover, the activities of vessel removal often create additional impacts to the damages due to manuever of the ship. Relative to the magnitude of the vessel grounding, damages to coral reef may range from superficial to the loss of coral structure where reef has been ruptured [4].

It was recorded that the cargo vessel, the Indinurmatalia07, ran aground on the evening of Saturday, August 7, 2021, at approximately 7:00 PM at cross-over reef after leaving Manokwari port, in middle part of Doreri Bay, Manokwari (Figure 1). The area has been known to host a variety of coral species such as *Pocillopora*, *Acropora*, *Coleoseris*, *Goniopora,* and *Millepora* [5]. In addition, five types of coral growth forms were identified in the area: *Acropora Branching* (ACB), *Acropora Submassive* (ACS), *Coral Branching* (CB), *Coral Massive* (CM), *Coral Submassive* (CS) and *Coral Mushroom* (CMR) [6]. The area has been known to have a relatively high biological diversity with multiple species of fish, invertebrates and many more and is a setting area for the use of artisanal fishing.

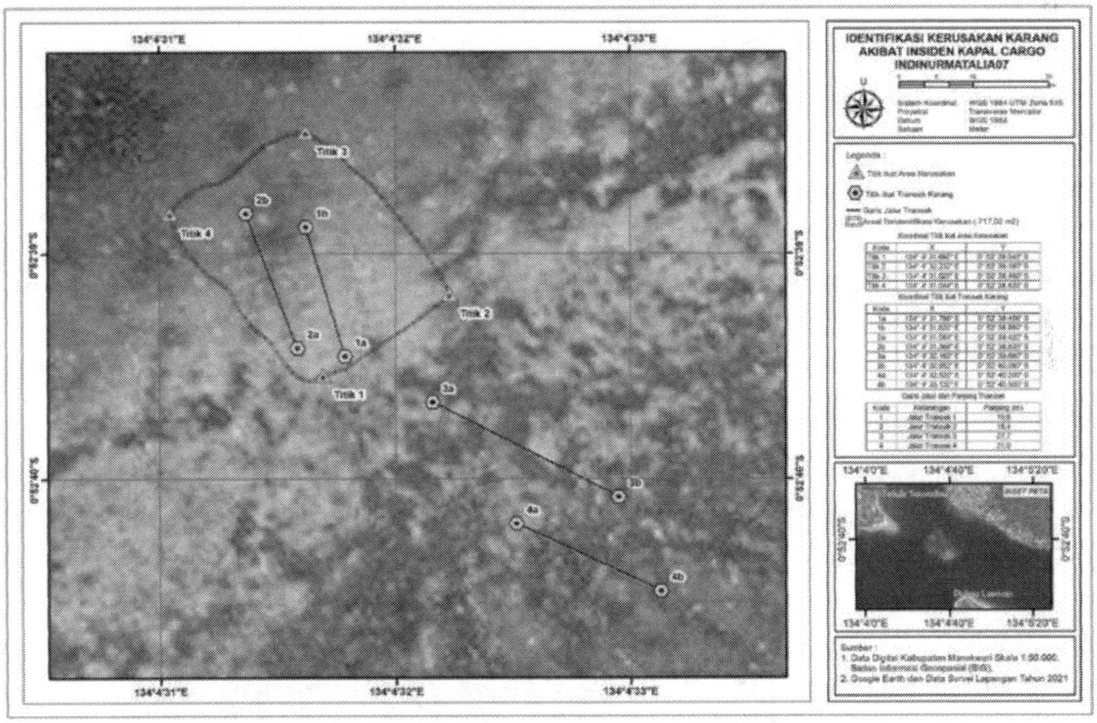

Figure 1. The location of grounded ship having damaged the coral reef at Doreri Bay, Manokwari.

The Ketapang Dive Center (KDC) took an initiative to conduct initial evaluation of the bottom part where the cargo ship ran aground and also any other damages that might be have been influenced by the anchor on the reef (Figure 2d).

KDC's initiative was to conduct the following roles:
1. Observation and presence during the vessel removal to evaluate the magnitude of the damage caused by the ship and initial observations of reef disruption (summary report available at KDC).

2. An initial quantitative evaluation once the cargo ship ran aground and after the vessel removal from the impacted area to assess the magnitude of impact of the vessel on the ground, and any other disruptions to the adjacent areas.
3. A list of component on the damaged area that need to be considered including the services that coral reefs provide and a charge of restoring the damaged reef.

2. Methods

KDC conducted task (2) on Sunday, August 8, 2021 and then Wednesday, August 11, 2021. The quantitative measurement of damaged corals was in the form of undertaking benthic surveys in the affected areas and using comparative quantitative studies at nearby locations of the same patchy reef as references (Figure 1).

The reference location was selected by diving approximately 100m away from the impacted area (Figure 1) as the location experienced similar physical and chemical conditions and showed a relatively uniform coral community resemblance. At both sites, a 25m line transect was laid across the bottom habitat. Coral cover was measured by taking one image per meter transect by including the position of the meter in each image (left side for odd meter and right side for even meter of the frame) using an underwater camera (Olympus TG6). Thus, 25 images are generated on each line transect covering a total area of $25m^2$. The images obtained were then analyzed at the Pacific Marine Resources Research Center – University of Papua using CPCe (https://www.nova.edu/ocean/software_cpce). Each photo frame was determined by 20 random points where each point was categorized in the lifeform category [6]. Surveys of percentage coverage and benthic conditions using the same method had been carried out in several other studies in Manokwari, Raja Ampat, Kaimana, South of Manokwari and Cenderawasih Bay and published in multiple peer-reviewed journals [6, 8, 9, 10]. The proportion of coral cover in each category was averaged between the two transects at each location.

The survey team used a measuring tape to measure several lengths and widths underwater and verified using GPS on the boat to estimate the impacted area by the shipwreck to the approximate actual shape. In addition, the survey team observed and measured the additional area caused by the shifting of the anchor and keel of the ship maneuvered after the removal. In addition, several approximate variables were compared between the affected benthic area and the reference at adjacent undamaged location using a t-test with the R program [11]. Prior to the t-test, the data was checked for normality using the Shapiro-Wilk test.

3. Results and Discussion

3.1. The area affected by the ran aground vessel

The main affected area was observed on coral reefs (Figure 2a) which consisted of the main part of the bottom location where the ship ran aground and a small part of the area on the left side caused by anchor placement. The total area of the affected area was $717.02m^2$.

Figure 2. (a). Damaged area by the ship and (b) Anchor dragging

3.2. Live coral coverage
Live coral cover was considerably higher (64.2%) at the reference location outside the damaged site compared to the damaged site (0.7%) (t-test, $p<0.01$). Undoubtedly, the mean proportion of damaged coral cover at the damaged site was 99.3% significantly different from that of the damaged coral at the reference location, 35.8% (t-test, $p<0.05$).

3.3. Diversity of coral genera
There were nine and four coral genera inventoried at the reference sites and damaged sites. *Acropora* (67%) and *Porites* (18%) were found dominating the reference location. Other coral genera make up less than 1% cover. The percentage cover of *Acropora* species was 2% at the damaged location and considerably lower than at the reference site (t-test, $p<0.01$). The percentage of *Porites* species cover was 1% at the damaged site and significantly different from the reference site (t-test, $p<0.05$).

3.4. Benthic composition
The benthic composition changed due to the considerable decrease in live corals replaced by broken corals (Figure 3a) at the damaged site. While the composition of benthic at reference location is still in good condition (Figure 3b).

Geographically, the cross-over reef is located adjacent to Lemon Island. Lemon Island and the cross-over reef are included in the waters of Doreri Bay. Almost all of the residents who live on this island make their main livelihood as traditional fishermen with very diverse patterns of utilization of coastal resources by people who catch fish using traditional fishing gear in the form of fishing rods and nets [6, 10, 13]. In addition, previously the community used fishing gear and materials that damaged coral reef and seagrass ecosystems and other biota, such as using bombs, potassium, and nets stuck in corals. Fishing activities are usually carried out around coral and cross-over reefs, seagrass, and waters around the island which have an impact on the condition of coral reefs as fish habitats and fishing locations, especially in coral reef areas. The catch of the community is often sold to the Sanggeng Fish Market in Manokwari and some of it is also consumed by the family. In addition to fishermen, there are also those who work in sea transportation services using traditional boats to take/pick up passengers from Manokwari City to Mansinam Island or Lemon or vice versa [14].

Figure 3. (a. Left). The composition of benthic at the damaged area (b. Right). The composition of benthic at the reference area

The grounding of the vessel, Indi Nurmatalia07, at crossover reef inside the Doreri Bay destroyed a total area of coral reef of 717.02m^2. The destruction of coral reef life form led to high loss of live corals and coral genera diversity. In contrast, the reference site had a high live coral cover and coral genera diversity and relative pristine reef life form (less of even absence of broken reefs). The waters of Lemon Island have various forms of coral reef growth. The growth forms found were *Acropora Branching* (ACB), *Acropora Submassive* (ACS), *Coral Branching* (CB), Coral Massive (CM), Coral Submassive (CS) and *Coral Mushroom* (CMR). Five types of growth forms were identified by [6] at station I adjacent to damaged area including *Acropora Branching* (ACB), *Acropora Submassive* (ACS), *Coral Branching* (CB), *Coral Massive* (CM), *Coral Submassive* (CS) and *Coral Mushroom* (CMR). Based on their observations at cross-over reef at a depth of 3m, it was found that the non-Acropora Coral Massive (CM) life coral category covered 22.78%, Coral Branching (CB) = 3.89%, Coral Submassive (CS) = 1.67%, Coral Mushroom (CMR) = 0.56%. There were also Acropora coral categories: *Acropora Submassive* (ACS) = 17.22% and *Acropora Branching* (ACB) = 2.22%. The total percentage of live coral cover found in the area of cross-over reef was 48.3% in the medium category. At this station, Rubble (R) was found of 23.33%, while Dead Coral (DC) was 12.78%.

The crushing effect of the ship's hull onto the coral-reef substrate led to the loss of live corals and structure complexity. Live coral cover at the affected location approached 0.7% which was 15.7 times lower than at the reference site. The destruction of coral reef life form led to more than approximately 75% decrease of coral genera diversity. *Acropora* reefs were the most impacted due to their typical fragile structure. *Acropora* reefs are a very diverse group (122 total species) [15]. *Acropora* spp. are the primary life form due to their common characteristics such as branching morphological structure and fast growth [16, 25]. Moreover, their typically branching provides habitat for other reef species (marine organisms) and complexity needed for coral diversity [17, 25]. Currently, an estimate of 45% of *Acropora* species are classified ranging from vulnerable to critically-endangered on the red list of IUCN [19]. In contrast, *Porites* reefs of massive morphology are also important reef life form. They were the least impacted, however, a few colonies have been known to be fragmented in damaged area (Figure 3a).

Observing the reference site located nearby, it is potential that the damaged location shifted ranging from a high-coral cover system to a broken-coral and dominated by rubble. Despite no data were collected on other ecological indicators (fish and reef invertebrates), the loss of coral habitat affected and may continue to influence reef species that have obligate association with live corals.

It has been widely known that the ship groundings are one of the most damaging on coral reefs [19, 20, 21] among threats from anthropogenic activities. Despite localized, the destruction of the system usually complete and critical to coral reef habitats. For example, multiple ship grounding case studies have demonstrated that the damaged grounds have been difficult to recover to its existing community structure without assistance of restoration initiatives [19, 21].

It has been very well known that coral reefs provide a variety of important goods and also services such as seafood, raw materials for medicines and cosmetics, live fish for aquarium trade, beach buffering, nitrogen fixation, sand generation, and support of tourism activities [10, 22]. To estimate the value of the loss coral-reef ecosystem, economics try to value these environmental goods and services from an anthropogenic perspective [23]. The Indi Nurmatalia07 ran aground in cross-over reef of Doreri bay in Manokwari, damages to coral reefs may still occur, in particular with the increase in tourism activities to Mansinam and Lemon Islands [13] and boat traffic in the area. The economic valuation of impacts is required to approximately set fines and penalties in relation with the level of impact as it has been known that coral reefs have a great value. In addition, coral reefs are protected under Indonesian laws and the World Heritage of UNESCO since 2012 [24]. We referred the value for coral reef loss based on Decree of Ministry of Environmental No. 07/2014. It is stated clearly in the decree that the setting of fines and penalties take into account essential components such as ecological system and integrity, economic and societies values, and ecosystem rehabilitation.

Acknowledgement
The authors would like to especially thank researchers from the Research Centre of Pacific Marine Resources at University of Papua (UNIPA) for their valuable advices on coral reef evaluation and valuation.

IOP Conf. Series: Earth and Environmental Science **989** (2022) 012030 doi:10.1088/1755-1315/989/1/012030

References

[1] Pusat Penelitian Oseanografi – Lembaga Ilmu Pengetahuan Indonesia (P2O-LIPI) (2018). Status Terkini Terumbu Karang Indonesia. http://lipi.go.id/siaranpress/lipi:-status--terkiniterumbu-karang-indonesia-2018-/21410.

[2] Asyawati, Y., Akliyah, L.S. (2014). Identifikasi Dampak Perubahan Fungsi Ekosistem Pesisir Terhadap Lingkungan Di Wilayah Pesisir Kecamatan Muaragembong. Perencanaan Wilayah dan Kota, Fakultas Teknik, Universitas Islam Bandung (UNISBA).

[3] Romadhon, A. (2014). Valuasi Ekonomi Manfaat Ekosistem Terumbu Karang Di Pulau Sapudi, Sumenep, Madura. Prodi Ilmu Kelautan, Fakultas Pertanian, Universitas Trunojoyo Madura.

[4] Lirman D, Gracias N, Gintert B, Gleason ACR, Deangelo G, Dick M, Martinez E, Reid RP. (2010) Damage and recovery assessment of vessel grounding injuries on coral reef habitats by use of georeferenced landscape video mosaics. Limnol. Oceanogr. Methods 8:88–97.

[5] Muljadi A, Tapilatu RF. 2019. Coral Biodiversity and colour visualization during night dive at Cross-over reef adjacent to Lemon Island, Manokwari. https://www.penyelaman.com/2019/01/29/misteri-fluorescence-terumbu-karang/. Accessed on 12 November 2021.

[6] Aglutomo D, Tapilatu RF, Kusuma AB. 2022. Visualization of Coral Reef Cover with Photogrammetry Method at Coastal Waters of Lemon Island, Manokwari. Indonesia. Ecology, Environment and Conservation (EEC) Vol 28: S87-S94.

[7] Kohler KE, Gill SM (2006) Coral Point Count with Excel extensions (CPCe): A Visual Basic program for the determination of coral and substrate coverage using random point count methodology. Comput. Geosci. 32:1259–1269.

[8] Pattiasina TF, Herawati EY, Semedi B, Sartimbul, A, Pelasula DD and Krey M. 2018. Detecting and visualizing potential multiple coral reef regimes in Doreri Bay, Manokwari Regency, Indonesia. *Aquaculture, Aquarium, Conservation & Legislation*, *11*(1), pp.118-131.

[9] Dasmasela YH, Pattiasina TF, Syafril S, Tapilatu RF. 2019. Evaluasi Kondisi Terumbu Karang Di Pulau Mansinam Menggunakan Aplikasi Metode Underwater Photo Transect (UPT). *Median: Jurnal Ilmu Ilmu Eksakta*, *11*(2), pp.1-12.

[10] Peck M, Tapilatu RF, Kurniati E, Rosado C. 2021. Rapid coral reef assessment using 3D modelling and acoustics: acoustic indices correlate to fish abundance, diversity and environmental indicators in West Papua, Indonesia. *PeerJ*, *9*, p.e10761.

[11] R Development Core Team. 2015. R: A language and environment for statistical computing. Vienna, Austria.

[12] Larsen SN, Leisher C, Mangubhai S, Muljadi A and Tapilatu RF. 2018. Fisher perceptions of threats and fisheries decline in the heart of the Coral Triangle. *Indo Pacific Journal of Ocean Life*, *2*(2), pp.41-46.

[13] Siburian RHS, Faidiban R, Tapilatu RF. 2021. The Climate Change Resilience of Lemon Island Community, West Papua, Indonesia. Ecology, Environment and Conservation (EEC) 27 (August 2021), S297-S302.

[14] Wallace CC, Done BJ, Muir PR. 2012. Revision and Catalogue of Worldwide Staghorn Corals *Acropora* and *Isopora* (Scleractina: Acroporidae) in the Museum of Tropical Queensland. Queensland Museum,

[15] Gladfelter EH. 1984. Skeletal development in *Acropora cervicornis*. III. A comparison of monthly rates of linear extension and calcium carbonate accretion measured over a year. Coral Reefs 3:51–57.

[16] Graham NAJ and Nash KL. 2013. The importance of structural complexity in coral reef ecosystems. Coral Reefs 32:315–326.

[17] Roff G and Mumby PJ. 2012. Global disparity in the resilience of coral reefs. Trends Ecol. Evol. 27:404–413.

[18] IUCN. 2021. IUCN Red List of Threatened Species. http://www.iucnredlist.org.

IOP Conf. Series: Earth and Environmental Science **989** (2022) 012030 doi:10.1088/1755-1315/989/1/012030

[19] Precht WF, Aronson RB, Swanson DW. 2001. Improving scientific decision-making in the restoration of ship- grounding sites on coral reefs. Bull. Mar. Sci. 69:1001–1012.

[20] Gouze M, Otto EI, Olsudong D, Mereb, G., 2015. Quantitative assessment of coral-reef damage caused by the Palau Siren grounding in Koror State Southern Lagoon. *Palau International Coral Reef Center Technical Report*, (15-16), pp.1-21.

[21] Tapilatu RF, Parenden D, Dimara R, Imbiri O, Hidayat NI, Noer A. 2017. The assessment of coral reef damage caused by the Caledonian sky Grounding in Cross-over reef, Dampier Strait of Raja Ampat Papua Barat.

[22] Moberg F and Folke C. 1999. Ecological goods and services of coral reef ecosystems. Ecol. Econ. 29:215–233.

[23] Barbier EB, Burgess JC, Folke C, Trzyna TC, Osborn JK, Munasinghe M, Shearer W, Wagner MR, Cobbinah JR, Kirmase RD. 1995. Paradise Lost?: the ecological economics of biodiversity.

[24] UNESCO. 2012. Rocks Islands Southern Lagoon. http://whc.unesco.org/en/list/1386.

[25] Tuhumena JR, Tapilatu RF, Boli P. 2019. The effect of type and duration of substrate collector placement to the coral genus recruitment in Saleo Beach Area, Dampier Strait, Raja Ampat, Indonesia. Aquaculture, Aquarium, Conservation & Legislation 12 (5), 1643-1652.

IOP Conf. Series: Earth and Environmental Science **989** (2022) 012031 doi:10.1088/1755-1315/989/1/012031

Hexa Helix, participatory concept to support landslide early warning system

Z Zakaria [1, 2, *], N Sulaksana [2, 3, 4], R I Sophian [1,2], and BYCSSS Alam [2,3]

[1] Laboratory of Engineering Geology, Faculty of Geological Engineering, Unpad (Universitas Padjadjaran), Jl. Raya Bandung-Sumedang, KM-21, Jatinangor-Sumedang 45363, Indonesia

[2] Researcher Association at Disaster Research Center - Universitas Padjadjaran, Jl. Raya Bandung-Sumedang, KM-21, Jatinangor- Sumedang 45363, Indonesia

[3] Laboratory of Geomorphologi, Faculty of Geological Engineering, Universitas Padjadjaran, Jl. Raya Bandung-Sumedang, KM-21, Jatinangor- Sumedang 45363, Indonesia

[4] Head of Department of Applied Geology, Faculty of Geological Engineering, Universitas Padjadjaran, Jl. Raya Bandung-Sumedang, KM-21, Jatinangor- Sumedang 45363, Indonesia Sumedang, KM-21, Jatinangor-45363, Indonesia

*Email: zufialdi.zakaria@unpad.ac.id

Abstract. The Indonesian archipelago lies at triple junction between the world's tectonic plates that have many natural disasters. Furthermore, Indonesia lies accross the equator which result in a tropical climate (dry and rainy seasons). During the rainy season, rainfall often triggers landslides. Landslides can be identified by the material type (soil, rock, or debris) and movement (slide, fall, flow, or combination). Landslide-prone areas can be influenced by lithology (rock type and weathering), slope (geomorphology), seismic vibrations (tectonic aspects), and human activities. In order to anticipate landslides, both structural and non-structural mitigation are required. Non-structural mitigation is required to determine the condition of rainfall, and the type of landslides. RAFLEWS (Rainfall Apparatus for Landslide Early Warning System) design model is a part of non-structural mitigation. Non-structural mitigation is also required to anticipate the possibility of landslides. Proposal of Hexa Helix participatory concept is a part for early warning system which will be useful in supporting non-structural mitigation. Hexa Helix consists of six participant elements, namely: Academia, Government, Society, Industry, Mass Media, and Laws/Regulations/Legislations. Combination of RAFLEWS model and Hexa Helix concept is expected to be influential in dealing with landslide disasters in the future.

1. Introduction

Based on the tectonic aspect, the Indonesian Archipelago is found at the juncture of three of the world's major structural plates. Indonesia is encompassed by subduction zones which are the cause of the strongly seismicity and volcanic action within the locale [1] (Figure 1). Three major plates – the northward moving Indo-Australian Plate, the south-eastward moving SE-Asian Plate, and the westward-moving Pacific Plate - meet at a plate triple-junction situated in the south of New Guinea's Bird's Head. The narrow North-Moluccan plate is interposed between the Asia and Pacific [1, 2].

Due to the position of the Indonesian Archipelago which is found at the juncture of three world structural plates, there are numerous common calamities. The meeting between these plates gives geological disaster zones, such as the Earthquake Zone, Volcanic Eruption Zone, Tsunami Zone, and Mass Movement Zone (Landslide Zone).

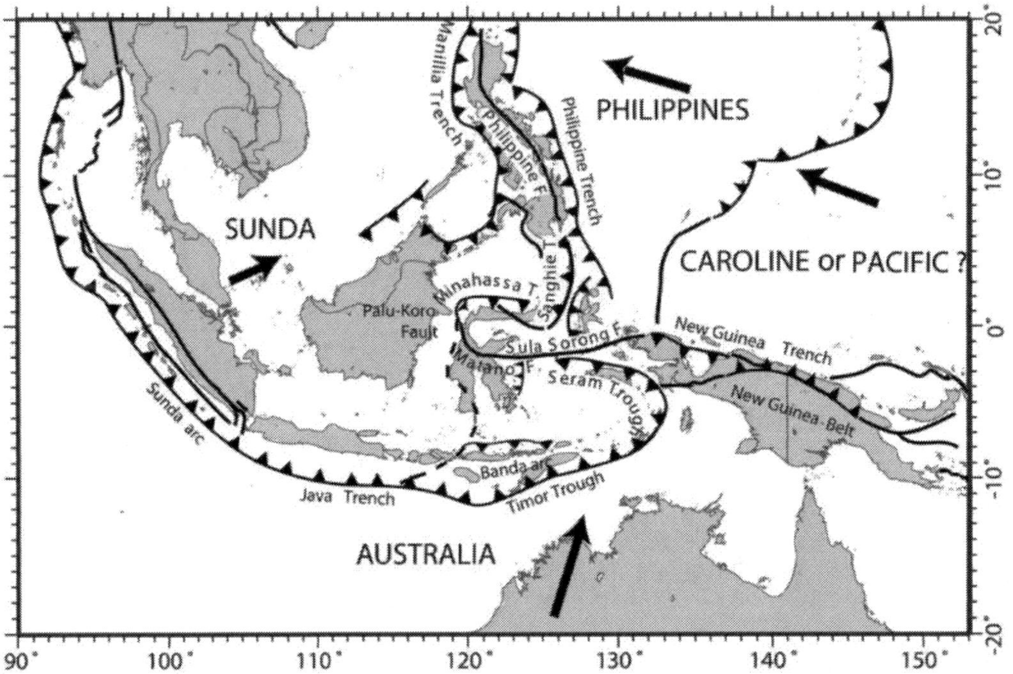

Figure 1. Dynamic subduction zones and major fault zones of the Indonesian region, the intersection of three major structural plates: (1) generally steady Sunda (Eurasia), (2) North-moving Australia- Indian Sea and (3) West-moving Philippine Ocean- West Pacific Plates (Socquet *et al.* 2006, from [3]).

Based on climate, Indonesia is located on the equator with a tropical climate. There are two seasons, namely dry season and stormy season. In the middle of the rainy season, rainfall often triggers landslides. Several mass media detailed that in Indonesia there are many landslide-prone areas. Illustration of landslide-prone areas, especially in West Java. This area is the most vulnerable to landslides. Throughout 2012, in Indonesia there were 116 cases of landslides that caused 128 deaths. In that year, the largest number of cases was recorded in West Java. There were 58 landslides with 33 deaths [4].

For this landslide disaster problem, it is required to Disaster Risk Reduction through mitigation and expand knowledge to the community as part of the early warning system. This paper discusses 1) how efforts to reduce disaster risk through an early warning system, 2) how community participation in an early warning system for landslides, 3) how the Hexa Helix as innovation of participatory concept.

2. Methods

2.1. Landslides and their problems

Landslides is defined as the mass transfer of material from one place to another that is lower. Landslides can be identified by the type of material (soil, rock, or debris) and their movement (slide, fall, flow, or a combination thereof). Varnes in 1978 gave a classification of landslides based on the type of material and its movement [5]. The condition of landslide-prone areas can be influenced by lithology (rock types and weathering), slope (geomorphology), seismic vibrations (tectonic aspects), and human activities. During the rainy season, rainfall often triggers landslides. Rainfall is related to slope stability [6]. High rainfall will reduce the safety factor (FK) of the slopes, thereby reducing the level of slope stability (Figure 2). Landslide prediction, especially during the rainy season, can be estimated based on the slope, type of relief, and rainfall.

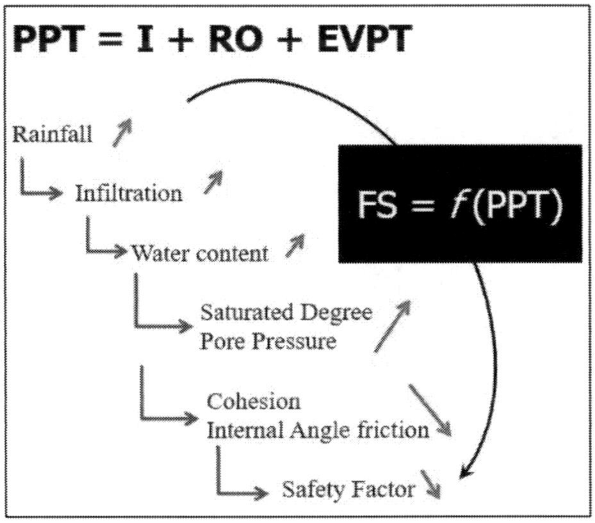

Figure 2. Relationship of rainfall with safety factor (FS)

2.2. Mitigation

Structural and non-structural mitigation is required to reduce the landslide hazard. In non-structural mitigation for landslide problems, data related to landslide problems is required, for example, it is necessary to know the conditions of rainfall and soil types, as well as the possibility of landslides. In addition, efforts to increase community capacity in dealing with landslide disasters are also required. These efforts can be done through counseling and training so that the community knows things related to the landslide problem. Community participation supports non-structural mitigation. Helix's participation concept is a proposal that will be useful in supporting non-structural mitigation, as well as in early warning systems.

2.3. RAFLEWS

Anticipation of landslide crisis is approached through a tool with the RAFLEWS design model (Rainfall Apparatus for Landslide Early Warning System). This apparatus can support landslide mitigation efforts. In addition, cooperation or participation from the community is also required. The community is required to maintain the RAFLEWS tool, measure the amount of rainfall on safety slope conditions, and respond if there are critical indications of landslides. Landslides can be triggered by extreme rainfall. If rainfall causes a landslide crisis slope, then RAFLEWS will provide a warning through the sound of a siren or an online report sent via smartphone to the surrounding community or government officials [7].

2.4. Helix Innovation

The helix concept is the concept of support from different parties in a venture or movement. This concept proceeds to advance concurring to the changing society and is an increment in development and innovation [8]. A few concepts have been created, extending from Triple Helix, Quadruple Helix, Penta Helix, to Hexa Helix [9]. The Triple Helix concept centers on the relationship between the scholarly community or academia, industry or business, and government (Figure 3) [10]. The

IOP Conf. Series: Earth and Environmental Science **989** (2022) 012031 doi:10.1088/1755-1315/989/1/012031

Quadruple Helix completes the Triple Helix by including “media-and-culture-based public” and "civil society" as the fourth helix [9][11]. The Quadruple Helix model (Triple Helix plus community – Non Government Organization) is also used in the concept of the *Starlet* Model (*Stabilisasi dan Rancangbangun Lereng Terpadu*, or stabilization and stable slope design), to be specific the participation relationship between academia (scientists), government, community (NGO), and industry (business) within the environmental management and monitoring of engineering slopes (Figure 4) [9][12].The Quintuple Helix innovation model is even more extensive and comprehensive by contextualizing the Quadruple Helix and by adding one helix, namely the natural environment of society [13] or mass media (Figure 5) [14]. This helix innovation is known as the Penta Helix (Academia, Business, Society, Government, and Mass Media) [14]. Penta Helix also developed into a more complete Hexa Helix by adding laws and regulations [15].

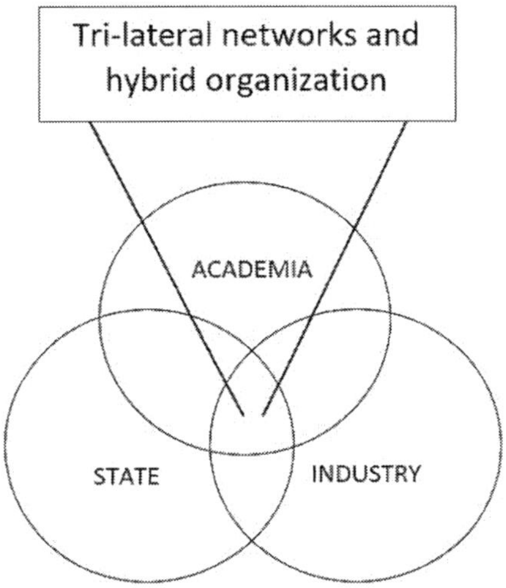

Figure 3. Triple Helix [10]

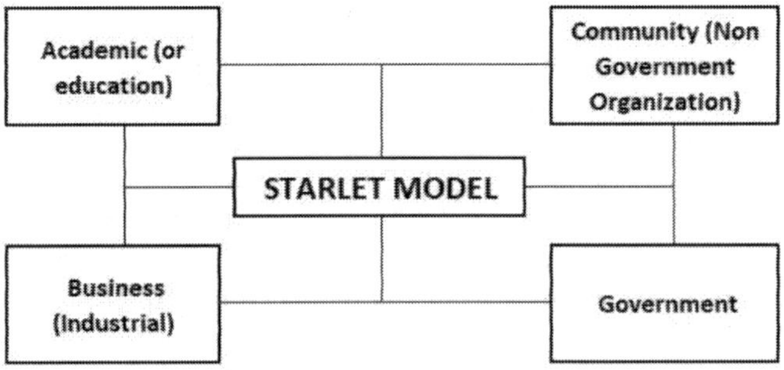

Figure 4. Quadruple Helix in the *Starlet* model [9][12]

IOP Conf. Series: Earth and Environmental Science **989** (2022) 012031 doi:10.1088/1755-1315/989/1/012031

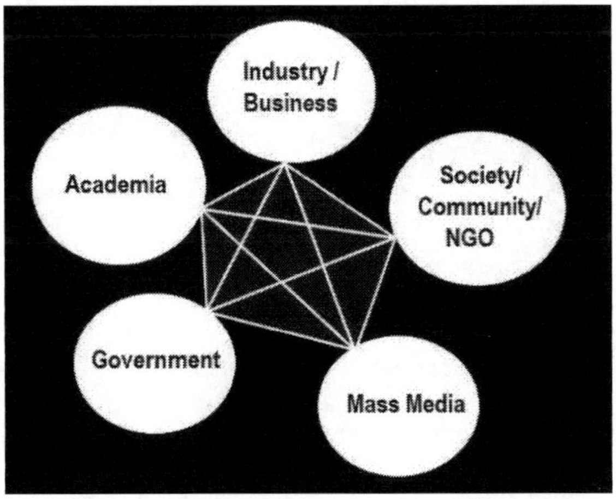

Figure 5. Penta Helix concept in the *Starlet* model [14]

3. Results and Discussion

3.1. RAFLEWS Work System

RAFLEWS is an early warning tool for landslides triggered by rainfall. It is known that in many avalanches, rain is the main trigger. The RAFLEWS tool provides rainfall signals that will cause critical slopes or unstable slopes. Installation of the RAFLEWS tool can be placed around landslide prone areas. Every rain that falls and the duration of rainfall will be recorded by the Field Officer. Any rainfall that indicates a possible critical slope will be connected directly to the Smartphone via an SMS signal and/or a signal connected to a siren. Sirens are installed in residential areas or areas that are suspected to be affected by landslides (Figure 6), so that residents can evacuate immediately.

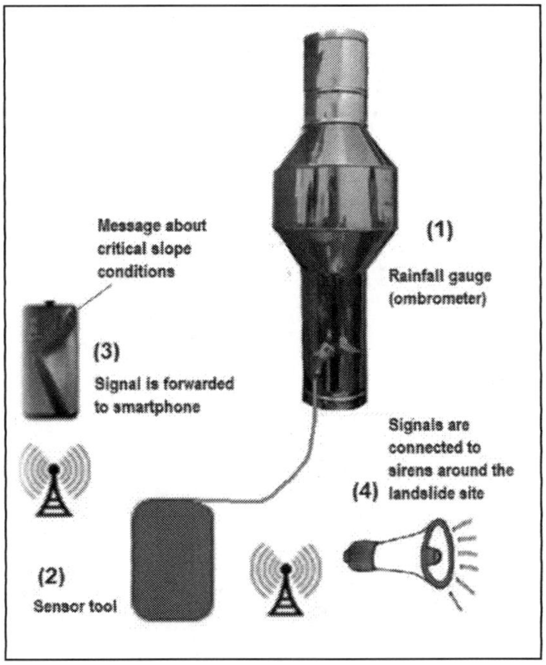

Figure 6. Early Warning System through RAFLEWS

3.2. Hexa Helix

The Hexa Helix concept was developed from the penta-helix by adding legal and regulatory aspects [9]. The six components of the Hexa Helix are Academia, Business (Industry), Community (Society, Non-Governmental Organizations), Government (State), Mass Media (and communication) and and Laws/Regulations/Legislations (Figure 7) with the following details:

- Government (State), consisting of the central government and regional governments
- Industry, include entrepreneur, business, service, and trade
- Community, consisting of society or people from various professions and NGOs
- Academia or scientists, including the world of general education and higher education
- Mass media and communication media, including communication devices such as Smartphone.
- and Laws/Regulations/Legislations, including local government regulations.

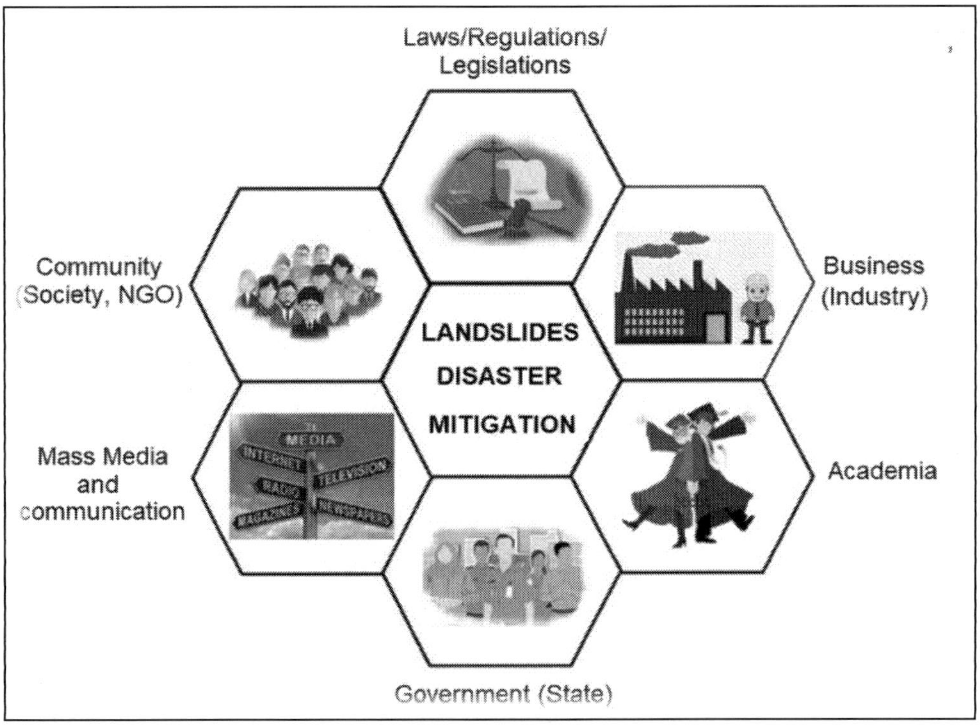

Figure 7. The six hexa-helix components

The six components share a common view of regional potentials and constraints, as well as disasters that damage infrastructure or cause casualties. The six components also have a role in the early warning system. disaster. Special attention is required for monitoring environmental damage around slopes or infrastructure [9].

The support or participation of each helix in Hexa Helix is required for the success of the Early Warning System and through the RAFLEWS (Rainfall Apparatus for Landslide Early Warning System) tool. The role of each helix is as follows (Table 1):

Table 1. Role of Hexa-helix Concept in Landslide Early Warning System using RAFLEWS

Agent	Management
Academia, scientists, education	Designing the RAFEWS tool. Placing RAFLEWS in a safe position in landslide prone locations. Placing the siren in the right location to be heard by the public. Provide direction in the management and monitoring process. Provide counseling and teach how RAFLEWS works, including how to measure rainfall.
Business, entrepreneurship, industry, service, trade	Reproduce RAFLEWS equipment. Build an early warning system for landslides. Build siren poles in landslide-prone settlements.
Community, Society, Non-Governmental Organizations (NGO)	Monitor the RAFLEWS tool for malfunctions. Measure rainfall regularly. If it rains heavily, be aware of the early warning of landslide hazards. Get counseling about the landslide early warning system.
Central government, local government	Coordinate with scientists, entrepreneurs, and the public. Deliver a work allow. Provide counseling on landslide-prone slope issues. Provide information data on indications of landslide-prone slopes. Provides climate estimate information.
Mass media, communication	Provide information on landslide events and indications of landslides. Disseminate information on landslide prone areas. Spreading information the weather forecast.
Laws, regulations, legislations	Arrange building permits. Environmental regulations. Regulation on Directions for Environmental Management and Monitoring Plans. Regulating Land Use. Oversee patents or copyrights

4. Conclusions

The landslide early warning system model through RAFLEWS (Rainfall Apparatus for Landslide Early Warning System) is one alternative in Disaster Risk Reduction. The community can be involved in disaster risk reduction efforts through the Hexa Helix concept. Hexa Helix consists of six participant elements, namely: Academics, Government, Society, Industry, Mass Media, and Laws/Regulations/Legislations. The combination of the RAFLEWS model and the Hexa Helix concept is anticipated to have an impact on taking care of landslide disasters within the future.

Acknowledgements

Thanks to the Director of *DRPM* (*Direktorat Riset, dan Pengabdian pada Masyarakat* or Directorate of Research and Community Service),UniversitasPadjadjaran. This paper is a small part of research which is funded by Universitas Padjadjaran through ALG (Academic Leadership Grant) - *RKDU (Riset Kompetensi Dosen Unpad* or Competency Research of Unpad Lecturer) year 2021.

References

[1] Hall R 2014 Indonesia tectonics: Subduction, extension, pronevance and more *Proceedings, Indonesia Petroleum Association (IPA) Thirty-Eighth Annual Convention & Exhibition, May 2014 Thirty-Eighth Annual Convention & Exhibition, May 2014*

[2] Verstappen H Th 2007 Indonesian Landforms and Plate Tectonics *Jurnal Geologi Indonesia Vol. 5 No. 3 September 2010: 197-207*

[3] Van Gorsel J T 2018 Bibliography of the geology of Indonesia and surrounding areas Edition 7.0 accessed from *https://vangorselslist.com/pdf/BIG_I_Regional_7.pdf*

[4] Zakaria Z 2020 *Slope Stability* (in Bahasa: Kestabilan Lereng) Unpad Press 204 pp.

[5] Varnes D J 1978 Slope movement types and processes. In: Special Report 176: *Landslides: Analysis and Control (Eds: Schuster, R. L. & Krizek, R. J.)* Transportation and Road Research Board, National Academy of Science, Washington D. C. 11-33

[6] Zakaria Z, Jihadi, L H, Sabila Z S & Oscar A W 2015 2015 Simulation of Slope Stability in the Dry and Rainy Season at Jatinangor, District of Sumedang, West Java, *Proceedings of The 2nd International Conference Hanoigeo ISBN: 978-604-913-418-0 "Engineering Geology in Respond to Climate Change and Sustainable Development of Infrastructure" 27-28 November 2015 pp. 365-367*

[7] Zakaria Z, Sophian RI, & Alam, BYCSSS 2020 *Starlet-Perdana* Model and Hexa Helix Concept for Landslide Disaster Mitigation. (Case Study: Citatah Avalanche Mitigation with Engineering Geology and Forensic Geology Approach), Unpad Lecturer Competency Research, 134 pages, unpublished (In Bahasa)

[8] Praswati A N 2017 Development of the helix model to enhance innovation, *Proceedings of the National Seminar on Management & Business Research 2017 "Development of E-Business Concepts and Research in Indonesia" pp. 690-705* (in Bahasa). *https://publikasiilmiah.ums.ac.id/bitstream/handle/11617/9022/sansetmab2017_15.pdf?sequence=1&isAllowed=y,* accessed on September 19, 2021, at 08.30

[9] Zakaria Z, Sophian R I, Muljana B, Gusriani N, Zakaria S 2019 The Hexa-Helix Concept for Supporting Sustainable Regional Development (Case Study: Citatah Area, Padalarang Subdistrict, West Java, Indonesia) IOP Conf. Series: Earth and Environmental Science 396 (2019) 012040

[10] Lindmark A, Roos M N and Sturesson E 2009 *Difficulties of Collaboration for Innovation, A Study in the Öresund Region*, Department of Business Administration, Lund University, Swedish, from*https://lup.lub.lu.se/student-papers/search/publication/1437850.*

[11] Carayannis E G and Campbell D F J 2009 Mode 3 and 'Quadruple Helix': toward a 21st century fractal innovation ecosystem *International Journal of Technology Management 46 (3) 201-234.*

[12] Zakaria Z 2010 Model Starlet: A proposal for landslide mitigation with land genetic approach (Case study: Citatah Landslide, Padalarang), West Java *Indonesian Journal on Geosciences - Jurnal Geologi Indonesia Vol 5 Number 2 Juni 2010: 93-112* (in Bahasa)

[13] Carayannis E G, Barth T D and Campbell D F J 2012 The Quintuple Helix innovation model: global warming as a challenge and driver for innovation, *Journal of Innovation and Entrepreneurship 2012, 1-12.*

[14] Zakaria, Z.,, Sophian, R.I., & Khoirullah, N., 2018, Modification of the Participatory Rural Appraisal Concept for debriefing of Community Service Program in West Java, Indonesia, (in bahasa: Modifikasi konsep Participatory Rural Appraisal untuk pembekalan Kuliah Kerja Nyata Mahasiswa di Jawa Barat, Indonesia), *Dharmakarya: Jurnal Aplikasi Ipteks untuk Masyarakat Vol. 7, No. 1 Maret 2018, p. 38-45*

[15] Zakaria Z, Sophian R I, and Gusriani N 2019 Environmental Aspect in Infrastructure Planning Using Starlet-Perdana Model, *IOP Conf. Ser.: Earth Environ. Sci. 248 012013, https://iopscience.iop.org/article/10.1088/1755-1315/248/1/012013/pdf*

IOP Conf. Series: Earth and Environmental Science **989** (2022) 012032 doi:10.1088/1755-1315/989/1/012032

Factor Affecting and Damage Mapping of Flood Inundation in Japan: A Case Study of Hitoyoshi City, Kumamoto Prefecture, Japan

Arbi Surya Satria Ridwan[1*], Riken Homma[2], and Hang Liu[1]

1 Graduate School of Science and Technology, Kumamoto University, 2-39-1 Kurokami, Chuo-ku, Kumamoto City, Japan
2 Faculty of Advanced Science and Technology, Kumamoto University, 2-39-1 Kurokami, Chuo-ku, Kumamoto City, Japan

*Email: arbissr94@gmail.com

Abstract. One of the most water-related disasters is flooding induced by heavy rain. According to several model experiments and an analysis of historical data, such natural disasters will become more common in the future. In recent years, torrential rain and flooding have caused significant damage to agricultural land, homes, and human life in numerous Japanese towns. We used the Hitoyoshi City in Kumamoto Prefecture, Japan, as a case study for the area's flood event in July 2020. This study uses machine learning to study spatial analysis of the city of Hitoyoshi after the 2020 flood disaster. Assessments such as flood hazards, exposure, and susceptibility are critical components of total flood risk assessment. In addition, this study also uses several driving factors for mapping the scenario to the evacuation center, how the impact of flooding on the residential environment, and changes to land use in the Hitoyoshi area. For land-use planners, estimating the quickly flooded region using distinct land use profiles is quite important. It aids in the analysis of flood losses and the allocation of resources for recovery and restoration.

1. Introduction

Japan is a country that is very prone to many natural disasters due to its geographical location, topographical and meteorological condition [1]. Most cities in Japan are located on alluvial plains which have a relatively high risk of flooding. Some areas in Japan often experience annual flooding[2]. However, Japan is one of the countries that has a good disaster management and evacuation procedures that have been prepared carefully [3]. Residents in Japan have a handling procedure where before or after a disaster occurs, residents are required to go to shelters designated by the government to become temporary residences such as schools, sports halls, and gymnasium[4].

The risks of natural disasters such as floods are very diverse and complex[5][6]. Many phenomena are directly related to the flood disaster, one of which is rainfall and the ongoing damage of the flood itself. Many impacts caused by floods such as the impact on nature, disruption of human activities in it, economic losses in the region, to the loss of lives [7][8].

Japan has a long history, especially of floods. The first recorded flood disaster in Japan was in the mid-6th century and the flood is still one of the greatest disasters in modern Japanese history. In 1896, the Japan River Law was enacted to make a regulation that diverts floodwaters to the sea as quickly as possible[9]. Events such as floods in Japan are purely natural phenomena and social conditions at that

IOP Conf. Series: Earth and Environmental Science **989** (2022) 012032 doi:10.1088/1755-1315/989/1/012032

time also play an important role which will vary from one area to another and develop from one time to another[10].

2. Case Study Area

2.1. 2020 Flood in Southern Kumamoto and Disaster Risk

Hitoyoshi as one of the case studies in this research is because at the beginning of July 2020, Hitoyoshi experienced a heavy flood that destroyed the city of Hitoyoshi, especially the city center of the small town of Hitoyoshi in Kumamoto Prefecture, Japan [11]. The central part of Hitoyoshi is located close to the Kuma River which was flooded by the heavy rain in July 2020.

Floods in Hitoyoshi were caused by heavy rains that lasted for days. In addition to flooding, several points in Hitoyoshi also experienced landslides[12]. The overflowing river also swept across roads and bridges, the rapidity of the flood disaster made it difficult for rescue workers to find people who were still trapped and had not had time to evacuate[13]. This research is expected to help for the development of better disaster mitigation for small cities and efficiency of evacuation facilities in the future. In the future, this dataset usage status can be used for various spatial information, especially in emergency response and allocation of impact estimates on the city of Hitoyoshi[14].

Based on Hitoyoshi City Hall data, the city's population reached 31,558 at the end of October 2020 with a total city area of 210.5 km^2[15]. The Kuma River is the only major river that passes through the city of Hitoyoshi and it runs through the residential and urban areas of the city of Hitoyoshi in Kumamoto prefecture. The Kuma River, commonly nicknamed by the public as Aberagawa (in Japanese means a raging river) is one of the 3 fastest flowing rivers in Japan after the Fuji River in Nagano, Shizuoka, and Yamanashi prefectures as well as the Mogami River in Yamanashi prefecture.

Figure 1. Location of Hitoyoshi City, Kumamoto Prefecture, Japan
Source: http://www.qsr.mlit.go.jp/yatusiro/river/kouzui/index.html

2.2. Flood History in Hitoyoshi

The Kuma River has historically experienced several floods, but one of the biggest flood impacts occurred in 1965. The geographical location of the city of Hitoyoshi causes this area to experience frequent flooding during the season with high rainfall. The Kuma River that passes through the town of Hitoyoshi passes through 3 areas, namely Yatsushiro City, Hitoyoshi City and Kuma Village before reaching the Yatsushiro Sea. The following are some of the floods that have occurred in the Hitoyoshi area[16].

Tabel 1. Major floods that have occurred on the Kuma river since 1960

Year Record	Flood Situation
July 1965	The flood submerged two thirds of the town of Hitoyoshi and destroyed the Hagiwara Dam in the town of Yatsushiro.
July 1982	The flood caused 5000 houses to be affected by the flood and 47 to be swept away by the flood
Aug. 2004	The typhoon brought heavy rains which caused the Kuma River to overflow and forced residents to evacuate to safer places.
Sept. 2005	119 houses were submerged by the overflowing river and displaced 750 families.
July 2006	80 houses were submerged and over 900 families needed to be evacuated due to the overflowing Kuma River.
June 2008	33 houses were inundated and 1000 families were affected in Hitoyoshi, Yatsushiro, and Ashikita areas.
June 2011	Due to heavy rain, the river overflowed beyond the limit and forced residents to evacuate themselves.

Source: http://www.qsr.mlit.go.jp/yatusiro/river/kouzui/index.html

2.3. *Building Type in Hitoyoshi area*

The Hitoyoshi city center is one of the more inhabited by the town's residents than some of the surrounding villages. .In this study, the northern part of Hitoyoshi became one of the considerations in this study because the flood caused casualties, the number of residential areas, infrastructure, and several other important facilities[17]. Based on several visits to the Hitoyoshi area, most of the buildings in Hitoyoshi, especially residential ones, are wooden construction building.

In this study also adapted to the data set of Hitoyoshi city, buildings are distinguished based on the type of material into 4 types, namely: wooden construction, fire-resistant wooden construction, reinforced concrete and other types of materials. Because during the visit, many wooden buildings were damaged, destroyed, and abandoned by their residents due to the flood in Hitoyoshi in 2020.

Figure 2. Building Type (Wooden Construction) in Hitoyoshi

3. Tools of Assessment

3.1. *Data Source*

Several processes were carried out to collect research data, especially for the Hitoyoshi area. The first step is to collect data that is considered important for mapping, especially the areas affected by the flood in Hitoyoshi. The data source will be attached to table number 1 in this paper.

The system in Japan, some data can be accessed directly by the public, but some data need permission from the Hitoyoshi city center government office. For data of a Hitoyoshi city basic map, accessed directly through the Basic Map Information at the Geospatial Information Authority of Japan (GSI) for research purposes so that it can be accessed by the public. Some other supporting data such as population data, land use data, and building information are data obtained directly through the Hitoyoshi city government. The type of land use metadata is vector and then converted into raster data of 30 m x 30 m by ArcGis polygons to raster.

Table 2. Data Set source

Category	Data	Year	Source
Land Use	Land Use Data	2017	Hitoyoshi City Basic Survey
Population	Population census data	2020	Basic Resident Register population of Hitoyoshi City
Building Information Data	Building Data	2017	Hitoyoshi City Basic Survey
Disaster Data	Flood Area	2020	Japan National Land Numerical Website: https://nlftp.mlit.go.jp/ksj/index.html

3.2. Building Status in Flood Inundation Depth

The flood hazard is calculated based on the depth of the water to the building and area. According to The Ministry of Land, Infrastructure, Transport, and Tourism, Japan categorized level of flood hazard[18][19].

Table 3. Flood Hazard Level according to MLIT, 2020

Warning level	Flood Inundation depth (meters)	Hazard
Level 0	0	Not affected by flood
Level 1	≤ 0.5 m	Very low hazard
Level 2	0.5 – 1 m	Low Hazard
Level 3	1 – 2 m	Medium Hazard
Level 4	2 – 5 m	High Hazard
Level 5	≥ 5 m	Extreme Hazard

Source: MLIT. (2020): About inundation depth and evacuation behavior. (https://www.river.go.jp/kawabou/reference/ index05.html, 2020.10.2) (in Japanese).

In Japan there are 5 classifications of vulnerability to flooding[20], namely when the depth of the flood inundation is at level 1 (depth less than 0.5 m) when people can still evacuate using feet and buildings are also classified as safe. Level 2 (depth 0.5 - 1 m) is categorized as low level, in this zone evacuation becomes quite difficult for adults and babies, pets have also entered the danger zone. Level 3 (1-2 m depth) belongs to the medium vulnerability zone, where humans will start to sink if forced to evacuate, but if they are on the 2nd floor of their house, they will remain safe for the time being. Furthermore, this level 4 (depth reaches 2-5 m) has already entered the Danger Zone, when people are not safe and they can only have the possibility of being safe when they are on their roofs. And this happened a lot during the flood disaster in Hitoyoshi in 2020. Level 5 (a depth of more than 5 m), this zone is classified as extreme conditions and even residents who save themselves on the roofs of their houses will no longer be safe.

3.3. Data Set Arrangement

This Hitoyoshi city boundary is also one of the data that is considered important as a boundary for the Data Set. Based on data from Google Earth and basic survey data from the city of Hitoyoshi, this study tries to identify which areas are affected by flooding and the types of buildings in it for the validation process[21]. Because this study aims to evaluate each building's impact with the visual interpretation method on flood exposure in the Hitoyoshi area. Then some data from polygons such as Building Information and Land Use are put together. The spatial data resolution which is a collection of several raster data will be used for population calculations and the distribution of buildings and population estimates in the Hitoyoshi area[22]. Then this flood inundation map is also masked and developed in the form of polygons to provide area limits for land use and building data in calculating the impact of floods in the Hitoyoshi area[23]. ArcGis 10.6 is needed to determine the category of the area based on the level of flooding and the impact on the floods that occurred in Hitoyoshi (Figure 3).

IOP Conf. Series: Earth and Environmental Science **989** (2022) 012032 doi:10.1088/1755-1315/989/1/012032

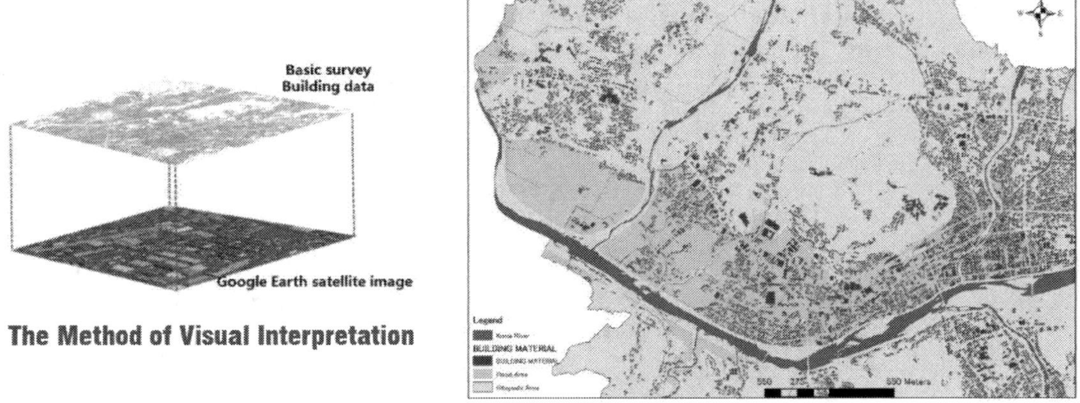

Figure 3. Building Material and Flood Inundation Map in Hitoyoshi city

4. Results and Discussion

4.1. Population Estimation Mapping

In mapping the flood exposure in the Hitoyoshi area, the first thing to do is to calculate the estimated population in the Hitoyoshi area. The results of the cross between Hitoyoshi city block database and Hitoyoshi city land use data are needed to calculate the estimated population ratio in Hitoyoshi city. Then, each area based on the residential area from the land use data is used as a base block unit calculation in independent variable units in mapping the estimated population in Hitoyoshi. Then from the results of the data, it can be seen where the location of the population in Hitoyoshi and the distribution of the population in it is according to Figure 4.

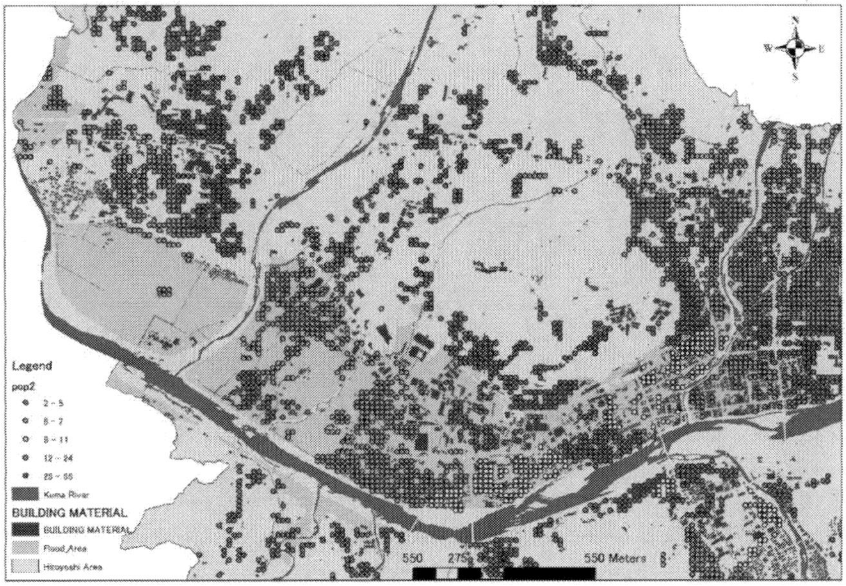

Figure 4. Population estimation result in Hitoyoshi city

IOP Conf. Series: Earth and Environmental Science **989** (2022) 012032 doi:10.1088/1755-1315/989/1/012032

4.2. Flood Inundation Mapping and Damage Mapping

Based on the results of spatial mapping research for the Hitoyoshi area, there are about 20947 building data recorded from the Hitoyoshi government data set. Through the simulation results for flood-inundation areas, it was found that almost most of the buildings directly affected by the flood were at level 5 flood as many as 3026 buildings (14.4%). Then followed by 927 buildings (4.42%) located at level 4. The buildings that were submerged at level 3 were 493 buildings (2.35%). And the rest are at level 2 as many as 132 buildings (0.63%) and level 1 only 13 buildings (0.06%).

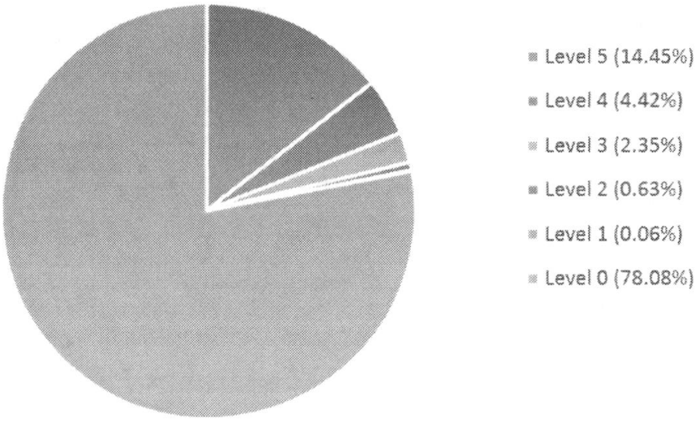

Figure 5. Building Percentage in Inundation Area, Hitoyoshi, Kumamoto

s

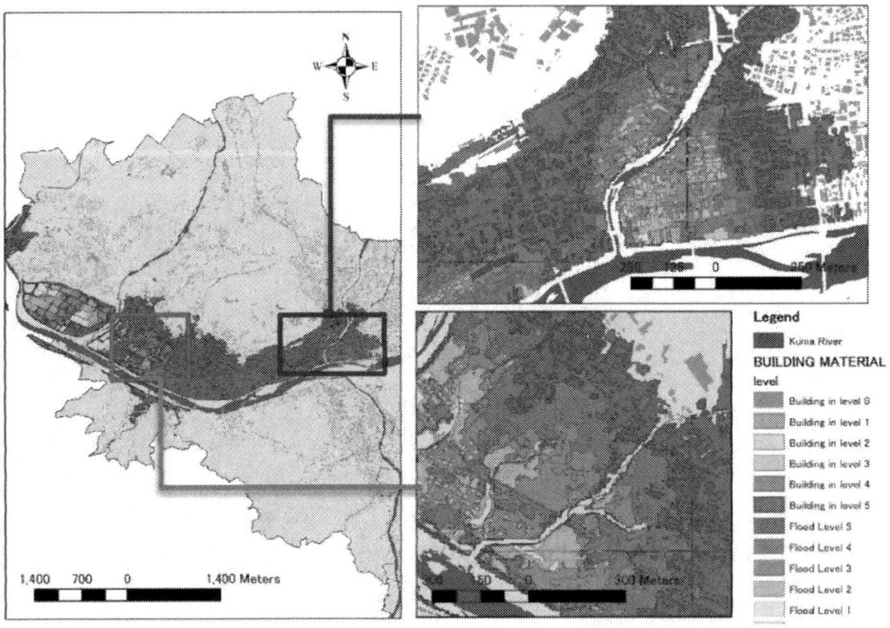

Figure 6. Population estimation result in Hitoyoshi city

4.3. Percentage of Building in Inundation Area
Based on the data set from Hitoyoshi City, the buildings in Hitoyoshi are divided into 4 types, namely wooden construction, fire-resistant wood construction, reinforced concrete, and other materials. However, if a flood disaster occurs, usually a building that has a wooden construction is one of the buildings at risk of having the most severe damage and after the disaster occurs it cannot be reused.

It is known from the data that up to 2434 buildings made of wood are in a level 5 inundation area. Meanwhile, at level 4 there are 795 buildings that use wood construction which have the same risk of damage caused by flooding. And mostly in the Hitoyoshi area, these wooden construction buildings are traditional Japanese houses. Meanwhile, in the level 3 inundation area, there are about 407 wooden buildings. This research is expected to be a guide for the Hitoyoshi government to provide a rebuilding proposal for the city of Hitoyoshi to rebuild the economy and tourism in Hitoyoshi.

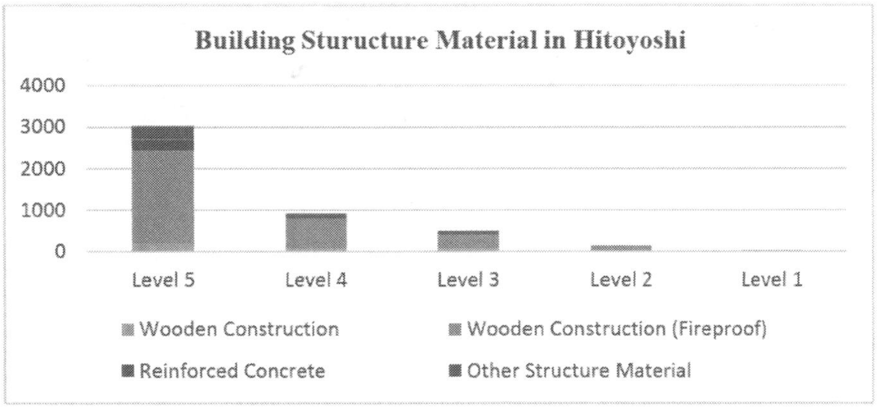

Figure 7. Number of Building Structure Material Damage in Hitoyoshi

Table 4. Number of Building Construction Material in each level of inundation flood

Flood Level	Wooden Construction	Wooden Construction (Fireproof)	Reinforced Concrete	Other Structure Material
Level 5	173	2261	276	316
Level 4	78	717	44	88
Level 3	69	338	38	48
Level 2	34	80	5	13
Level 1	6	2	3	2

4.4. Building Construction Material
Based on the survey results, it also proved and validated after the flood that some wooden buildings in Hitoyoshi were damaged which were quite severe and could not even be used anymore and were abandoned by the previous owners. As for houses made of reinforced concrete or other materials that are more resistant to water, until now some can still be used and only need a few renovations. This situation indirectly has an impact on the number of empty buildings in Hitoyoshi and slowly Hitoyoshi has become a small town in a suburban area that was abandoned by its residents due to flooding in 2020. So that some considerations for future Hitoyoshi city planning are also needed some solutions so that in the future, Hitoyoshi city does not become a city abandoned by its residents due to natural disasters such as floods.

IOP Conf. Series: Earth and Environmental Science **989** (2022) 012032 doi:10.1088/1755-1315/989/1/012032

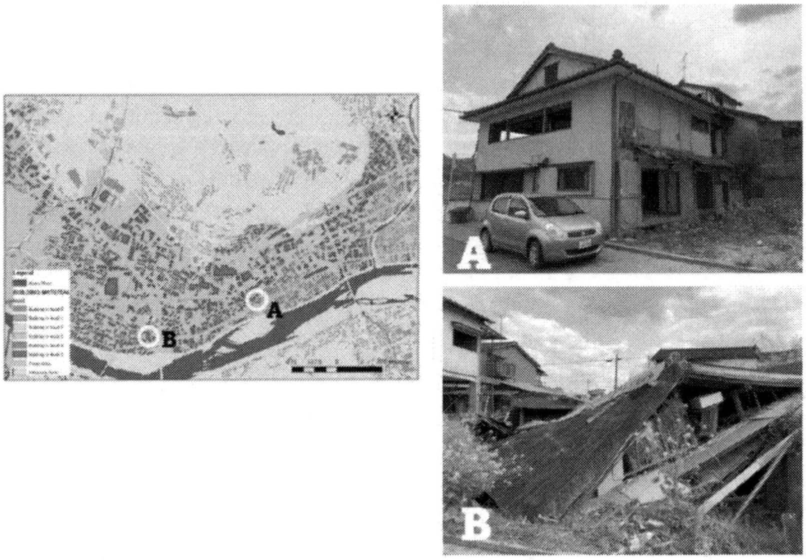

Figure 8. Damage Map and damage buildings in Hitoyoshi city

5. Conclusion

Based on the simulation above, it proves that most affected areas by the Hitoyoshi flood in 2020 are in areas that have a dense building density. Almost most of the area that is inhabited by many residents is in the level 5 flood area. Even most of sthe population is in areas that experience flood inundation between level 3 to level 5 which shows that this area is experiencing extreme flood impacts and tends to be dangerous for the lives of the Hitoyoshi people themselves. Based on direct observation of the environment, the buildings in Hitoyoshi, especially houses made of wood, have the greatest risk of damage due to flooding caused by flood inundation. The geographical location of Hitoyoshi which is a mountainous area forces the residents to stay close to the Kuma River which has a vulnerability to overflowing at certain times, especially high rainfall.

Japan already has an evacuation system and a warning system that is quite good to anticipate a natural disaster before it actually occurs. However, there are still losses to life, economy loss, and society problems caused by unavoidable natural disasters. Until now, even the economic and tourism conditions in Hitoyoshi have not fully recovered due to the 2020 floods due to the damage to infrastructure such as trains and roads leading to locations that make people still feel the impact of the floods in Hitoyoshi in 2020. There needs to be participation from the Hitoyoshi government to rebuild the city to be safer from the natural disasters that occurred in Hitoyoshi with direct participation with the community and society. There is still a lot of room for improvement as seen from the high number of flood impacts in Hitoyoshi for better urban planning in the future.

References

[1] Ministry of Foreign Affairs Japan, "Disaster Prevention," 2021. https://www.mofa.go.jp/policy/disaster/21st/2.html (accessed Sep. 16, 2021).

[2] A. Rimba, M. Setiawati, A. Sambah, and F. Miura, "Physical Flood Vulnerability Mapping Applying Geospatial Techniques in Okazaki City, Aichi Prefecture, Japan," *Urban Sci.*, vol. 1, no. 1, p. 7, 2017, doi: 10.3390/urbansci1010007.

[3] Kyung-Su Choo, D.-H. Kang, and B.-S. Kim, "Impact Assessment of Urban Flood on Tra ffi c Disruption using Rainfall – Depth – Vehicle," *Water*, vol. 12, no. 926, 2020.

[4] M. G. Wilson, A. M. Lane, C. J. Beedie, and A. Farooq, *7 Risk Returns*. 2011.

[5] J. De Boer, W. J. Wouter Botzen, and T. Terpstra, "Improving flood risk communication by focusing on prevention-focused motivation," *Risk Anal.*, vol. 34, no. 2, pp. 309–322, 2014, doi: 10.1111/risa.12091.
[6] P. Hudson, "Evaluating the effectiveness of flood damage mitigation measures by the application of Propensity Score Matching," *Nat. Hazards Earth Syst. Sci. Discuss.*, vol. 2, no. 1, pp. 681–723, 2014, doi: 10.5194/nhessd-2-681-2014.
[7] S. S. Lin, N. Zhang, Y. S. Xu, and T. Hino, "Lesson learned from catastrophic floods in Western Japan in 2018: Sustainable perspective analysis," *Water (Switzerland)*, vol. 12, no. 9, 2020, doi: 10.3390/w12092489.
[8] S. L. Cutter, C. T. Emrich, D. P. Morath, and C. M. Dunning, "Integrating social vulnerability into federal flood risk management planning," *J. Flood Risk Manag.*, vol. 6, no. 4, pp. 332–344, 2013, doi: 10.1111/jfr3.12018.
[9] J. Fan and G. Huang, "Evaluation of flood risk management in Japan through a recent case," *Sustain.*, vol. 12, no. 13, 2020, doi: 10.3390/su12135357.
[10] E. warning S.-C. of the I.-M. C. on I. C. of D. Reduction, "Japan ' s Natural Disaster Early Warning Systems and International Cooperative Efforts March 2006 Early Warning Sub-Committee of the Inter-Ministerial Committee on International Cooperation for Disaster Reduction Government of Japan," no. Marzo, pp. 1–29, 2006.
[11] T. Mukunoki *et al.*, "Reconnaissance report on geotechnical damage caused by a localized torrential downpour with emergency warning level in Kyushu, Japan," *Soils Found.*, vol. 61, no. 2, pp. 600–620, 2021, doi: 10.1016/j.sandf.2021.01.008.
[12] K. Prefecture, P. C. Shakti, K. Hirano, S. Iizuka, and R. Misumi, "Quick Exposure Assessment of Flood Inundation : A Case Study of Hitoyoshi Quick Exposure Assessment of Flood Inundation : A Case Study of Hitoyoshi City in Kumamoto Prefecture , Japan," no. December 2020, 2021, doi: 10.24732/NIED.00002409.
[13] W. E. Highfield, W. G. Peacock, and S. Van Zandt, "Mitigation Planning: Why Hazard Exposure, Structural Vulnerability, and Social Vulnerability Matter," *J. Plan. Educ. Res.*, vol. 34, no. 3, pp. 287–300, 2014, doi: 10.1177/0739456X14531828.
[14] P. C. Shakti *et al.*, "Assessing flood risk of the chao phraya river basin based on statistical rainfall analysis," *J. Disaster Res.*, vol. 15, no. 7, pp. 1025–1039, 2020, doi: 10.20965/jdr.2020.p1025.
[15] Hitoyoshi City Government, "Hitoyoshi City," 2020. https://www.city.hitoyoshi.lg.jp/q/list/750.html (accessed Sep. 20, 2021).
[16] Y. R. and N. H. Office, "Past floods," 2021. http://www.qsr.mlit.go.jp/yatusiro/river/kouzui/index.html (accessed Sep. 16, 2021).
[17] S. Das, R. Shaw, S. Kanbara, M. Abe, and T. Komino, "DISASTER DURING A PANDEMIC," 2020.
[18] T. and T. Ministry of Land, Infrastructure, "National Flood Risk," *2021*. https://www.river.go.jp/kawabou/pc/ov?zm=5&clat=35.67514743608467&clon=138.47167968750003&mapType=0&viewGrpStg=0&viewRd=1&viewRW=1&viewRiver=1&viewPoint=1 (accessed Sep. 13, 2021).
[19] A. Ishiguro and E. Yano, "Tsunami inundation after the Great East Japan Earthquake and mortality of affected communities," *Public Health*, vol. 129, no. 10, pp. 1390–1397, 2015, doi: 10.1016/j.puhe.2015.06.016.
[20] S. J. Kang, S. J. Lee, and K. H. Lee, "A study on the implementation of non-structural measures to reduce urban flood damage -Focused on the survey results of the experts-," *J. Asian Archit. Build. Eng.*, vol. 8, no. 2, pp. 385–392, 2009, doi: 10.3130/jaabe.8.385.
[21] J. Mallick *et al.*, "Risk assessment of resources exposed to rainfall induced landslide with the development of gis and rs based ensemble metaheuristic machine learning algorithms," *Sustain.*, vol. 13, no. 2, pp. 1–30, 2021, doi: 10.3390/su13020457.
[22] J. Dabbeek, H. Crowley, V. Silva, G. Weatherill, N. Paul, and C. I. Nievas, "Impact of exposure

spatial resolution on seismic loss estimates in regional portfolios," *Bull. Earthq. Eng.*, no. 0123456789, 2021, doi: 10.1007/s10518-021-01194-x.

[23] H. Liu, R. Homma, and K. Iki, "Case Study of the 2016 Kumamoto Earthquake: The Disaster Response Capability of Kumamoto Compact City, Japan," *MATEC Web Conf.*, vol. 331, no. December, p. 01002, 2020, doi: 10.1051/matecconf/202033101002.

AUTHOR INDEX